Bow Ties in Process Safety and Environmental Management

Bow Ties in Process Safety and Environmental Management

Current Trends and Future Perspectives

Edited by

Anjani Ravi Kiran Gollakota, Sneha Gautam, and Chi-Min Shu

CRC Press
Taylor & Francis Group
Boca Raton London New York

CRC Press is an imprint of the
Taylor & Francis Group, an **Informa** business

First edition published 2022
by CRC Press
6000 Broken Sound Parkway NW, Suite 300, Boca Raton, FL 33487-2742

and by CRC Press
4 Park Square, Milton Park, Abingdon, Oxon, OX14 4RN

ISBN: 9780367690885 (hbk)
ISBN: 9780367690915 (pbk)
ISBN: 9781003140382 (ebk)

DOI: 10.1201/9781003140382

Typeset in Times
by KnowledgeWorks Global Ltd.

Contents

Preface

A DYNAMIC APPROACH

Chemical process safety is an area of growing importance within chemical, oil and gas, and allied industries. Worldwide, the chemical sector alone represents a $5 Trillion industry and directly employee millions. The chemical sector is a fast-developing industry due to the rapid development of exploration and product technologies. The need for safe, well managed processes within the industry has been highlighted by several recent disasters including the Deepwater horizon oil spill in 2010 (caused by poor safety systems and cost cutting measures, according to White House report) and the Tianjin disaster in China in 2015. The disasters not only cost human life but had a huge impact on the local environment and represent major fiscal losses for the companies involved.

On the spin side the chemical, oil and gas, and allied industries are inherently risk-laden sectors. The continuous occurrence of major process incidents has increased the awareness within the process industry about the importance of making development and operational decisions based on a thorough assessment of the associated risks to identify measures that can be taken to prevent potential losses. This increased awareness has shaped and influenced process safety science. Chemical process safety is a rapidly evolving area and is moving to more dynamic and adaptive methods of design and management to improve health and safety across the industry.

Risk management is a vital systems activity across design, implementation and operational phases of process system. This integrative systems perspective is often missing or poorly emphasized in much of risk management literature. The purpose of this volume is to present a holistic approach to process risk management that is firmly grounded in systems engineering employing a life cycle. The framework is briefly segregated into two major thrust areas mingling the process safety, and environmental impacts across various industrial processes. The first section of the book is mainly devoted towards the alarming level of coal residual impacts and the mitigating approaches, whereas, the second section of the book is mainly devoted towards the neglected areas of environmental devastating factors through several tiny approaches.

http://www.routledge.com/9780367690885

Editors

Dr. Anjani Ravi Kiran Gollakota is an assistant professor in Department of Safety, Health and Environmental Engineering, National Yunlin University of Science and Technology, Taiwan. He has received his PhD (Chemical Engineering) from Indian Institute of Technology (IIG), Guwahati, and an MTech (Chemical and Engineering) from Andhra University College of Engineering, India. After his PhD, he joined University of Surrey, United Kingdom (May 2016 – July 2017), received Post Doc Fellowship from Ministry of Science and Technology Taiwan and joined Post Doc (August 2017–February 2018), and soon promoted as assistant professor (February 2018–To Date).

Dr. Gollakota's primary research is on 4th generation biofuels via hydrodeoxygenation, hydrothermal liquefaction. His research is primarily focused towards upgrading the biofuels generated from physico-chemical-thermo chemical methods. The prime objective of this research is to enhance the efficacy of the biofuels which are highly underrated due to their complex compositions. At present, Dr. Gollakota is working on another novel aspect of self-treatment procedures through green materials and green solvents. The present research will be a great boon to many of the coal industries struggling with their dumping systems.

His current research projects and works are focused in broad multidisciplinary areas of transmuted coal fly ash applications in geopolymers, biofuel upgrading catalysis, and leaching radioactive isotopes. Also Dr. Gollakota is working on green material synthesis on the basis of waste to energy generation.

Dr. Gollakota has published 18 articles in top-ranked journals (h-index 12; i10-index 12; citations >500), he has developed a network of collaborators across four continents. Prior experience as a software engineer for three years helped Gollakota to have a proficiency in traditional chemical and process engineering software such as Ansys and COMSOL.

Dr. Sneha Gautam is an assistant professor in Department of Civil Engineering at Karunya Institute of Technology and Sciences, Coimbatore, India. He has received his PhD (Engineering) from Indian Institute of Technology (IIT), Kharagpur, and an MTech (Environmental Science and Engineering) from Indian Institute of Technology, Indian School of Mines (IIT – ISM) Dhanbad. After his PhD, he joined INCLEN and University of California, Berkeley, CA (May 2015 – July 2016), received Postdoc Fellowship from Ministry of Science and Technology Taiwan and joined Postdoc (August 2016–May 2017), after which he joined Marwadi University, Rajkot Gujarat as an assistant professor and head (September 2017–April 2019).

His fundamental and application-oriented cross-disciplinary research is focused on the interfaces of clean air engineering/science, human health and smart/sustainable living in cities/megacities. His research builds an understanding of the formation and emission of particles, both from vehicle exhausts and non-vehicular sources.

He investigates their contribution to pollution, especially in megacity contexts. He is developing approaches to low-cost sensing and contributing to the development of exposure control technology and guidelines for policymakers to curtail pollution exposure in cities, with associated health benefits.

His current research projects and works are focused in broad multidisciplinary areas of air pollution monitoring/modelling, low-cost sensing, nature-based solutions, climate change mitigation, and developing innovative technological and passive (e.g. green infrastructure) solutions for air pollution exposure control for both developing and developed world.

He has received several prestigious awards (i.e., IEI Speical Appreciation Award 2021, IEI Young Scientist Award, Outstanding Reviewer Award – Elsevier, International Travel Grant – Government of India, etc.). With over 62 articles in top-ranked journals (h-index 23; i10-index 34; citations >1486), he has developed a network of collaborators across four continents, serving on editorial boards of several international journals. His research has featured in well-read media outlets such as the Aljajeera, Times of India, and Trinity.

Prof. Chi-Min Shu is Distinguished chair professor in Department of Safety, Health and Environmental Engineering at National Yunlin University of Science and Technology, Taiwan. Prof. Shu is a chemical engineering graduate and migrated towards the process safety upon the thrust towards the industrial safety. He initiated a prodigious impulse in the field of process safety research by establishing the Process Safety and Disaster Prevention Laboratory (PS&DPL), Taiwan. The focus of PS&DPL is primarily on loss prevention, thermokinetic hazard analysis, runaway reaction analysis, combustible gas or dust explosion analysis, risk assessment, and process-safety-related problems. In the past decade, he has received the North American Thermal Analysis Society (NATAS) Fellow Award in 2011 and in 2016 was elected as an American Institute of Chemical Engineers Fellow. In 2017, his continuous contribution was recognized by receiving the NATAS Mettler Toledo Award, the most honorable award in the field of Thermal Analysis since 1968.

Prof. Shu's expertise in process safety led him to receive invitation from various governments across the globe to be a part of investigation teams. For instance, the underground pipeline explosion that occurred in August 2014 in Kaohsiung, Taiwan, which is a well-known disaster, is one among them. Since 2005, Prof. Shu has been an advisory board member of fire accident investigation at (1) the National Fire Agency, Ministry of the Interior, (2) Taichung City, and (3) the Yunlin County Fire-Fighting Department. In 2015, he was the first scholar in the field of chemical process safety to assume an office at the Environmental Impact Assessment (EIA) Committee, accredited by Environmental Protection Administration of the Executive Yuan, Taiwan. To date, Prof. Shu has reviewed more than 500 EIA reports and attended more than 100 EIA meetings.

Prof. Shu is continuously striving to overcome challenges to industrial safety by intensively publishing papers in various respected international scientific journals.

His laboratory findings have been documented through publications (350+) in reputed science journals of loss prevention and process safety.

Prof. Shu is serving as an editorial board member in various respected scientific journals, including *Process Safety Progress (SCI)*, the *Journal of Safety Research (SSCI)*, and the *Journal of Loss Prevention in the Process Industries (SCI)*, and as an associate editor for the *Journal of Thermal Analysis and Calorimetry (SCI)* during the periods 2009–2010 and 2018–2020.

Contributors

Amarpreet Singh Arora
School of Chemical Engineering
Yeungnam University
Gyeongsan, South Korea

Hariharan B
CSIR-National Environmental
    Engineering Research Institute
    (CSIR-NEERI)
Nagpur, India

Hemant Bherwani
CSIR-National Environmental
    Engineering Research Institute
    (CSIR-NEERI)
Nagpur, India
and
Academy of Scientific and Innovative
    Research (AcSIR)
Ghaziabad, India

Rahil Changotra
School of Energy and Environment
Thapar Institute of Engineering and
    Technology
Patiala, India

Liang-Chao Chen
Department of Safety Engineering
College of Mechanical and Electrical
    Engineering
Beijing University of Chemical
    Technology
Beijing, People's Republic of China

Vikram Chopra
Department of Chemical Engineering
Indian Institute of Technology Roorkee
Roorkee, India

Alaka Das
CSIR-National Environmental
    Engineering Research Institute
    (CSIR-NEERI)
Nagpur, India

Jun Deng
School of Safety Science and
    Engineering
and
Shaanxi Key Laboratory of Prevention
    and Control of Coal Fire, Xi'an
    University of Science and
    Technology,
Xi'an, People's Republic of China

Amit Dhir
School of Energy and Environment
Thapar Institute of Engineering and
    Technology
Patiala, India

Zhan Dou
Department of Safety Engineering
College of Mechanical and Electrical
    Engineering
Beijing University of Chemical
    Technology
Beijing, People's Republic of China

Mohit Garg
Department of Chemical Engineering
Indian Institute of Technology Roorkee
Roorkee, India

Sneha Gautam
Department of Civil Engineering
Karunya Institute of Technology and
    Sciences
Coimbatore, India

Sarbani Ghosh
Department of Chemical Engineering
Birla Institute of Technology and
    Science (BITS)
Pilani Campus
Vidyavihar, India

Ankit Gupta
CSIR-National Environmental
    Engineering Research Institute
    (CSIR-NEERI)
Nagpur, India
and
Academy of Scientific and Innovative
Research (AcSIR), Ghaziabad, India

Mukesh Gupta
Environmental Science Program
Asian University for Women
and
Center for Climate Change and
    Environmental Health,
Asian University for Women,
    Chattogram, Bangladesh

Muhammad Anwar Iqbal
Department of Environmental Science
    and Disaster Management
Noakhali Science and Technology
    University
Noakhali, Bangladesh

Nafisa Islam
Environmental Science Program
Asian University for Women
Chattogram, Bangladesh

Brema Jayanarayanan
Department of Civil Engineering
Karunya Institute of Technology and
    Sciences
Coimbatore, India

Suneel Kumar Joshi
Research and Development Centre
Geo Climate Risk Solutions
Visakhapatnam, India

Md. Badiuzzaman Khan
Department of Environmental Science
Bangladesh Agricultural University
Mymensingh, Bangladesh

Amit Kumar
Department of Chemical
    Engineering
Institute of Technology
Nirma University
Ahmedabad, India

Rakesh Kumar
CSIR-National Environmental
    Engineering Research Institute
    (CSIR-NEERI)
Nagpur, India
and
Academy of Scientific and Innovative
Research (AcSIR), Ghaziabad, India

Li-Li Li
Department of Safety Engineering
College of Mechanical and Electrical
    Engineering
Beijing University of Chemical
    Technology
Beijing, Jiangsu
People's Republic of China

Maorui Li
Jiangsu Key Laboratory of
    Hazardous Chemicals Safety
    and Control
Nanjing Tech University
Nanjing, People's Republic of China

Qing-Wei Li
School of Safety Science and
 Engineering
and
Shaanxi Key Laboratory of Prevention
 and Control of Coal Fire
Xi'an University of Science and
 Technology
Xi'an, People's Republic of China

Ru Li
Department of Safety Engineering
College of Mechanical and Electrical
 Engineering
Beijing University of Chemical
 Technology
Beijing, People's Republic of China

Zhenbao Li
School of Petrochemical Engineering
Lanzhou University of Technology
Lanzhou, People's Republic of China

Horng-Jang Liaw
Department of Safety, Health, and
 Environmental Engineering
National Kaohsiung University of
Science and Technology, Taiwan,
 Republic of China

Kun-Hua Liu
School of Safety Science and
 Engineering
Xi'an University of Science and
 Technology
Xi'an, People's Republic of China

Huiping Liu
Department of Safety Engineering
Shanghai Institute of Technology
Shanghai, People's Republic of China

Zhangrui Liu
Department of Safety Engineering
Shanghai Institute of Technology
Shanghai, People's Republic of China

Tingguang Ma
Department of Safety Engineering
Shanghai Institute of Technology
Shanghai, People's Republic of China

Indra Deo Mall
Department of Chemical Engineering
Indian Institute of Technology
 Roorkee
Roorkee, India

Atsumi Miyake
Institute of Advanced Sciences/Center
 for Creation of Symbiosis Society
 with Risk
Yokohama National University
Yokohaama, Japan

Jo Nakayama
Institute of Advanced Sciences/Center
 for Creation of Symbiosis Society
 with Risk
Yokohama National University
Yokohaama, Japan

Trent Parker
Department of Chemical
 Engineering
Texas A&M University
College Station, Texas

Aditya Kumar Patra
Department of Mining Engineering
Indian Institute of Technology
 Kharagpur
Kharagpur, India

Bruno Pavoni
Dipartimento di Scienze Ambientali
Informatica e Statistica, Università
    Ca' Foscari
Venezia, Mestre, Italy

Dhasarathan R
Department of Civil Engineering
Karunya Institute of Technology and
    Sciences
Coimbatore, India

Nazifa Rafa
Environmental Science Program
Asian University for Women
Chattogram, Bangladesh

Himadri Rajput
School of Energy and Environment
Thapar Institute of Engineering and
    Technology
Patiala, India

Satya Prakash Sahu
Mahanadi Coalfields Limited
Coal India Limited
Kolkata, India

Mohammed Abdus Salam
Department of Environmental Science
    and Disaster Management
Noakhali Science and Technology
    University
Noakhali, Bangladesh

Mahmud Hossain Sumon
Department of Soil Science
Bangladesh Agricultural University
Mymensingh, Bangladesh

Ryan Shen
Department of Chemical Engineering
Texas A&M University
College Station, Texas

Zeyang Song
Department of Safety
Xi'an University of Science and
    Technology
Xi'an, People's Republic of China

Guan-Yu Suo
Department of Safety Engineering
College of Mechanical and Electrical
    Engineering
Beijing University of Chemical
    Technology
Beijing, People's Republic of China

Jin-Fu Tao
Department of Safety Engineering
College of Mechanical and Electrical
    Engineering
Beijing University of Chemical
    Technology
Beijing, People's Republic of China

Sayed Mohammad Nazim Uddin
Environmental Science Program
Asian University for Women
and
Center for Climate Change and
    Environmental Health
Asian University for Women
Chattogram, Bangladesh

Qingsheng Wang
Department of Chemical
    Engineering
Texas A&M University
College Station, Texas

Yang Xiao
School of Safety Science and
    Engineering
and
Shaanxi Key Laboratory of Prevention
    and Control of Coal Fire
Xi'an University of Science and
    Technology
Xi'an, People's Republic of China

Shen-Qing Xuan
Department of Safety Engineering
College of Mechanical and Electrical
    Engineering
Beijing University of Chemical
    Technology
Beijing, People's Republic of China

Jian-Feng Yang
Department of Safety Engineering
College of Mechanical and Electrical
    Engineering
Beijing University of Chemical
    Technology
Beijing, People's Republic of China

Zhijin Yu
Department of Safety
Xi'an University of Science and
    Technology
Xi'an, People's Republic of China

Hui Zhan
Department of Safety Engineering
College of Mechanical and Electrical
    Engineering
Beijing University of Chemical
    Technology
Beijing, People's Republic of China

Hao Zhang
School of Safety Science and
    Engineering
Xi'an University of Science and
    Technology
Xi'an, People's Republic of China

Xiaoliang Zhang
Department of Safety Engineering
Shanghai Institute of Technology
Shanghai, People's Republic of China

Jingyu Zhao
Department of Safety
Xi'an University of Science and
    Technology
Xi'an, People's Republic of China

Mei-Chen Zhou
Department of Safety Engineering
College of Mechanical and Electrical
    Engineering
Beijing University of Chemical
    Technology
Beijing, People's Republic of China

.

Introduction

The journey towards the safe, sustained and excellent process safety requires a continual effort that includes individual participation, industrial support, and policy frameworks. The key personnel in an industry ensuring the process safety includes (Nwanko et al., 2020)

- Senior and middle level managers help to provide the men and resource
- Technical department responsible for designing and evaluation of process
- Operating personal to operate instruments and troubleshoot in case of faulty functioning
- Mechanical personal providing the reliable machinery
- R&D for their continued efforts for the betterment of the operations and scalability.

The process safety programs are beyond the traditional ones such as wearing mask, safety glass, hard hats, gloves etc. The effective process safety addresses the functional errors that may lead to toxic releases, fire accidents, oil spilling, runaway reactions and so on resulting the catastrophic consequences to human, environment, and equipment (Klein and Vaughen, 2017).

A thorough knowledge on organization's process safety, goals and systems helps to achieve the safe and reliable operations thereby contributing both exceptional standards of safety performance and optimized manufacturing of product streams.

WHAT IS A PROCESS AND WHAT IS SAFETY?

It is a sequence of steps and decisions involved in the way to accomplish the task. In other words "Process" is the actions happening while something is happening or being done. Process is the term which is not only confined to the chemical process industries, rather used in daily life. For instance, the cleaning of kitchen requires a series of steps which is defined as a cleaning process. On a large scale, granting permission to establish an industry with prior concerns of different departments.

On the other hand, "safety" is a state of being protected from potential harm or something that has been designed to protect and prevent them. Similar to the process safety exists everywhere, for example wearing a seat belt while driving helps to save a life in case of any accident. Another example related to industries is using safety jackets, gloves, hard hats, and googles at the work environment prevents the possible dangers.

A combination of these things i.e., "process", "safety" is quite important when it comes to industry. The technical definition of process safety in terms of

TABLE 1

Series of Accidents in Process Industries

Incident	Country	Year	Fatalities
Flixborough	England	1974	28
Roland Mill	Germany	1979	14
Bhopal	India	1984	2500+
Chernobyl	Ukraine	1986	30+
Piper Alpha	North Sea, Scotland	1988	167
Pasadena	USA	1989	24
Toulouse	France	2001	30
Texas City	USA	2005	15
Imperial Sugar	USA	2008	13
Deepwater Horizon	Gulf of Mexico	2010	11
Amuay Refinery	Venezuela	2012	47
West Fertilizer	USA	2013	14
Kaohsiung gas explosion	Taiwan	2014	32
Binhai	China	2015	173
Chittagong Urea Fertiliser Limited	Bangladesh	2016	-
Philadelphia Refinery Explosion	USA	2019	-
LG Polymers	India	2020	12

industry is that managing potential hazardous materials, energies and process to prevent serious incidents, and injuries. To ensure the process safety, one requires a clear determination, continuous dedication, and strong commitment towards the organization. Unfortunately, majority of nations and the industries lack the process safety frameworks resulting frequent incidents now and then. A detailed list of major disasters is presented in Table 1.

The aforementioned incidents are samples and countable, while there are plenty of disasters and incidents which caused catastrophic damages. Despite of the stringent regulations and the established frameworks, law enactments, the industrial accidents are becoming quite common causing serious damage to both health, economy, and nature. The prime reason for these repeated incidents is lack of process safety culture in industrialization, in case of presence but not followed to the fuller extent.

ROLE OF CHEMICAL INDUSTRIES IN WORLD ECONOMY

Nowadays, each nation is striving to improve the economic status via rapid industrialization. Among which the massive populous nations, where the demands are diversified and complex to cater the needs. This rapid surge in growing needs and probes swift industrialization especially the chemical industries. According to the reports of UNEA – four chemical industries contribute enormously to the global economy (ICCA, 2019). Some of the key facts are

- Chemical industries generated an astonishing revenue of $5.7 trillion dollars through direct, indirect and induced impacts which is equivalent to 7% of the global GDP and 120 million jobs.
- Specifically, the processing industries added $1.1 trillion to world GDP and employed 15 million people, fostering as fifth largest global manufacturing sector.
- Further, every penny generated by the chemical industry, generated $4.2 across the globe.
- Another stunning factor related to the chemical process industries estimated to spend $3 trillion in buying raw material and services to manufacture the products. This supply chain is contributing roughly $2.6 trillion to global GDP, and also supporting 60 million jobs.
- The global chemical industry investments touch an all-time high of $51 trillion in R&D, supporting 1.7 million jobs and $92 billion USD in economic activity.

NEED OF THE HOUR

In order to overcome this potential endangers, effective process safety programs are highly necessary. Implementing these programs benefited many industries time to time in controlling the risks associated with process hazards, prevention of serious incidents or mitigating the potential consequences to avoid more severe accidents. The basis for this process safety program is classified into two sections namely materials and processes derived from the standard occupational safety health programs (Knegtering and Pasman, 2009). Further, three key parameters are identified for the process safety, and the strength of interlinking between these parameters defines how safe the process is. However, every successful organization should have three baseline requirements to reach the objective of safety and they are as follows:

EFFECTIVE MENTORING

The success of any organization is majorly reliant on its underlying safety culture and its leadership capabilities. In detail, a strong safety culture assures deep commitment of safety, health and environment in line with the organizational goals; however, the weak culture dooms the safety aspect due to conflict of interest in production schedules and lowering operational costs (Vinodkumar and Bhasi, 2010). The inherent safety cultures are strongly associated to the leadership, personal commitment that helps the industry to be sustainable and accountable. This involves the commitment of management at levels starting from senior level managers to the front-line employees.

PROCESS SAFETY SYSTEM

This is the second most important thing that contains the integrated management framework to ensure the identification of potential hazards and risks, thoroughly evaluated and controlled. It is always better to start with the existing system and make sure that it has the minimum essential level of practice (Zhang et al., 2018). But, with these base line systems many of the significant process hazards are covered up and requires further evaluation and addition to the existing systems. In some instances, additional requirements beyond the regulatory systems may be necessary to attain adequate risk reduction due to process hazards.

OPERATIONAL DISCIPLINE

This defines the human behavior inside the organization i.e., whether the personal is following the guidelines, standard procedures, systems in a right manner each time for a safe and reliable operation or not (Bitar et al., 2018). Regardless of how strong, comprehensive and well-designed safety programs may be, day-to-day discipline and the individual's anticipation always matter for the successful transformation of written policies from concept to reality. The key refrainment of the process safety program is lack of organizational discipline, lack of knowledge, lack of personal commitment, and lack of awareness in daily activities, which would lead to the failure of a system.

In summary, the coordination among these three tools can make or weak the organization. Besides, the scope and depth of each tool varies and highly process dependent.

WHY IS IT IMPORTANT?

In general, every company has a standard framework of the process safety programs that solely meets the regulatory requirement. It is mandatory, but may be ineffective because the regulations may not adequately manage the process risks at the preset facility. In other words, the process safety program should be applicable to every point in the industrial facility that includes the life cycle assessment which is hard to manage but if implemented produce great results in preventing serious incidents and losses (Zwetsloot et al., 2020). Unfortunately, no organization is steeping towards the comprehensive process safety regulation framework in line with the existing policies. They are mainly concerned towards meeting the basic requirements which is resulting into some or other incidents every now and then. For instance, in case of catastrophic accident at the operational facility is very stressful even for an established emergency response team (Cournoyer et al., 2013). In case of any loss of lives, it may seriously affect the company's survival, or in case of property loss there may be a huge opposition from the people and they oppose the cooperation and support to the facility.

BUSINESS IMPACT

The entire life of organizational business is relied upon the safe and sustainable process operations. A successful organization with amicable process safety programs may hit the following targets (Ismael et al., 2019).

QUESTION OF SUSTAINABILITY AND REPUTATION

Reputation holds the key for any successful business or organization especially having reputation in the market and the surrounding community is invaluable. An organization with high reputation will have a healthy support, cordial relations among the employees, neighbor communities, customers, suppliers, regulators etc. In case of one failure incident, it will lead to severely damage the connection chain of relationships thereby leading to regulatory fines, hampered production capacity, interruption in product supply, declining market share values, most importantly ecological imbalances and so on (Cornejo et al., 2019). Hence, there is a definite need of effective process safety programs for any company or industry to be in limelight and sustain for a longer duration. Further, some critical factors are identified how the business if benefited through process safety.

EMPLOYEE PROTECTION

The primitive object is to provide a safer working conditions to employee to attract and extract high level of engagement (Singh and Mishra, 2020). Effective process safety programs play a crucial role for ensuring the safety in the work environment and avoids fatalities, serious injuries, but high turnover rates of employees are associated with high training expenditure.

ADVANCED MANUFACTURING METHODOLOGIES AND LEVERAGING COSTS

A safer work environment requires huge investment and also enhance the production capabilities. Further the reliable operations are essential to high quality products, asset productivity, low capital, operating costs, and maintenance costs (Fernandez and Perez, 2015). Finally, in case any unforeseen dangers incur incident costs such as property damage, emergency response and recovery, investigation, regulatory penalization, high insurance premiums can be avoided with effective process safety programs.

ENVIRONMENT CONCERNS

On the spin side everything we do, starting from what we eat to the energy we use, has a solid impact on globe – but it wasn't always that way. However, the industrial revolution turned the tables for the process industries and raised serious unanswered questions like how did the industrialization impacted the ecological imbalances and to what extent the environment can be restored to the normal position (Gobbo et al., 2018). It is not possible that a single day passes by without newspaper and television is not exposed to a litany of global environmental information: Acid rain resulting from atmosphere SO_2 and NO_x, global warming in relation to increased atmospheric CO_2, toxic nuclear waste disposal, contamination of oceans, etc., which are more prevalent (Ahmad et al., 2020). Localized disclosures include a leaky underground fuel tank, increased urban air pollution, the flow of nitrates, nitrites, pesticides, and industrial solvents into ground water supplies, and the poisoning of soil, water, and air by toxic industrial effluents and gases. It then comes as no surprise that our generation of mankind has become afflicted with the pervasive and acute diseases. There are copious indications, though, most of the chemicals are tainted or dissoluted into a very fragile ecology, despite efforts by environmental ethicists and the media to convince us otherwise. But for most scientists involved in reduction of environmental contaminants, there is indeed room for improvement in virtually all spheres.

IMPACT ON ECO-SYSTEM

Each time an impact assessment is made, one has to look for the ascribing parameters and assess the possible damage. Accidents are quite unwarned, whereas the damage is outside the box mainly because of vulnerability. For instance, a fire accident in one section of the industry may blowout to various sections if not handled properly at the initial stages by the plant managers. All this prevention mindset comes from preparedness to the catastrophic situations which is again a function of strong training skill set. Further, the outset fire may release toxins to the environment that may seriously affect air, water, and soil to such an extent where there won't be any living species possible for generations all together. In addition to this, the fall out gaseous clouds or toxins migrate to the neighboring places or effect the local fauna or flora by interacting with the living systems. For instance, the caustic gases cause wilting, necrosis, and chlorosis, or forbid the photosynthesis mechanism of the plants. Also, these toxic gases accompany acid rains, which are another serious threat to the soil, air, and water bodies. Similar suffering is seen in Canada, one of most advanced nations worldwide whose lakes, pools, and puddles receive acid rain due to heavy industrial activity from the Unites States along Canada's Southern border. Similarly, cloud smog in most populous nation, China, degraded the visibility levels causing serious distress to the breathable oxygen content (Halim and Mannan, 2018). Another instance is in 2019 in India, the major cosmopolitan cities had the threatening level of air quality and breathable oxygen content due to excessive industrial pollution and the vehicle pollution. Furthermore, European nation experience the trans-continental migration of pollutants by the wide spread of nucleoside in the European nations due to Chernobyl accident. On the other hand, it is very well known that single gram of soil has trillions of bacteria that participate in bio-geo-chemical cycles in degrading the organic matter to form simple compounds otherwise known as nutrients (Altmann, 2008). In case of mine failures or especially coal-based thermal plants waste product coal fly ash upon storage in ponds contains several heavy metals will join the aquifer contaminating the ground water bodies.

Summarizing these facts, it is clear that the process safety and the ecology are closely related to each other. However, proper coordination among these needs to be established by the human activities such as controlled operations, full dedication toward the safer and optimized process releasing less toxicants, preventing catastrophic accidents, and avoiding unnecessary leakages to ground systems for the safer, sustainable, and balanced ecology.

REFERENCES

Ahmad, S.I., Yunus, N.A., Akbar, Md.R., Hashim, H., Rashid, R. Solvent design and inherent safety assessment of solvent alternative for palm oil recovery. Journal of Loss Prevention in the Process Industries. 2020, 65, 104120.

Altmann, S. Geochemical research: A key building block for nuclear waste disposal safety cases. Journal of Contaminant Hydrolygy. 2008, 102, 174–179.

Bitar, F.K., Chadwick-Jones, D., Lawrie, M., Nazaruk, M., Boodhai, C. Empirical validation of operating discipline as a leading indicator of safety outputs and plant performance. Safety Science. 2018, 104, 144–156.

Cournoyer, M.E., Garcia, V.E., Sandavol, A.N., Peabody, M.C., Schreiber, S. A behavior-based observation program's contribution to a nuclear facility operational safety. Journal of Chemical Health and Safety. 2013, 20, 23–29.

Cornejo, C.P., Puente, E.Q., Bautista, J., Garcia, D. How to manage corporate reputation? The effect of enterprise risk management systems and audit committees on corporate reputation. European Management Journal. 2019, 37, 505–515.

Fernandez, F.B., Perez, M.A.S. Analysis and modeling of new and emerging occupational risks in the context of advanced manufacturing processes. Procedia Engineering. 2015, 100, 1150–1159.

Gobbo, J.A., Busso, C.M., Gobbo, S.C.O., Carreao, H. Making the links among environmental protection, process safety, and industry 4.0. Process Safety and Environmental Protection. 2018, 117, 372–382.

Halim, S.A., Mannan, M.S. A journey to excellence in process safety management. Journal of Loss Prevention in Process Industries. 2018, 55, 71–79.

International Congress and Convention Association 2019. 58th ICCA Congress, Houston, USA.

Ismael, S., Herrera, S., Donate, M.J. Occupational safety and health (OSH) and business strategy: The role of the OSH professional in Spain. Safety Science. 2019, 120, 206–225.

Klein, J.A., Vaughen, B.K. Process Safety. Key Concepts and Practical Applications. 2017. CRC Press, Taylor and Francis, Boca Raton, USA.

Knegtering, B., Pasman, H.J. Safety of the process industries in the 21st century: A changing need of process safety management for a changing industry. Journal of Loss Prevention in the Process Industries. 2009, 22, 162–168.

Nwanko, C.D., Theophilus, S. C., Arewa, A.O. A comparative analysis of process safety management (PSM) systems in the process industry. Journal of Loss Prevention in the Process Industries. (In Press). 104171.

Singh, A., Misra, S.C. A dominance based rough set analysis for investigating employee perception of safety at workplace and safety compliance. Safety Science. 2020, 127, 104702.

Vinodkumar, M.N., Bhasi, M. Safety management practices and safety behavior: Assessing the mediating role of safety knowledge and motivation. Accident Analysis & Prevention. 2010, 42, 2082–2093.

Zhang, Z., Wu, Z., Durand, H., Albalawi, F., Christofides, P.D. On integration of feedback control and safety systems: Analyzing two chemical process applications. Chemical Engineering Research and Design. 2018, 132, 616–626.

Zwetsloot, G., Kampen, J.V., Steijn, W., Post, S. Ranking of process safety cultures for risk-based inspections using indicative safety culture assessments. Journal of Loss Prevention in the Process Industries. 2020, 64, 104065.

1 Bow Ties of Process Safety and Environmental Impact Transition from Theory to Practice

Hui Zhan, Zhan Dou, Guan-Yu Suo, Shen-Qing Xuan, Jin-Fu Tao, Jian-Feng Yang, Liang-Chao Chen, Li-Li Li, Ru Li, and Mei-Chen Zhou
Department of Safety Engineering, College of Mechanical and Electrical Engineering, Beijing University of Chemical Technology, Beijing, People's Republic of China

CONTENTS

1.1 INTRODUCTION

Bow-tie analysis (Bow-tie, referred to as BT) has this name because the shape of the diagram resembles a bow tie. It is a method of risk analysis that shows the relationship between the accident (top event), cause of the accident, path leading to the accident, consequence of the accident, and also the measures to prevent the accident[1].

The prototype of the bow-tie analysis was formed in the early 1970s[2]. The original method was composed of an accident tree and an event tree (causal diagram), and the key events that connected the two. This could be regarded as the earliest

DOI: 10.1201/9781003140382-1

diagram of bow-tie analysis. At the same time, the concept of using barriers to control risks had gradually formed. In the developing process of bow-tie analysis, the combination of an accident tree, an event tree, and the concept of barriers had constructed a more complete bow-tie analysis we have been seeing so far.

After the emergence of simple diagrams of bow-tie analysis, the academia gradually established different bow-tie analysis models, including qualitative (causal relationship model), quantitative (complete accident tree, event tree model, occupational risk model), and some other models. Although the quantitative model can be found in many research literatures, it is not widely used at present due to insufficient historical data used in the bow-tie analysis.

In the early 1990s, the Royal Dutch Shell Group established the Hazard and Impact Management Program (HEMP), and the bow-tie analysis was the core of this management program[3]. Since then, the bow-tie analysis has received more attention and development in the oil and gas industry and has gradually been applied in chemical and pharmaceutical fields[4].

The main purpose of the bow-tie analysis is to provide safety experts with a means to communicate with operation experts on safety issues discovered. Therefore, operation experts can find "protective measures" and "emergency measures" while safety experts can maintain in a neutral position and analyze whether there is a high-value approach for the organization to manage the risks. At the same time, the bow-tie diagram also demonstrates the effectiveness of the organization's HSE system in a concise graphic, allowing employees and managers to clearly understand how major hazards may occur and what protective measures can be used to prevent disasters.

Because it is a combination of an accident tree and an event tree, it not only eliminates the shortcomings of the two diagrams but also clearly shows the scene analysis of the accident. The bow-tie diagram takes the top event as the central node, which is also called energy/material leakage (Loss of Containments, LOCs in short). The fan-shaped structures on the left and on the right sides of the bow-tie diagram are the accident tree and the event tree. The accident tree on the left shows how "preventive measures" can block the "threats" from developing toward the direction of LOCs and leakage accidents. The event tree on the right shows how "emergency measures" can eliminate or reduce the subsequent negative effects caused by LOCs so that they will not develop into "disasters". Therefore, the top event should be used as a core to deduct the left and the right sides of the bow-tie diagram. To the left, the threats are identified to cause the top event. To the right, the final possible consequences of the top event are determined.

1.2 KEY ELEMENTS

The bow-tie diagram takes the top event as the central node, and the fan-shaped structures on the left and the right sides are the accident tree and the event tree, respectively. The key elements of the accident tree on the left are "hazards" and "preventive barriers" that intercept "threats". The key elements of the event tree on the right are "passive barriers". A summary of the specific terms in the bow-tie diagram is shown in Table 1.1.

TABLE 1.1
Summary of the Specific Terms in the Bow-Tie Diagram

Term	Definition
Hazard	It may harm people and the environment, cause property damage or loss, and adversely affect reputation. The beginning of any bow-tie diagram is "hazard". Hazards can cause damage. They are something inside, around, or part of an organization.
Top event	The selected reliable scenarios related to the hazardous release.
Threat	The possible reasons that lead to the release of the hazard and the occurrence of the top event.
Consequences	The event or a chain of the events due to the release of a hazard, usually resulting in some forms of damage.
Barriers	Any measures taken against certain undesired forces or intentions to achieve or maintain the desired state. A proactive barrier will terminate certain scenarios so that the threat will not lead to a top event. Reactive barriers are passive and can prevent top events from causing more serious consequences.
Escalation Factor	Conditions that make preventive measures/emergency mitigation measures ineffective or surpassed. These conditions may increase risks.
Escalation Factor Barriers	A barrier set up to manage conditions (factors) that lead to the failure of the active preventive barrier or the reactive barrier.

1.3 ANALYSIS STEPS

The bow-tie analysis can be conducted in the following steps:

1. Identify hazards: identify potential hazards in the chemical process, such as flammable and explosive substances, toxic substances, radioactivity, other dangerous chemicals, rotating machinery, high-altitude operations, hoisting operations, etc.
2. Identify top events: identify top events caused by hazardous release.
3. Identify threats: after determining the top event, brainstorming is used to find out all possible causes of the top event. Then place them on the far left side of the bow-tie diagram and connect them to the top event.
4. Identify the consequences: a top event usually causes multiple consequences, such as: fire, explosion, environmental pollution, toxic gas leakage, casualties, and downtime.
5. Identify active preventive barriers: in order to prevent and reduce the possibility of threats triggering the release of hazards, effective barriers (active preventive barriers) should be set up in the path of threats

reaching the top event. The number of preventive barriers will depend on how effective they are to control threats. When the effect of preventive barriers is limited, more preventive barriers should be added. When the types of preventive barriers are the same, other types of barriers should be considered.

6. Determine passive response barriers: after a top event occurs, it is particularly important to determine all measures to reduce the consequences. These measures minimize the impacts of these consequences and are called passive response barriers.

7. Determine escalation factors: in certain circumstances, specific reasons or conditions can make the active prevention barrier and passive response barrier fail. These factors become escalation factors, and the determination of escalation factors will affect the effectiveness of the safety barrier.

8. Determine the barrier of upgrade factors: a series of measures will be taken to prevent the failure of upgrade factors.

Through the bow-tie analysis, we can intuitively see the weak links of the system, which is the chain of events with missing or insufficient barriers. The organization can take effective measures to ensure the safety of the system. For human factors, the bow-tie analysis can generate a list of all safety-critical tasks required to manage "soft" barriers, and assign each task to a person or a job role, thereby creating a "safety-critical role" within the organization. Identify those who will assume safety-critical roles, and the organization should consider providing appropriate training for them. For equipment factors, the bow-tie analysis can generate a "hard" barrier list of HSE key equipment and systems, and organize it into several "HSE key equipment and system groups". Topics covered by these groups include structural integrity, process leak prevention, monitoring, shutdown, etc. The performance requirements of each HSE key equipment and system group or subgroup can be developed in the performance standard document, which will determine any necessary inspection, maintenance, and performance verification tasks. The practical way to determine the key HSE equipment and systems is to match them with the equipment belonging to the groups or the subgroups in the factory/facility. In this way, inspection and maintenance requirements can be set at the equipment level, and each key HSE equipment and system can be well maintained.

1.4   ADVANCES IN RESEARCH AND APPLICATION

All walks of life are striving to identify and control the inherent risks in their own fields. They try to understand the methods of systematic analysis and risk assessment in order to reduce risks. The bow-tie approach is a popular evaluation method in process industries, such as oil, natural gas, and mining. Scholars in related fields have used the bow-tie approach to conduct a series of studies on issues in the chemical and petroleum industries. From the viewpoint of process safety and environmental impact, the accident tree on the left side of the bow-tie diagram is designed to analyze a series of basic events in process safety, and the event tree on the right side is designed to analyze the consequences to people and the environment after the top event occurs. The bow-tie approach can qualitatively analyze the system and identify the entire chain of events, the security barriers, and the weak links of the system. It can prevent dangerous events from happening and mitigate the consequences when top events do occur. Wilday et al.[5] used the bow-tie approach for the purpose of hazard identification in the carbon capture and storage process. Chen et al.[6] used the bow-tie approach to identify the sources of environmental risks. Wengang et al.[7] established a bow-tie approach to comprehensively analyze the causes and consequences of oil spills during tanker berthing and loading operations, and the risk factors in the entire process were identified. Bilal et al.[8] analyzed the causes and consequences of explosion using the fault tree method and the bow-tie approach. Baker et al.[9] used the bow-tie approach in seven hazardous operational activities of the oil sands tailings operations as a case study to visually identify unwanted events potential threats, consequences, and the controls used to prevent incidents from occurring. The bow-tie approach was used to conduct risk assessment of a wastewater treatment plant in Moorchekhort Industrial Park (MIC). Multilayered bow-tie approach was adopted to explore the causes of the benzene tank explosion, along with the corresponding consequences, and preventive measures[10]. Wang et al.[11] proposed prevention and control measures for oil and gas leakage in floating production, storage, and offloading. They also discussed the impacts of oil and gas leakage on the marine environment. Based on the bow-tie approach, a generic framework for quantitative risk management of storage fires was established via the analysis of previous storage fires[12]. The bow-tie approach combining the accident tree and the event tree can also be quantitatively analyzed. The probability (or frequency) of basic events can be obtained from the historical fault data. As for quantitative risk assessment of dust explosions, Yuan et al.[13] brought forward a bow-tie-approach-based framework via the review and the analysis of previous major dust explosions. Aziz et al.[14] obtained failure data of different elements from ship maintenance logs and accident records, which can estimate the frequency of incidents and the failure rate of safety fences. They also conducted quantitative risk assessment using the proposed bow-tie approach. However, historical failure data of basic events in some scenarios were uncertain and not easy to obtain. In order to solve the problem of insufficient original probability data in probabilistic risk assessment, the corresponding parametric uncertainty was dealt with in a Monte Carlo simulation following the bow-tie approach, which was the technique mostly used in PRA studies[15]. Fuzzy Set Theory (FST) was applied to address the uncertain risk data for the occurrence probability (OP) and consequence severity (CS) in the risk analysis process. Tang et al.[16] and Li et al.[17] established a fuzzy bow-tie model (FBTM) to obtain a quantitative risk assessment result according to the analysis of the FFTA and FETA. Fuzzy logic was used to deal with vagueness and imprecision of expert judgment[18]. In the research of Ferdous et al.[19], fuzzy set and evidence theory were incorporated to assess data uncertainties. In the fault tree analysis, it is assumed that each basic event is independent, but in reality, each basic event may have mutual influence, which will result in inaccurate quantitative analysis. To solve this problem, many researchers have introduced the Bayesian network model into the bow-tie approach. Badreddine and Amor[20] developed a new Bayesian approach to construct the bow-tie approach from real data, and they also demonstrated that by adding a new numerical, this bow-tie approach could be improved to help the implementation of appropriate preventive and protective barriers in a dynamic manner. A Bayesian bow-tie approach was proposed by mapping and connecting the bow-tie approach to Bayesian networks[21]. Yuan et al.[22] proposed a dust explosion risk analysis method based on Bayesian networks, using the bow-tie approach to better depict the logical relationship between the factors and consequences of dust explosions. Bow-tie and Bayesian network models were established to conduct risk assessments of a cotton storage fire[23]. Based on some case studies and experts' experience of typical natural gas pipeline accidents, Fang et al.[24] established a bow-tie approach combined with Bayesian networks. The bow-tie (BT) approach was used to establish a systematic well integrity failure model, this was then combined with a Bayesian Network (BN) to assess risks[25]. When the Bayesian network was used to establish a bow-tie approach, there was also a situation where the failure data was uncertain. A Bayesian network (BN) for the leak safety evaluation of heavy oil gathering pipelines was established via mapping from a bow-tie (BT) approach[26], and information diffusion theory was combined with FST to obtain the failure probability. Focusing on the fuel-liquid road transportation system, Li et al.[27] established a bow-tie approach based on the Bayesian Network (BN), and combined expert judgment and FST to estimate the probability. A Bayesian Network (BN) model for gasholder leakage was established by converting BT to BN (BTBN), and the fuzzy logic based on expert judgment was applied to cope with the uncertainty of the failure data[28]. A hybrid method consisting of bow-tie-Bayesian network (BT-BN) analysis and fuzzy theory was proposed, and a probabilistic safety assessment (PSA) was run for a tunnel section in the Wuhan metro system[29].

Combining bow-tie approach with Bayesian network could overcome the limitation of the traditional bow-tie approach in the dynamic safety analysis. In the study of Khakzad et al.[30], the failure probability of safety barriers of the bow-tie approach was periodically updated using Bayes' theorem. As new information became available over time, an updated bow-tie approach was used to estimate the posterior probability of the consequences which in turn could result in an updated risk profile. Chang et al.[31] proposed a bow-tie approach combined with Bayesian network to conduct dynamic risk analysis on the fracture failure of deepwater drilling risers. Gao et al.[32] proposed a quantitative dynamic risk assessment model for gasifier systems based on dynamic bow-tie (DBT) and dynamic Bayesian networks (DBN). At the same time, in order to deal with the uncertainty of the failure data, fuzzy numbers and defuzzification methods were used to convert the language of experts into failure rates. Khakzad et al.[33] combined the bow-tie approach with Bayesian networks to conduct dynamic security analysis of multiple accident scenarios.

1.5   ADVANTAGES AND DISADVANTAGES

The bow-tie analysis has become a hotspot of application and research, and is closely related to the change of risk management methods. Early risk management methods are based on the standards proposed by the regulatory authorities, and they are also known as regulatory-based safety management methods. Its main content is to identify system risks, mainly in qualitative evaluation. The bow-tie analysis based on direct causality diagram can be applied to this type of method. However, this method is based on regulatory standards and always regards the insufficiency of the management system as the main cause of the accident; in addition, in the qualitative evaluation, all risk factors have the same importance, and it is difficult to guide the optimal allocation of management resources. Therefore, a risk-based safety management method has come into being. It uses the risk itself as an indicator and combines quantitative evaluation methods with risk acceptance criteria in order to rank the hazards and urgency of risks. It also provides support for the formulation of accurate, efficient, and economical risk management measures.

In this safety management framework, the advantages of the bow-tie analysis begin to emerge:

1. The bow-tie analysis puts risk as a top event in the center of the analysis chart, which fully embodies the risk-based evaluation concept in form.
2. The bow-tie analysis constructed by the accident tree and the event tree is very quantifiable, and it can meet the needs of quantitative risk assessment.
3. Compared to a single analysis of an accident tree or an event tree, the bow-tie method has the ability to identify risks and describe the entire process of risk events from the causes to the consequences; the bow-tie method has considered measures to prevent and mitigate the consequences of accidents in the evaluation process, namely, the role of barriers. Combined with quantitative evaluation methods, it can also provide a more accurate guidance for the formulation of safety measures.
4. The method of visualizing the bow-tie diagram allows the system and barriers to be presented intuitively, which is convenient to identify the weak links of the system.

In summary, in addition to qualitative and quantitative evaluation, the bow-tie method is superior to other commonly used risk evaluation methods. Based on its basic functions, it also effectively covers all the main contents of the risk management process, including risk identification, risk evaluation, and risk prevention. The bow-tie approach not only can serve as a consequence mitigation but it is also a quantitative risk assessment technology that is most compatible with risk-based safety management methods. Research projects on risk assessment methods funded by organizations and countries, such as the European Union and Norway, have successively used the bow-tie approach integrated with the accident tree and the event tree as the basis of evaluation. These projects reflected the recognition of bow-tie approach among domestic and foreign scholars in the risk-based safety management system. Scholars were expanding the scopes of applications of bow-tie approach by making it suitable for cutting-edge issues in other research fields, such as missing data and dynamic evauation. When applying the traditional bow-tie method in the quantitative risk assessment, insufficiency of historical data, non-independence of data, and the limitations of expert knowledge might generate uncertainties in the evaluation results. In order to solve this problem, the existing research combined the bow-tie method with fuzzy theory, using fuzzy data to replace accurate probability data and avoiding the uncertainties produced by data and the experts. In addition, the bow-tie method could be combined with the Bayesian network, and a dynamic risk assessment method would be established through the calculation of failure probability. This method could further optimize the results of the risk analysis.

1.6   CASE STUDY

In the case of a particularly serious explosion of Jiangsu Xiangshui Tianjiayi Chemical Co., Ltd. "3.21", the accident was analyzed in the bow-tie analysis. This analysis method mapped the relationship among hazard sources, harmful factors, preventive control measures, overhead incidents, mitigation measures, and consequences in the diagram of a bow tie, which was extremely useful for device risk control, grassroots safety management, and field operations. With strong applicability and operability, the bow-tie analysis had set an example of the transition from theory to practice.

Examining the production, collection, and storage of nitrification waste in Tianjiayi Company, it was understood that the company's nitrification waste had a certain degree of defects and hidden dangers in the process.

As mentioned above, the bow-tie diagram was composed of the central event and its connected parts, which were the accident tree on the left and the event tree on the right. Therefore, when analyzing an actual case, the accident tree and the event tree were combined in order to comprehensively analyze and evaluate the accident, following the key analysis steps of the bow-tie analysis.

1.6.1   Determine the Top Event and Identify of Harmful Factors

Through accident analysis, it was identified that "nitrification spontaneous combustion" was the top event. The main reasons for spontaneous combustion of nitrate waste were:

1. **Human unsafe behaviors**
   a.   Failure to label as required.
   b.   Failure to declare nitrification waste in time.
   c.   Deliberately conceal the situation of nitrification waste.

2. **The unsafe state of the object**
   a. Storage violations of nitrification waste
      - Large storage of nitrification waste.
      - Poor ventilation in the old solid waste warehouse.
      - Long storage time of nitrification waste.
      - The release of heat from the self-dissociation of nitrification waste.
   b. Illegal disposal of nitrification waste
      - Illegal burial and transfer of solid waste.
      - Sneaky discharge of nitrification wastewater.
      - Illegal smuggling of waste materials.
      - Illegal incineration of nitrification waste.
3. **Management defects**
   a. The person in charge has not been assessed.
   b. The person in charge has no corresponding management ability.
   c. Confusion in safety production management.
   d. The intermediary agency issues false documents.

1.6.2 ACTIVE PREVENTIVE MEASURES

1. Set temperature alarm for reaction material temperature.
2. Regularly inspect the materials in the solid waste warehouse.
3. Strictly conduct daily safety management inspections of process safety.
4. Strictly review the process flow and the design of occupational health and safety.

Figure 1.1 shows a fault tree analysis with "nitrate spontaneous combustion" as the top event.

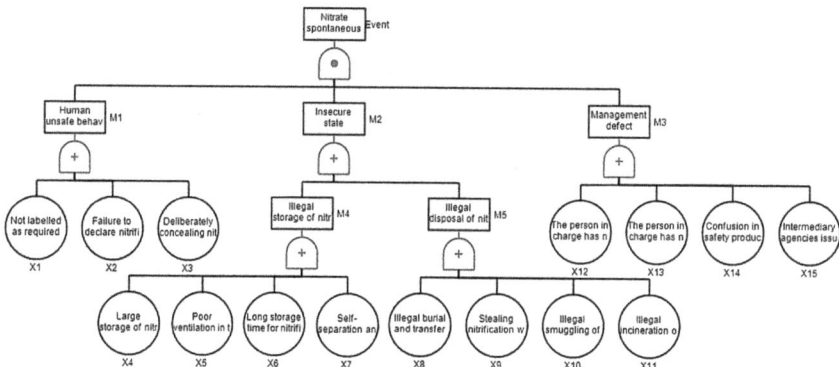

FIGURE 1.1 Fault tree analysis with "nitrate spontaneous combustion" as the top incident.

1.6.3 Analysis of Accident Consequences

The consequences of the accidents caused by the loss of control after the spontaneous combustion of nitrates can be analyzed in the event tree in following aspects:

1. Water quality: the leakage of benzene, p-phenylenediamine and other materials will pollute the water source.
2. To the air: dense smoke caused by fires and explosions has a greater impact on air quality; benzene is volatile and will produce volatile toxic and harmful gases after the leakage.
3. To ecology: the concentration of sulfur dioxide and nitrogen oxide threatens ecological security.

1.6.4 Passive Response Measures

The main mitigation measures against the consequences of spontaneous combustion of nitrification waste are:

1. Extinguish the spontaneous fires in the facilities.
2. Implement precipitation plans and emergency monitoring.
3. Conduct concentration detection of leaking toxic substances.
4. Contain and intercept of sewage and firefighting wastewater.
5. Activate sewage technology connection in the park.
6. Set up sewage plant disposal facilities.

Figure 1.2 shows the event tree analysis with "nitrate spontaneous combustion" as the initial event.

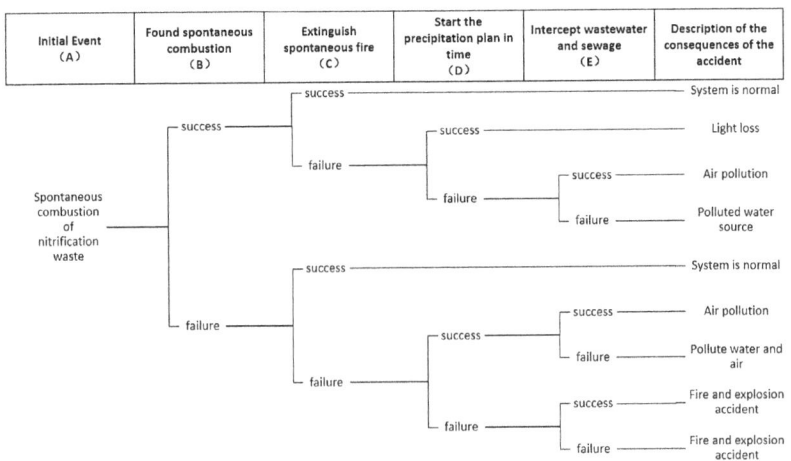

FIGURE 1.2   Event tree analysis with "nitrate spontaneous combustion" as the initial event.

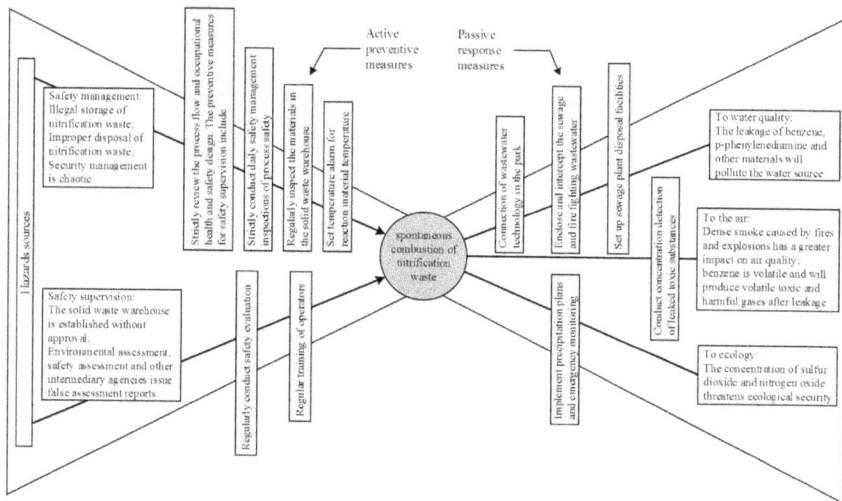

FIGURE 1.3 Bow-tie analysis of the "nitrate spontaneous combustion" event.

The accident tree and the event tree are combined into a bow-tie diagram, as shown in Figure 1.3. Through the above-mentioned bow-tie analysis, we can identify the causes and consequences of the accident in great details, and we can use the bow-tie diagram to intuitively illustrate the entire process of the accident. It can also help organizations establish effective measures to prevent and control accidents before and after the occurrence.

1.7 CONCLUSIONS

Through the bow-tie analysis, the accident tree and the event tree are used to analyze the hazards that may cause the central event in a safety process, as well as the impacts on the environment after the central event occurs. This method has a clear structure, and it can intuitively display the whole of the hazard situation, including the accident, and consequences. Other related qualitative analysis can be shown in the diagram. In the process of analyzing the accident chain, active protective barriers against hazards and passive protective barriers against the consequences of accidents are given. This process can help people establish effective measures to prevent and control the accident before and after its occurrence. Before an accident happens, use barriers to implement measures that prevent the occurrence of the accident. When an accident does happen, the impacts of the accident will not be widespread or cataclysmic. After using the bow-tie approach to analyze the case of the "3.21" explosion accident in Jiangsu Xiangshui Tianjiayi Chemical Co., Ltd., it was shown that bow-tie analysis is useful in both theory and practice.

REFERENCES

1. ima Khakzad, Faisal Khan, Paul Amyotte. Dynamic safety analysis of process systems by mapping bow-tie into Bayesian network. *Process Safety & Environmental Protection*, 2013, 91(1–2): 46–53.
2. Frank Lees. Lees' loss prevention in the process industries: *hazard identification, assessment and control*. Butterworths, USA, 2012.
3. C. Zuijderduijn. Risk management at Shell Pernis refinery/chemicals – implementation of SEVESO-II based on build up experiences, using a hazards and effects management process. *Proceedings of the Seveso 2000 European Conference*, 1999, 10–12 November, 1999.
4. Zhendong Guo, Jinlin Li, Kewei Wang, Boying Li, Chihua Ma. Bowtie analysis on pig launcher system safety management. *Chemical Industry and Engineering Progress*, 2019, 38(1): 2018–2228.
5. Jill Wilday, Mike Wardman, Michael Johnson, Mike Haines. Hazards from carbon dioxide capture, transport and storage. *Process Safety and Environmental Protection*, 2011, 89(6): 482–491.
6. Chen Qinqin, Qian Jia, Zengwei Yuan, Lei Huang. Environmental risk source management system for the petrochemical industry. *Process Safety and Environmental Protection*, 2014, 92(3): 251–260.
7. Cui Wengang, Fan Houming, Yao Xi, Yang Yu, Ma Mengzhi. Oil Spill Risk Analysis of Tanker Berthing and Handling Operation Based on Fuzzy Bow-Tie Model. 2016, *IEEE: NEW YORK*. 72–77.
8. Zerouali Bilal, Kara Mohammed, Hamaidi Brahim. Bayesian network and bow tie to analyze the risk of fire and explosion of pipelines. *Process Safety Progress*, 2017, 36(2): 202–212.
9. Kathleen E. Baker, Renato Macciotta Pulisci, Michael T. Hendry, Lianne M. Lefsrud. Combining safety approaches to bring hazards into focus: An oil sands tailings case study. *Canadian Journal of Chemical Engineering*, 2020, 98(11): 2330–2341.
10. Hanhai Dong, Yi Liu, Dongfeng Zhao, Meng Qi, Jian Chen, Yanchun Wang, Wei Wang. Lessons learned from analyzing an explosion at Shanghai SECCO petrochemical plant. *Process Safety Progress*, 2020, 39(e120941): 1–11.
11. Longting Wang, Liping Sun, Jichuan Kang, Baoping Cai, Yanfu Wang, Yaonan Wu. Risk identification and control of oil and gas leakage in the marine environment. *Journal of Coastal Research*, 2019, 98: 50–54.
12. Jun Zhang, Haifeng Bian, Huanhuan Zhao, Xuexue Wang, Linlin Zhang, Yiping Bai. Bayesian network-based risk assessment of single-phase grounding accidents of power transmission lines. *International Journal of Environmental Research and Public Health*, 2020, 17(18416): 1–17.
13. Zhi Yuan, Nima Khakzad, Faisal Khan, Paul Amyotte, Genserik Reniers. Risk-based design of safety measures to prevent and mitigate dust explosion hazards. *Industrial & Engineering Chemistry Research*, 2013, 52(50): 18095–18108.
14. Abdul Aziz, Salim Ahmed, Faisal Khan, Chris Stack, Annes Lind. Operational risk assessment model for marine vessels. *Reliability Engineering & System Safety*, 2019, 185: 348–361.
15. Fares Innal, Chebila Mourad, Mouloud Bourareche, Antar Si Mohamed. Treatment of Uncertainty in Probabilistic Risk Assessment Using Monte Carlo Analysis, *In 3rd International Conference on Systems and Control* (763–768).
16. Yang Tang, Jiajia Jing, Zhidong Zhang, Yan Yang. A quantitative risk analysis method for the high hazard mechanical system in petroleum and petrochemical industry. *Energies*, 2018, 11(141): 1–18.

17. Jishuoli Li, Kaili Xu, Bingjie Fan, Liyan Geng. Risk assessment of oxygen lance burning loss using bow-tie analysis based on fuzzy theory. *Mathematical Problems in Engineering*, 2020: 1–12. https://doi.org/10.1155/2020/7172184.

18. Emre Akyuz, Ozcan Arslan, Osman Turan. Application of fuzzy logic to fault tree and event tree analysis of the risk for cargo liquefaction on board ship. *Applied Ocean Research*, 2020, 101: 1–21. https://doi.org/10.1371/journal.pone.0160045.

19. Refaul Ferdous, Faisal Khan, Rehan Sadiq, Paul Amyotte, Brian Veitch. Analyzing system safety and risks under uncertainty using a bow-tie diagram: An innovative approach. *Process Safety and Environmental Protection*, 2013, 91(1–2): 1–18.

20. Ahmed Badreddine, Nahla Ben Amor. A Bayesian approach to construct bow tie diagrams for risk evaluation. *Process Safety and Environmental Protection*, 2013, 91(3): 159–171.

21. Fang Yan, Kaili Xu, Xiwen Yao, Yang Li. Fuzzy Bayesian network-bow-tie analysis of gas leakage during biomass gasification. *PLoS One*, 2016, 11(7): e0160045. https://doi:10.1371/journal.pone.0160045.

22. Zhi Yuan, Nima Khakzad, Faisal Khan, Paul Amyotte. Risk analysis of dust explosion scenarios using Bayesian networks. *Risk Analysis*, 2015, 35(2): 278–291.

23. Long Ding, Jie Ji, Faisal Khan, Xiaohua Li, Shaoan Wan. Quantitative fire risk assessment of cotton storage and a criticality analysis of risk control strategies. *Fire and Materials*, 2020, 44(2):165–179.

24. Weipeng Fang, Jiansong Wu, Yiping Bai, Laobing Zhang, Genserik Reniers. Quantitative risk assessment of a natural gas pipeline in an underground utility tunnel. *Process Safety Progress*, 2019, 38(4): 1–13.

25. Long Zhao, Yifei Yan, Peng Wang, Xiangzhen Yan. A risk analysis model for underground gas storage well integrity failure. *Journal of Loss Prevention in the Process Industries*, 2019, 62: 103951. https://reader.elsevier.com/reader/sd/pii/S09 50423018308490?token=E17717C9A2409C01C25076828F6DA7EB58A9DA18F2 C07F3782DF3470EBD5699482A4FDC132DD9365829CF21E107D55A0&originR egion=us-east-1&originCreation=20211001085431.

26. Peng Zhang, Xiangsu Chen, Chaohai Fan. Research on a safety assessment method for leakage in a heavy oil gathering pipeline. *Engergies*, 2020, 13(13406): 1–19. https://doi.org/10.3390/en13061340.

27. Yuntao Li, Doudou Xu, Jian Shuai. Real-time risk analysis of road tanker containing flammable liquid based on fuzzy Bayesian network. *Process Safety and Environmental Protection*, 2020, 134: 36–46.

28. Mostafa Mirzaei Aliabadi, Afshin Pourhasan, Iraj Mohammadfam. Risk modelling of a hydrogen gasholder using Fuzzy Bayesian Network (FBN). *International Journal of Hydrogen Energy*, 2020, 45(1): 1177–1186.

29. Wen Liu, Shihong Zhai, Wenli Liu. Predictive analysis of settlement risk in tunnel construction: A bow-tie-Bayesian network approach. *Advances in Civil Engineering*, 2019, 2019(2045125): 1–20. https://doi.org/10.1155/2019/2045125.

30. Nima Khakzad, Faisal I. Khan, Paul Amyotte. Dynamic risk analysis using bow-tie approach. *Reliability Engineering & System Safety*, 2012, 104: 36–44.

31. Yuanjiang Chang, Changshuai Zhang, Xiangfei Wu, Jihao Shi, Guoming Chen, Jihua Ye, Liangbin Xu, Anti Xue. A Bayesian Network model for risk analysis of deepwater drilling riser fracture failure. *Ocean Engineering*, 2019, 181: 1–12.

32. Han Gao, Yili Duo, Tie Sun, Xuefeng Yang. Dynamic safety management on the key equipment of coal gasification based on Dbt-Dbn method. *Mathematical Problems in Engineering*, 2020, 2020(7469470): 1–14. https://doi.org/10.1155/2020/7469470.

33. Nima Khakzad, Faisal Khan, Paul Amyotte. Dynamic safety analysis of process systems by mapping bow-tie into Bayesian network. *Process Safety and Environmental Protection*, 2013, 91(1–2): 46–53.

2 Fire and Explosion Hazards of Liquid Mixtures

Horng-Jang Liaw
Department of Safety, Health, and Environmental
Engineering, National Kaohsiung University of
Science and Technology, Taiwan, Republic of China

CONTENTS

2.1 FLASH POINT

On August 6, 2018, a tanker carrying flammable liquid experienced a collision on a motorway in northern Italy, Bologna, causing a major fire and explosion (F&E) that killed two people and injured more than 60 (BBC News, 2018). It was reported that more than 60% of hazardous chemical-tanker accidents involve flammable liquids (MHIDAS, 2004). Flash points are one of the most important quantities used in characterizing the F&E hazard risk of flammable liquids. In addition, a series of F&E accidents have been attributed to the leakage of flammable liquids and gases in central Taiwan between 2010 and 2012 (Liaw and Tsai, 2014). This highlights the importance of flash point calculations and their effect on the operational safety parameters of flammable liquids. In 2000, Taiwan experienced a major incident, now referred to as the Shengli event (Liaw *et al.*, 2002; Liaw and Chiu, 2003; 2006), that highlighted storage safety problems relating to flammable liquids, particularly with regard to the importance of flash points. Among other things, it was shown that combustion of liquids can occur in

DOI: 10.1201/9781003140382-2

a gaseous state, not just a liquid state. The flash point of a substance is defined in literature as the lowest temperature at which the concentration of that fuel in air can be ignited by a fire source. At the flash point, the concentration of flammable material in the gas phase is equivalent to its lower flammability limit (LFL). This leads to the derivation of Eq. (2.1):

$$LFL_i = \frac{P_{i,fp}^{sat}}{P} \tag{2.1}$$

where LFL_i is the lower flammability limit of a substance i, $P_{i,fp}^{sat}$ and P are the saturated vapor pressures of the flammable substance i at its flash point and ambient pressure, respectively. The lower the flash point value, the higher the F&E hazard. Current flash point test methods are either closed cup or open cup. The standard test methods, ASTM D56 (2001) and ASTM D93 (2000), belong to the closed cup test type, and ASTM D92 (1998) is the open type. An image of the Tag closed cup tester used in this study is presented in Figure 2.1. Since flammable vapor would escape into the local environment during an open cup test, the concentration of flammable vapor would be lower than when tested under closed cup conditions; closed cup conditions would lead to saturated vapor pressure. Thus, measured values under closed cup experiments are lower than those measured under open cup experiments. The flash point values measured by the closed cup method are always used for calculation of values for safety tests.

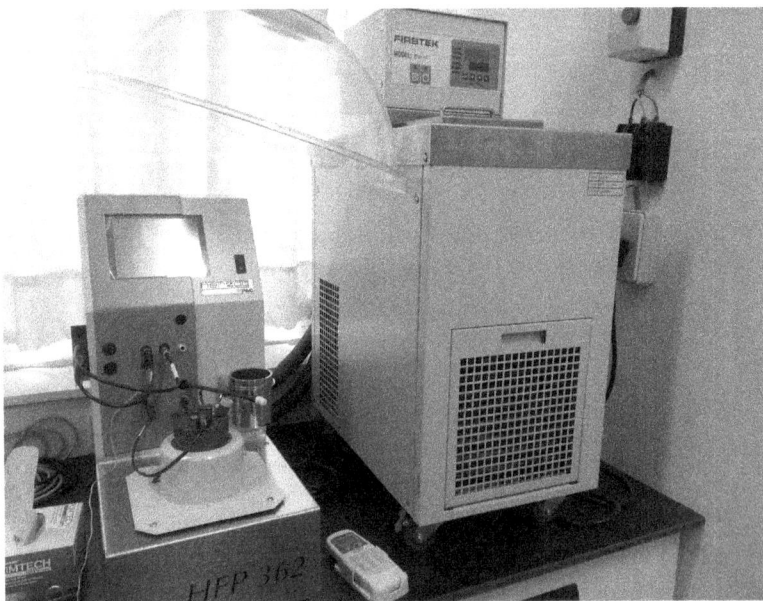

FIGURE 2.1   Photograph of the Tag closed cup test device.

2.2   FLASH POINT BEHAVIOR OF MIXTURES

In application, pure substances are far less frequent than mixtures, both in industry and daily life. One example is alcohol drinks. During flash point tests of aqueous-organic solutions, the flammable material and water evaporate simultaneously. Steam, being non-flammable, makes no contribution to the flammability of the vapor; rather, it serves as a diluent. LFL of the flammable material remains almost constant as the proportion of steam in the gas phase increases; thus, the flash point of the aqueous-organic solution increases with an increase in the liquid phase concentration of water. The typical flash point variation curve for aqueous-organic solutions is shown by ethanol + water, presented in Figure 2.2.

For an ideal solution of flammable solvents, such as octane + heptane, the flash point of the mixture increases as the quantity of the high flash point component, octane, increases. For reference, Figure 2.3 shows the flash point variation of octane + heptane. For most mixtures that include flammable solvents, the flash point values of such mixtures, as ideal solutions, lie between those of the pure components with the highest and lowest flash points. Ellis (1976) first reported that the flash point of a mixture may be lower than those of its constituent components.

The authors indicated that the flash point of the mixture that he was testing, butyl alcohol + mineral spirits, was as low as 38°C, although the open cup flash points independently were 42°C and 46°C, respectively (Ellis, 1976). Ellis

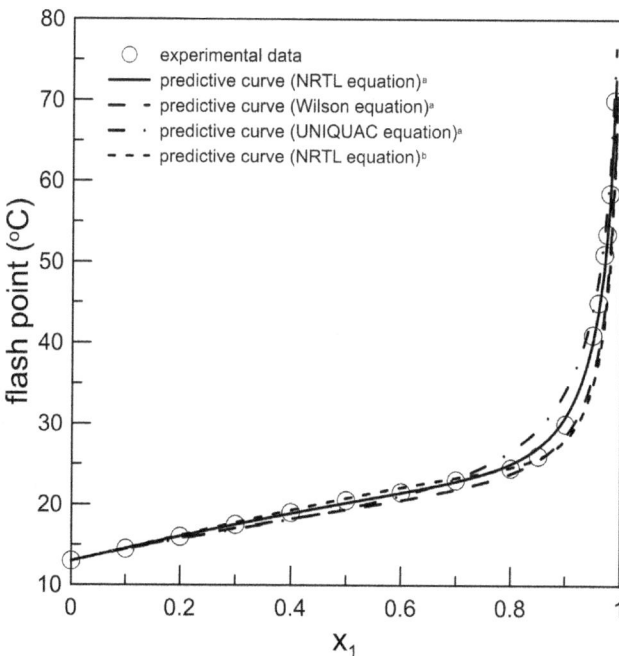

FIGURE 2.2   Flash point variation of water (1) + ethanol (2).

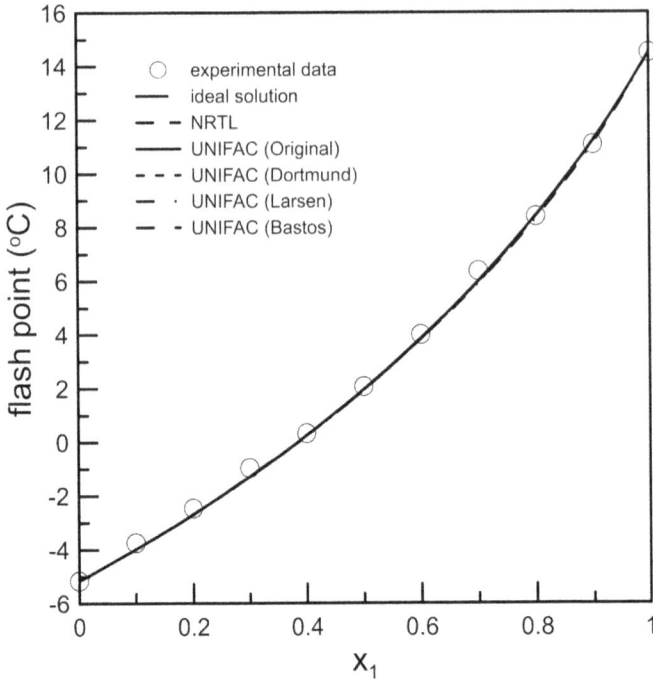

FIGURE 2.3   Flash point variation of octane (1) + heptane (2).

emphasized that the first impulse for any normal person would be to re-test the flash points; however, the measured values for the mixture was correct. This special behavior of flash points, where a minimum flash point value of a mixture is lower than those of its pure components, is referred to as Minimum Flash Point Behavior (MinFPB, formerly MFPB). A mixture exhibiting MinFPB is more hazardous than those of its pure components. Later in 1982, Gmehling and Rasmussen reported that isobutanol + toluene also exhibited MinFPB properties (Gmehling and Rasmussen, 1982). Anderson and Magyari (1984) then observed that the flash point of methanol mixed with alkane was lower than those of the individual components.

The discovery of MinFPB's for mixtures is very important. Unfortunately, Ellis' discovery was not valued by academia and industry in the previous century. For example, Martel and Cassidy (1997) stated that flash point estimation methods of mixtures are not needed because knowledge of the flash points of the pure components in the mixtures is sufficient. This argument of Martel and Cassidy (1997) is based on the hypothesis that the flash point of mixtures should lie between those of the pure components which delineate the lowest and highest flash points. Catoire *et al.* (2006) point out that the statement made by Martel and Cassidy (1997) is incorrect. Furthermore, they state that the idea the flash point of a mixture cannot be lower than the lowest flash point of the substituent components, or that the flash point of a mixture must lie somewhere between the lowest and highest flash points of the components, may be entirely wrong.

Ellis had explained why the flash point of butyl alcohol + mineral spirits is lower than those of its individual components qualitatively. However, Ellis did not predict or simulate the MinFPB of mixtures quantitatively. As mentioned previously, although the MinFPB observations were made by Ellis as early as 1976, this important finding was seemingly overlooked by the research and industrial community. As such, studies on the MinFPB of mixtures were rare in the 20th century. The field of research on MinFPB lay almost untouched until 2002, where Liaw *et al.* (2002) proposed a model to predict the flash point of binary miscible mixtures of flammable solvents. The MinFPB of ethanol + octane was accurately described and since then MinFPB research has begun to gain interest in academia. At the time of writing, 2021, there is now a growing body of research on the matter. Currently, the miscible mixture MinFPB's have been reported for the following combinations: ethylbenzene + *n*-propanol; methanol + methyl acrylate; decane + propanol, octane + ethanol; octane + propanol; octane + isopropanol; octane + 2-butanol; isopropanol + toluene; butanol + *p*-xylene; acetone + *n*-heptane; octane + methyl butyrate; ethanol + methyl butyrate; octane + propyl acetate; ethanol + propyl acetate and isopropanol + propyl acetate. Figure 2.4 shows the MinFPB of ethanol + octane.

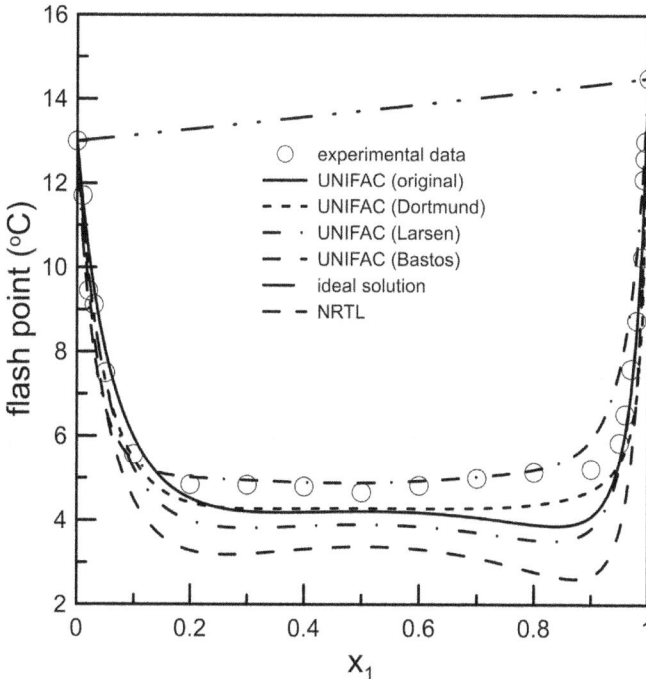

FIGURE 2.4 Minimum flash point behavior of octane (1) + ethanol (2).

The MinFPB of mixtures can be attributed to the repulsive force between unlike molecules being greater than the attractive force between same type molecules. More molecules escape during the gas phase than in an ideal solution, which results in an elevation of vapor pressure of the mixture. Therefore, the concentration of flammable materials reach their LFL at a lower temperature, leading to a lowered flash point. As the repulsive force between unlike molecules becomes more pronounced, the liquid phase mixture becomes partially miscible. Thus, it can be observed that partially miscible mixtures exhibit MinFPB. The reported partially miscible mixtures showing MinFPB include but are not limited to methanol + octane; methanol + 2,2,4-trimethyl pentane; methanol + nonane; methanol + decane; methanol + 1-dodecane; ethanol + tetradecane; and acetone + decane (Liaw *et al.*, 2008; 2010; 2014). Figure 2.5 shows the flash point behavior for the partially miscible mixture of methanol + octane.

Relative to the MinFPB, Liaw and Lin (2007) first reported that the binary mixtures phenol + cyclohexanol, phenol + cyclohexanone, phenol + acetophenone, and phenol + *p*-picoline all demonstrated maximum flash point behavior (MaxFPB), i.e., where the flash point of the mixture was higher than that of any substituent component. Later, Ha and Lee (2013; 2015) showed that both acetic acid + *n*-butanol and acetic acid + 2-pentanol exhibited MaxFPB properties.

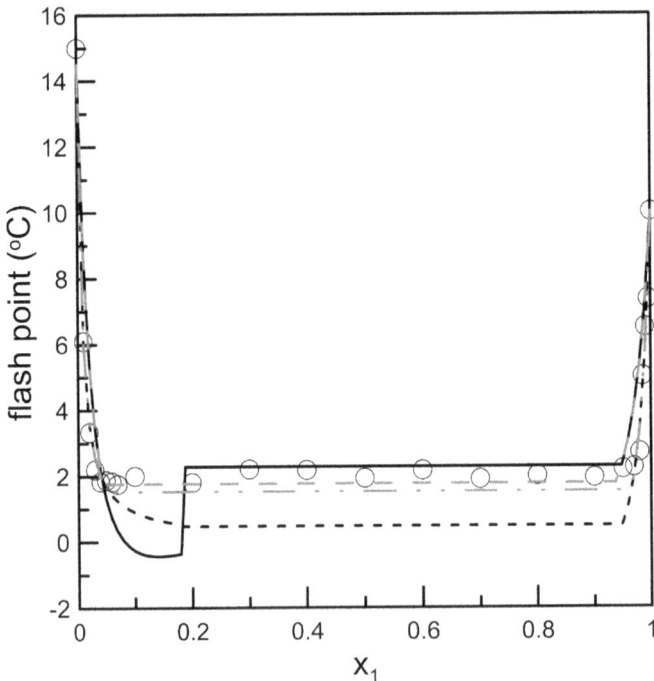

FIGURE 2.5   Flash point variation for the partially miscible mixture of methanol (1) + octane (2).

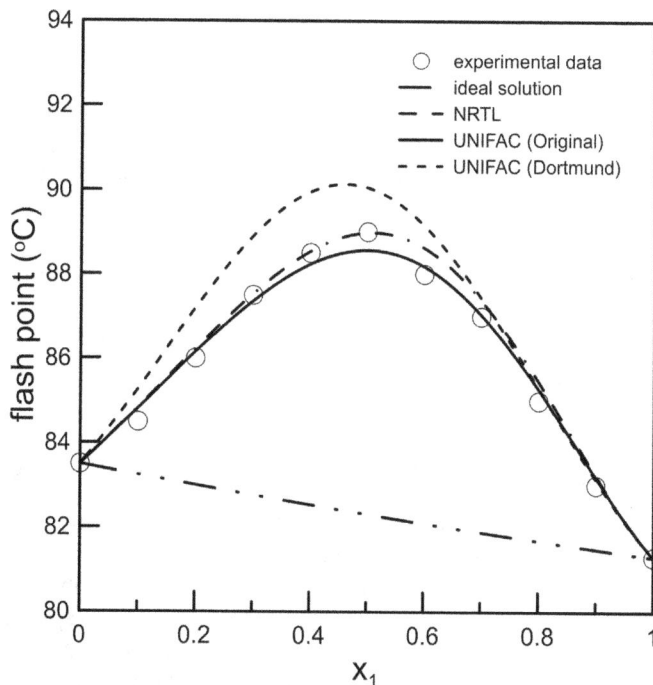

FIGURE 2.6 Maximum flash point behavior of phenol (1) + acetophenone (2).

MaxFPB can be attributed to the attractive force between unlike molecules being greater than the repulsive force of molecules of the same type; thus, less molecules escape during the vapor phase, reducing vapor pressure in the gas phase, in turn increasing the flash point of the mixture. The MaxFPB of certain mixtures is a unique property that has the potential to be exploitable with respect to F&E hazard reduction of flammable liquids. Mixtures that exhibit MaxFPB behave in a manner which is considered a highly negative deviation from an ideal solution. Since these mixtures are scarce, only a few demonstrate MaxFPB, the research showing MaxFPB values and combinations are accordingly scarce, far fewer in number than those exhibiting MinFPB. The MaxFPB of phenol + acetophenone is provided in Figure 2.6.

2.3 FLASH POINT PREDICTION OF MIXTURES

Due to the limitation of data concerning flash point prediction for mixtures, Crowl and Louvar (2002) recommended obtaining the flash points of mixtures that contain more than one flammable component by individual experimentation. Now there are some models available for effectively predicting the flash points of mixtures. The reviewed papers for flash point prediction can now be found in several journals (Vidal *et al.*, 2004; Liu and Liu, 2010; Phoon *et al.*, 2014).

However, some of the developed models have not been used in recent years. Liu and Liu (2010) stated that Liaw *et al.*'s models were the only ones that can describe the occurrence of MinFPB and MaxFPB. Phoon *et al.* (2014) further recommended estimation of mixture flash points by the use of Liaw *et al.*'s models as they predict the flash point of both ideal and non-ideal miscible/partially miscible mixtures, and they are more reliable than other models. Moreover, Liaw *et al.*'s models are the most commonly used (Phoon *et al.*, 2014). Thus, this manuscript introduces Liaw *et al.*'s models for the estimation of mixture flash points. These models are vapor pressure-based, meaning that they rely on the vapor pressure of the flammable substances involved. Since models that assume ideal solutions are limited in application to non-ideal solutions, these models can be adapted to take non-ideality of mixtures into account.

2.3.1   GENERAL FLASH POINT PREDICTION MODEL FOR MISCIBLE MIXTURES

At the flash point of a pure flammable liquid, the vapor pressure of such a liquid in the air is equivalent to form a concentration that corresponds to its LFL, as described above. This definition of flash point for a pure substance can be extended to mixtures. For a liquid mixture at its flash point, the vapor concentration of flammables is equal to LFL of its gas mixture. At the lower flammability limit of a gas mixture in air, the Le Chatelier rule for a flammable multiple component vapor-air is expressed as:

$$1 = \sum \frac{y_i}{LFL_i} \tag{2.2}$$

where y_i is the composition of component i in the vapor phase, and LFL_i is the lower flammability limit of component i. From the definition of flash point, the lower flammability of a pure liquid is described by Eq. (2.3):

$$LFL_i = \frac{P_{i,fp}^{sat}}{P} \tag{2.3}$$

Under the vapor liquid equilibrium assumption, the relationship between the vapor phase composition, y_i, and liquid phase composition, x_i, is described by:

$$y_i P = x_i \gamma_i P_i^{sat} \tag{2.4}$$

In the derivation of the above equation, the gas phase is assumed to be that of an ideal gas and the fugacity of the pure substance i is assumed to be the saturated vapor pressure of substance i, P_i^{sat}. In real-world applications the flash point is always measured at atmospheric pressure, where the vapor phase can be assumed to be an ideal gas and the fugacity of the pure substance i approaches to its saturated vapor pressure. Thus Eq. (2.4) can be re-written:

$$y_i = \frac{x_i \gamma_i P_i^{sat}}{P}$$ (2.5)

The LFL of component i can be expressed by Eq. (2.1). Then, when Eqs. (2.1) and (2.5) are substituted into Eq. (2.2), this results in:

$$1 = \sum \frac{x_i \gamma_i P_i^{sat}}{P_{i,fp}^{sat}}$$ (2.6)

Since only the flammable components can be burned in the gas phase, Eq. (2.6) is modified to produce:

$$1 = \sum_{i \neq k_l} \frac{x_i \gamma_i P_i^{sat}}{P_{i,fp}^{sat}}$$ (2.7)

where k_l are the non-flammable components. In the estimation of mixture flash points, the saturated vapor pressures of the pure substances, P_i^{sat}, are necessary, but can be estimated by means of the Antoine equation:

$$\log P_i^{sat} = A_i - \frac{B_i}{T + C_i}$$ (2.8)

The saturated vapor pressure of component i at its flash point, $P_{i,fp}^{sat}$, presented in Eq. (2.7), can be estimated by substituting its flash point into Eq. (2.8). In addition to the saturated vapor pressures of the pure substances, the activity coefficients of the solution components, γ_i, describing the non-ideality of the mixtures are also required. The activity coefficient values can be estimated by several models that describe the non-ideality of the solution components concerned. The nonrandom two liquid (NRTL), Wilson, and universal quasichemical (UNIQUAC) models are the most frequently used in the current literature. Table 2.1 displays these activity coefficient models.

In summary, the general flash point prediction model for miscible mixtures is described by the modified Le Chatelier's equation, Antoine equation, and any model describing the activity coefficients (Liaw and Chiu, 2006):

$$1 = \sum_{i \neq k_l} \frac{x_i \gamma_i P_i^{sat}}{P_{i,fp}^{sat}}$$ (2.9)

$$\log P_i^{sat} = A_i - \frac{B_i}{T + C_i}$$ (2.10)

TABLE 2.1

Some Frequently Used Models for Estimating Activity Coefficients of Multi-Component Systems

Name	Activity Coefficient for Component i
Wilson	$$\ln \gamma_i = -\ln\left(\sum_j^N x_j \Lambda_{ij}\right) + 1 - \sum_k^N \frac{x_k \Lambda_{ki}}{\sum_j^N x_j \Lambda_{kj}}$$ where $\Lambda_{ij} = \frac{v_j^l}{v_i^l} \exp\left(-\frac{\lambda_{ij} - \lambda_{ii}}{RT}\right)$
NRTL	$$\ln \gamma_i = \frac{\sum_j^N \tau_{ji} G_{ji} x_j}{\sum_k^N G_{ki} x_k} + \sum_j^N \frac{x_j G_{ij}}{\sum_k^N G_{kj} x_k}\left(\tau_{ij} - \frac{\sum_k^N x_k \tau_{kj} G_{kj}}{\sum_k^N G_{kj} x_k}\right)$$ where $\tau_{ij} = \frac{g_{ij} - g_{jj}}{RT}$, $\ln G_{ij} = -\alpha_{ij} \tau_{ij}$
UNIQUAC	$$\ln \gamma_i = \ln \frac{\Phi_i}{x_i} + \frac{z}{2} q_i \ln \frac{\theta_i}{\Phi_i} + l_i - \frac{\Phi_i}{x_i}\sum_j^N x_j l_j$$ $$- q_i \ln\left(\sum_j^N \theta_j \tau_{ji}\right) + q_i - q_i \sum_j^N \frac{\theta_j \tau_{ji}}{\sum_k^N \theta_k \tau_{kj}}$$ where $$\ln \tau_{ij} = -\frac{u_{ij} - u_{jj}}{RT}, \Phi_i = \frac{x_i r_i}{\sum_k^N x_k r_k},$$ $$\theta_i = \frac{x_i q_i}{\sum_k^N x_k q_k}, l_i = \frac{z}{2}(r_i - q_i) - (r_i - 1), z = 10$$

2.3.2 FLASH POINT PREDICTION FOR MISCIBLE MIXTURES OF FLAMMABLE SOLVENTS

Miscible mixtures of flammable solvents are the most commonly used mixtures in industry. For the mixtures of flammable solvents, all the individual components of the mixtures are flammable, Eq. (2.9) can be described as:

$$1 = \sum \frac{x_i \gamma_i P_i^{sat}}{P_{i,fp}^{sat}} \tag{2.11}$$

Thus, the flash point prediction model is reducible to:

$$1 = \sum \frac{x_i \gamma_i P_i^{sat}}{P_{i,fp}^{sat}} \tag{2.12}$$

$$\log P_i^{sat} = A_i - \frac{B_i}{T + C_i} \tag{2.13}$$

with the activity coefficient model (Liaw *et al.*, 2004). For the ternary mixtures, Eq. (2.12) reduces to:

$$1 = \sum \frac{x_i \gamma_i P_i^{sat}}{P_{i,fp}^{sat}} = \frac{x_1 \gamma_1 P_1^{sat}}{P_{1,fp}^{sat}} + \frac{x_2 \gamma_2 P_2^{sat}}{P_{2,fp}^{sat}} + \frac{x_3 \gamma_3 P_3^{sat}}{P_{3,fp}^{sat}} \tag{2.14}$$

Eqs. (2.10) and (2.7) can be combined with any model that describes the non-ideality of the mixtures to produce the flash point prediction model for ternary mixtures of flammable solvents. For binary mixtures of flammable solvents, Eq. (2.9) reduces to (Liaw *et al.*, 2002):

$$1 = \sum \frac{x_i \gamma_i P_i^{sat}}{P_{i,fp}^{sat}} = \frac{x_1 \gamma_1 P_1^{sat}}{P_{1,fp}^{sat}} + \frac{x_2 \gamma_2 P_2^{sat}}{P_{2,fp}^{sat}} \tag{2.15}$$

Thus, the flash point of the binary mixtures of flammable solvents can be estimated by Eqs. (2.13 and 2.15) and any activity coefficient model.

Figures 2.3, 2.4, and 2.6 compare the flash point prediction curves with the experimental flash point data for binary miscible mixtures, octane + heptane, octane + ethanol, and phenol + acetophenone. The mixture, octane + heptane, is an ideal solution. The other two mixtures behave highly non-ideally. The mixture ethanol + octane, shows MinFPB, and is a minimum flash point solution. The other mixture, phenol + acetophenone, exhibits MaxFPB, and is a maximum flash point mixture. By contrast, the mixture, octane + heptane, demonstrates neither MinFPB nor MaxFPB properties, and its flash points are between those of the pure components. It is shown in Figures 2.3, 2.4, and 2.6 that the flash point prediction model for binary mixtures of flammable solvents describes the experimental data well, and the prediction capability of the model depends upon which activity coefficient model is in use.

Figure 2.7 compares experimental flash point data with the prediction surface based on the flash point prediction model of ternary mixtures of flammable solvents for methanol + ethanol + acetone. Figure 2.7 shows that the model predicted the experimental data well.

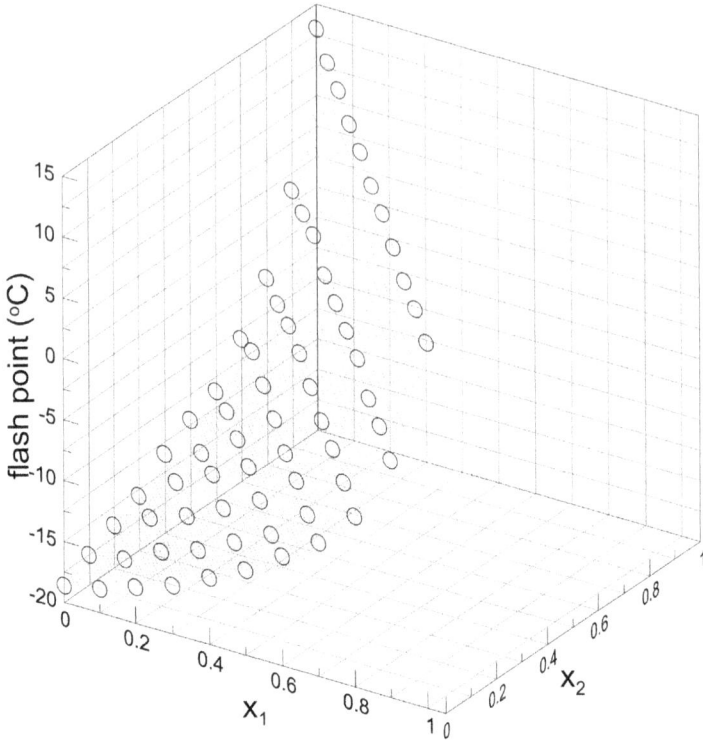

FIGURE 2.7 Comparison of flash point prediction surface and experimental data for methanol (1) + ethanol (2) +acetone (3).

2.3.3 FLASH POINT PREDICTION FOR MISCIBLE AQUEOUS-ORGANIC MIXTURES

For aqueous-organic solutions, steam is non-flammable, and the other organic compounds are flammable. Thus, Eqs. (2.9 and 2.10) and any model to estimate the activity coefficients of the solution components constitute the flash point prediction model for miscible aqueous-organic mixtures. Ternary aqueous-organic mixtures following Eq. (2.9) reduce to:

$$1 = \frac{x_2 \gamma_2 P_2^{sat}}{P_{2,fp}^{sat}} + \frac{x_3 \gamma_3 P_3^{sat}}{P_{3,fp}^{sat}} \tag{2.16}$$

where steam is denoted as component 1, and the other flammable components of the ternary mixtures are denoted as components 2 and 3. Thus, the flash point prediction model for the ternary aqueous-organic mixtures is constituted by Eqs. (2.10 and 2.16) and any model which can describe the activity coefficient of the solution components (Liaw and Chiu, 2006).

For the binary aqueous-organic mixtures, Eq. (2.9) is reducible to:

$$1 = \frac{x_2 \gamma_2 P_2^{sat}}{P_{2,fp}^{sat}} \qquad (2.17)$$

or

$$P_2^{sat} = \frac{P_{2,fp}^{sat}}{x_2 \gamma_2} \qquad (2.18)$$

The above equation is the same as that derived by Liaw and Chiu (2003). Eq. (2.18) combined with the Antoine equation, Eq. (2.8), and models describing the non-ideality of the liquid solution can be used to describe the flash point of binary aqueous-organic mixtures. Figure 2.2 compares the prediction curves estimated by the flash point prediction model for binary aqueous-organic mixtures with the experimental data for water + ethanol. Figure 2.2 indicates that the flash point prediction model describes the experimental data well, irrespective of which activity coefficient model is used.

2.3.4 FLASH POINT PREDICTION BASED UPON UNIFAC-TYPE MODELS

As described above, the non-ideality of the liquid mixtures is taken into account in Liaw *et al.*'s models; thus, the model estimating the activity coefficient of liquid mixtures is necessary when using those models. The NRTL, Wilson, and UNIQUAC thermodynamic models are frequently used in the estimation of activity coefficients, and such models require the binary interaction parameters regressed on experimental vapor liquid equilibrium data. Unfortunately, the binary interaction parameters of the relevant mixtures are frequently unavailable in the literature due to the vast number of possible mixture combinations. The UNIFAC-type models, such as original UNIFAC (Fredenslund *et al.*, 1975; 1977), UNIFACT-Dortmund models (Weidlich and Gmehling, 1987; Gmehling *et al.*, 1993), do not require experimental binary interaction parameters. The binary interaction parameters of the UNIFAC-type models can be evaluated by the chemical group contributions, which can be obtained by regression from large database.

Table 2.2 lists the original UNIFAC and UNIFAC-Dortmund models. The group volume, surface area parameters, and interaction parameters for the UNIFAC group can be obtained from literature (Fredenslund *et al.*, 1975; 1977; Weidlich and Gmehling, 1987; Gmehling *et al.*, 1993). The prediction capability of Liaw *et al.*'s (2011) models based on UNIFAC-type models to estimate the solution components' activity coefficients can be found in literature (Moghaddam *et al.*, 2012).

TABLE 2.2

The Original UNIFAC and UNIFAC-Dortmund Models

Name	Activity Coefficient for Component i
Original UNIFAC	$\ln \gamma_i = \ln \gamma_i^C + \ln \gamma_i^R$

The combinatorial part: $\ln \gamma_i^C = \ln \dfrac{\phi_i}{x_i} + \dfrac{z}{2} q_i \ln \dfrac{\theta_i}{\phi_i} + l_i - \dfrac{\phi_i}{x_i} \sum_j^N x_j l_j$

where $l_i = \dfrac{z}{2}(r_i - q_i) - (r_i - 1); \quad z = 10$

$$\phi_i = \frac{x_i r_i}{\sum_j x_j r_j}, \ \theta_i = \frac{x_i q_i}{\sum_j x_j q_j}$$

The residual part:

$$\ln \gamma_i^R = \sum_k \nu_k^{(i)} \left(\ln \Gamma_k - \ln \Gamma_k^{(i)} \right)$$

where $\ln \Gamma_k = Q_k \left[1 - \ln \left(\sum_m \Theta_m \Psi_{mk} \right) - \sum_m \dfrac{\Theta_m \Psi_{km}}{\sum_n \Theta_n \Psi_{nm}} \right]$

$$\Theta_m = \frac{Q_m X_m}{\sum_n Q_n X_n} \quad X_m = \frac{\sum_i \nu_m^{(i)} x_i}{\sum_i \sum_n \nu_n^{(i)} x_i} \quad \Psi_{mn} = \exp \left(-\frac{a_{mn}}{T} \right)$$

UNIFAC-Dortmund	$\ln \gamma_i^C = \ln \dfrac{\phi_i'}{x_i} + 1 - \dfrac{\phi_i'}{x_i} - \dfrac{z}{2} q_i \left(\ln \dfrac{\phi_i}{\theta_i} + 1 - \dfrac{\phi_i}{\theta_i} \right)$

where

$$\phi_i' = \frac{x_i r_i^{3/4}}{\sum_j x_j r_j^{3/4}}, \ \Psi_{mn} = \exp \left(-\frac{a_{mn} + b_{mn}T + c_{mn}T^2}{T} \right)$$

REFERENCES

Anderson JE, Magyari MW. Flashpoint temperatures of methanol-hydrocarbon solutions. *Combust. Sci. Technol.* 37, 193–199, 1984.

ASTM D 56. Standard test method for flash point by Tag closed tester. West Conshohocken, PA: American Society for Testing and Materials, 2001.

ASTM D 92. Standard test method for flash and fire points by Cleveland open cup. West Conshohocken, PA: American Society for Testing and Materials, 1998.

ASTM D 93. Standard test methods for flash point by Pensky-Martens closed cup tester. West Conshohocken, PA: American Society for Testing and Materials, 2000.

BBC News, Bologna crash: Tanker truck fireball kills two and injures dozens, Available at: https://www.bbc.com/news/world-europe-45087884. Accessed on August 8, 2018.

Catoire L, Paulmier S, Naudet V. Experimental determination and estimation of closed cup flash points of mixtures of flammable solvents. *Process Saf. Prog.* 25, 33–39, 2006.

Crowl DA, Louvar JF. Chemical Process Safety – Fundamentals with Applications. Upper Saddle River, NJ: Prentice Hall PTR, 2002.

Ellis WH. Solvent flash points – expected and unexpected. *J. Coat. Technol.* 48, 44–57, 1976.

Fredenslund A, Gmehling J, Michelsen ML, Rasmussen P, Prausnitz JM. Computerized design of multicomponent distillation columns using the UNIFAC group contribution method for calculation of activity coefficients. *Ind. Eng. Chem. Process Des. Dev.* 16, 450–462, 1977.

Fredenslund A, Jones RL, Prausnitz JM. Croup-contribution estimation of activity coefficients in nonideal liquid mixtures. *AIChE J.* 21, 1086–1099, 1975.

Gmehling J, Li J, Schiller M. A modified UNIFAC model. 2. Present parameter matrix and results for different thermodynamic properties. *Ind. Eng. Chem. Res.* 32, 178–193, 1993.

Gmehling J, Rasmussen P. Flash points of flammable liquid mixtures using UNIFAC. *Ind. Eng. Chem. Fundam.* 21, 186–188, 1982.

Ha D-M, Lee S. The measurement and prediction of maximum flash point behavior for binary solution. *Fire Sci. Eng.* 27, 70–74, 2013.

Ha D-M, Lee S., Calculation and measurement of flash point for n-decane + n-octanol and acetic acid + n-butanol using a Tag-open-cup apparatus. *Fire Sci. Eng.* 29, 45–50, 2015.

Liaw H-J, Chiu Y-Y. A general model for predicting the flash point of miscible mixture. *J. Hazard. Mater.* 137, 38–46, 2006.

Liaw H-J, Chiu Y-Y. The prediction of the flash point for binary aqueous-organic solutions. *J. Hazard. Mater.* 101, 83–106, 2003.

Liaw H-J, Gerbaud V, Chiu C-Y. Flash point for ternary partially miscible mixtures of flammable solvents. *J. Chem. Eng. Data.* 55, 134–146, 2010.

Liaw H-J, Gerbaud V, Li YH. Prediction of miscible mixtures flash-point from UNIFAC group contribution methods. *Fluid Phase Equilibr.* 300, 70–82, 2011.

Liaw H-J, Lee Y-H, Tang C-L, Hsu H-H, Liu J-H. A mathematical model for predicting the flash point of binary solutions. *J. Loss Prevent. Proc.* 15, 429–438, 2002.

Liaw H-J, Lin S-C. Binary mixtures exhibiting maximum flash-point behavior. *J. Hazard. Mater.* 140, 155–164, 2007.

Liaw H-J, Lu W-H, Gerbaud V, Chen C-C. Flash-point prediction for binary partially miscible mixtures of flammable solvents. *J. Hazard. Mater.* 153 (3), 1165–1175, 2008.

Liaw H-J, Tang C-L, and Lai J-S. A model for predicting the flash point of ternary flammable solutions of liquid. *Combust. Flame.* 138 (4), 308–319, 2004.

Liaw H-J, Tsai T-P. Flash-point estimation for binary partially miscible mixtures of flammable solvents by UNIFAC group contribution methods. *Fluid Phase Equilibr.* 375, 275–285, 2014.

Liu X, Liu Z. Research progress on flash point prediction. *J. Chem. Eng. Data.* 55, 2943–2950, 2010.

Martel B, Cassidy K. Chemical risk analysis: A practical handbook. London: Taylor & Francis, 1997.

Moghaddam AZ, Rafiei A, Khalili T. Assessing prediction models on calculating the flash point of organic acid, ketone and alcohol mixtures. *Fluid Phase Equilibr.* 316, 117–121, 2012.

Phoon LY, Mustaffa AA, Hashim H, Mat R. A review of flash point prediction models for flammable liquid mixtures. *Ind. Eng. Chem. Res.* 53, 12553–12565, 2014.

UK Atomic Energy Authority, *Major Hazard Incident Data Service (MHIDAS)*, CD-ROM version; Silver Platter, 2004.

Vidal M, Rogers WJ, Holste JC, Mannan MS. A review of estimation methods for flash points and flammability limits. *Process Saf. Prog.* 23, 47–55, 2004.

Weidlich U, Gmehling J. A modified UNIFAC model. 1. Prediction of VLE, h^E, and γ^∞. *Ind. Eng. Chem. Res.* 26, 1372–1381, 1987.

3 Bow-Tie Analysis of Underground Coal-Fire Hazards and Mining Activities Using Hybrid Data

A Case Study of Wuda Coalfield in Inner Mongolia, China

Zeyang Song, Zhijin Yu, and Jingyu Zhao
Department of Safety, Xi'an University of Science and Technology, Xi'an, People's Republic of China

Maorui Li
Jiangsu Key Laboratory of Hazardous Chemicals Safety and Control Nanjing Tech University, Nanjing, Jiangsu, People's Republic of China

Jun Deng
School of Safety Science and Engineering, and Shaanxi Key Laboratory of Prevention and Control of Coal Fire, Xi'an University of Science and Technology, Xi'an, People's Republic of China

CONTENTS

DOI: 10.1201/9781003140382-3

3.1   INTRODUCTION

Underground coal fires (UCFs) are a global environmental catastrophe (Stracher, 2004), increasingly frequent in China (Song and Kuenzer, 2014), India (Prakash et al., 2013), the USA (Stracher, 2004), Australia (Stracher, 2004), Russia (Stracher, 2004), and South Africa (Stracher et al., 2013) among other coal producing countries. As these fires devour valuable coal resources, they also emit hazardous materials—CO, H_2S, PAHs, tar, and Hg, for example—that threaten local residents' health (Hower et al., 2009; Hower et al., 2011; Hower et al., 2013; Li et al., 2018a; Li et al., 2018b; Liang et al., 2018; O'Keefe et al., 2010). Not only are the UCFs a severe mining hazard, but the massive heat from UCFs is transferred to the surface and leads to (Wessling et al., 2008) the degradation of vegetation (Kuenzer et al., 2007; Zhang and Kuenzer, 2007; Zhang et al., 2007). Geohazards (Kuenzer and Stracher, 2012), including landslides, collapses, and subsidence, can also be induced by UCFs. Smoldering fires and toxic gases pose great threats to the safety of mine workers and infrastructures (Song and Kuenzer, 2014; Song et al., 2015).

UCFs have received more recent attention due to their catastrophic hazards and heavy pollution. To mitigate this global threat, it is crucial to understand the root causes of hazards created by UCFs. Although it is well recognized that mining is a significant trigger for UCFs (Song et al., 2015), the precise cause-and-effect relationship between mining activities and coal-fire hazards is not yet well understood. This challenge arises from the complexity of the relationship in both time and space. Coal-fire hazards often occur on the surface, but mining activities are conducted at tens or hundreds of meters below ground. To compound the problem, the hazards from UCFs typically occur long after the mining activities have ended.

In this chapter, we propose a novel technical solution to address this challenge. Field surveys were used to demarcate surface thermal anomalies induced by UCFs. Coal-fire data, extracted from satellite images, was merged with underground mining maps and borehole data. Then, the cause-and-effect relationship between mining activities and coal-fire hazards was revealed through this merged data and a bowtie analysis. This work provides a valuable reference for the application of bowtie analyses to underground mining hazards and offers a tool for the mitigation of coal-fire hazards.

3.2   UNDERGROUND COAL FIRES IN THE WUDA COALFIELD, INNER MONGOLIA

3.2.1   GEOGRAPHIC LOCATION AND GEOLOGICAL SETTING OF STUDIED AREA

The Wuda coalfield is one of the biggest areas in the world impacted by UCFs. Wuda is located in the south-central part of Inner Mongolia of North China (see Figure 3.1). The Wuda coalfield covers a 35 km^2 syncline that strikes north-south (Kuenzer et al., 2012; Song and Kuenzer, 2017; Song et al., 2015). The climate is continental and fully arid with annual evaporation of 3500 mm and precipitation of 168 mm (Kuenzer, 2005; Kuenzer et al., 2012). Vegetation density is very low in the coalfield.

The Wuda coalfield comprises 17 coal seams. This includes five stable and minable coal seams: the 9th, 10th, 12th, 13th, and 15th. The average eastern dip angle of the strata varies between 6° and 10° (Kuenzer et al., 2012). The coal-bearing strata of the Wuda coalfield originate from the Pennsylvanian and Permian ages (Dai et al., 2012). The Pennsylvanian coal-bearing strata contain the Pennsylvanian Benxi Formation (C2b), the Pennsylvanian Taiyuan Formation (C2t), the Early Permian Shanxi Formation (P1s), the Xiashihezi Formation (P1x), and the Late Permian Shangshihezi Formation (P2sh) (Dai et al., 2012). All the stable and minable coal seams belong to the Taiyuan Formation, which is the major coal-bearing strata in the Wuda coalfield (Dai et al., 2012).

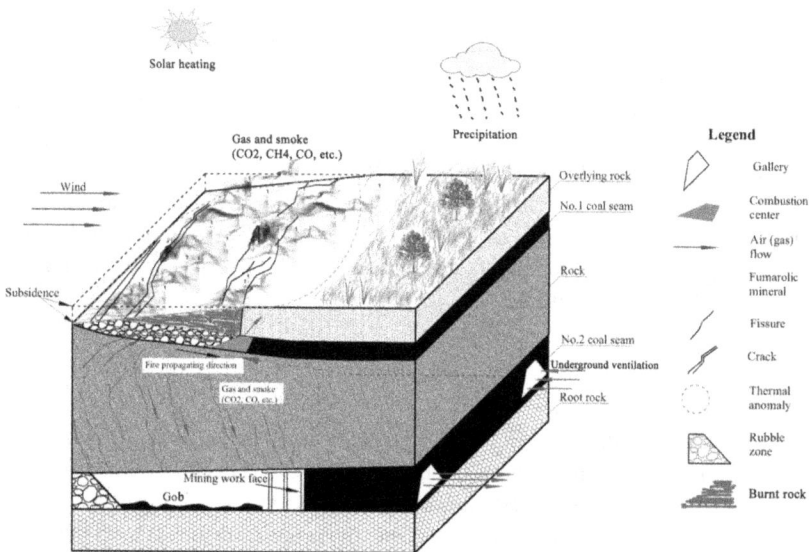

FIGURE 3.1   Schematic diagram of UCF and its hazards. (Song and Kuenzer, 2014.)

3.2.2 History and Current State of UCFs in the Wuda Coalfield

The first coal fires in the Wuda coalfield occurred in 1961 in the 9th and 10th coal seams. These UCFs have been attributed mainly to underground mining activities (Song et al., 2015). There were three state-owned coal mines—the southern Wuhushan Coal Mine, the northeastern Huangbaici Coal Mine, and the northwestern Su Hai Tu Coal Mine—as depicted in Figure 3.2. The long-wall mining technique was employed by state-owned coal mines to excavate the underground coal resources. This mining technique led to underground gobs and subsidence. Ventilation paths—cracks, fissures, and vents—were formed by subsidence which provided oxygen for the smoldering combustion in the underground coal seams. In addition to long-wall mining, illegal and unorganized mining activities were undertaken. This resulted in the exposure of coal to air which triggered UCFs.

As shown in Figure 3.3, the first UCF occurred in 1961 when the 9th and 10th coal seams in the Suhaitu Coal Mine, both with high sulfur contents, spontaneously combusted (Kuenzer and Stracher, 2012). In 1978, six coal fires near the surface occurred, creating a hazard for local miners and residents. According to satellite remote sensing, field surveys, and geologic analyses, by the year 2000, UCFs had spread to 16 zones (Kuenzer et al., 2012). Another two zones were identified in 2004 (Kuenzer et al., 2008). Coal fire areas shrunk from approximately 159.95 ha in 2004 to 112.96 ha in 2005, which was mainly attributed to the extinguishing of the No. 8 fire zone in the Wuda syncline (Kuenzer et al., 2012). In 2010, however, due to the accelerated extraction of coal and an increase in abandoned mines (Kuenzer et al., 2012), the total area of coal fires increased sharply to 227.08 ha.

Figure 3.3 shows the developments of coal production, burning area in hectares, and the number of fire zones in the Wuda coalfield from the 1950s to the 2010s. Coal fires began in 1961 when coal production was approximately 1.5 million tons annually. The total area of burning coal fires increased to ~100 ha, while the number of fire zones increased to eight in 1995 when coal production reached ~3.8 million tons annually. This indicates that there is a direct relationship between coal fires and coal mining activities. As coal production increases, so does the total burning area and the number of fire zones. Note that coal fire-fighting projects were launched in ~2000, leading to a decrease in both fire zones and burning area after 2000 as seen in Figure 3.3.

3.3 METHODS

The statistical data shown in Figure 3.3 presents a compelling but still vague clue about the causal relationship between coal mining activities and UCFs. However, the characteristics and detailed causal chains linking the mining activities and UCFs are a challenging puzzle to solve. One of the main reasons is both the lateral and vertical spatial separation between UCF hazards and underground mining activities. In this work, we propose a method that addresses this challenge. Hybrid data, including field surveys, stratigraphic data, underground mining galleries,

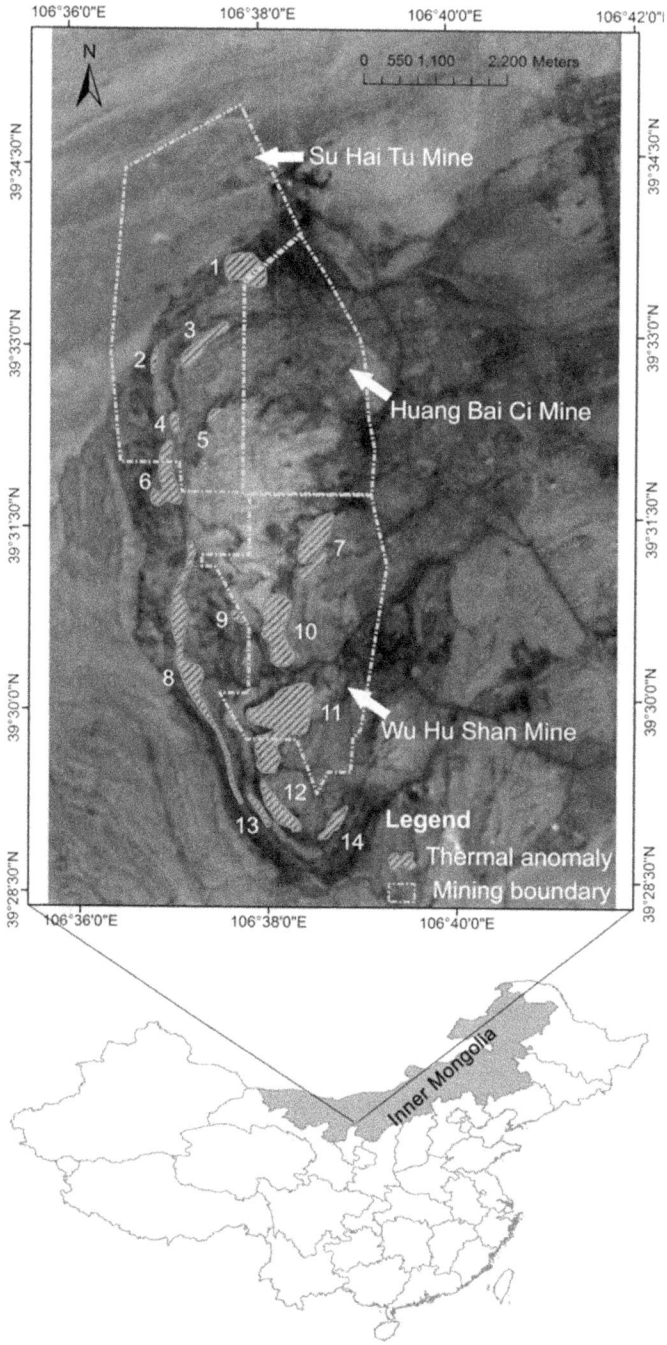

FIGURE 3.2 Geometric location, surface thermal anomalies, and coal mines of the Wuda coalfield in Inner Mongolia, China.

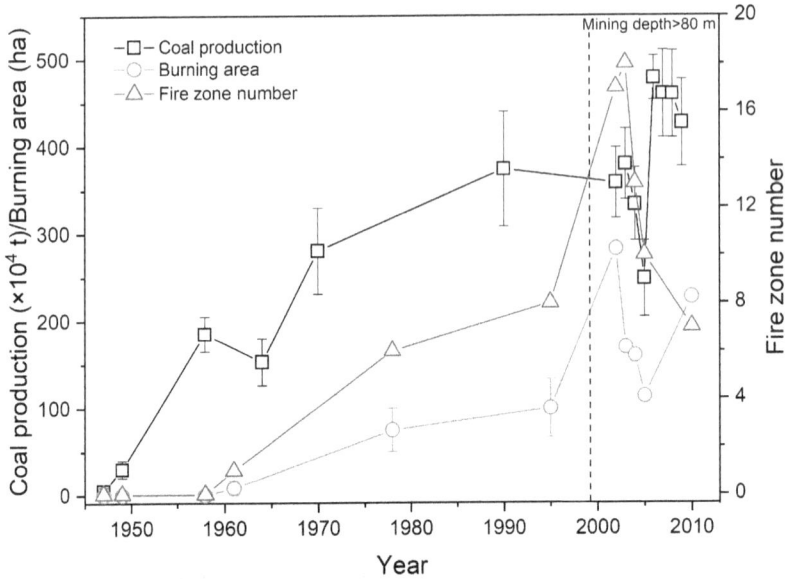

FIGURE 3.3 Developments of coal production, burning area, and fire zone number from 1950s to 2010s.

and goaf maps, were employed. Field surveys were conducted on the surface to investigate hazards induced by UCFs. In this work, thermal anomaly zones were considered to be the main indicator of the hazards. The thermal anomaly zones were demarcated as irregular polygons according to temperature data, as measured by handheld infrared thermometers. Solar radiation and background temperature influencing the interpreted thermal anomaly zones were also taken into consideration. The stratigraphic data was obtained to provide important information for coal seam depth and rock layers. Underground mining activities—such as coal and rock excavation—were determined by the maps of underground galleries and goafs. It is worth mentioning that field surveys, stratigraphic data, underground galleries, and goaf maps all describe different spatial dimensions. The integration of these data made possible the construction of a three-dimensional space, which helped to reveal the relationship between UCF hazards and underground mining activities. By unifying the coordinates and using geographic information systems, the field surveys, underground galleries, and goaf maps were fused into a single map of a fire zone. With the merged maps and the stratigraphic data, a bow-tie analysis was used to explore the causes and effects of UCFs.

3.4 RESULTS AND DISCUSSION

UCFs in the Wuda coalfield are mainly located in the Su Hai Tu Coal Mine and Wu Hu Shan Coal Mine. The coal seams strike towards the southeast (see Figures 3.4 and 3.6). UCFs are rare in the Huang Bai Ci Coal Mine, and this could

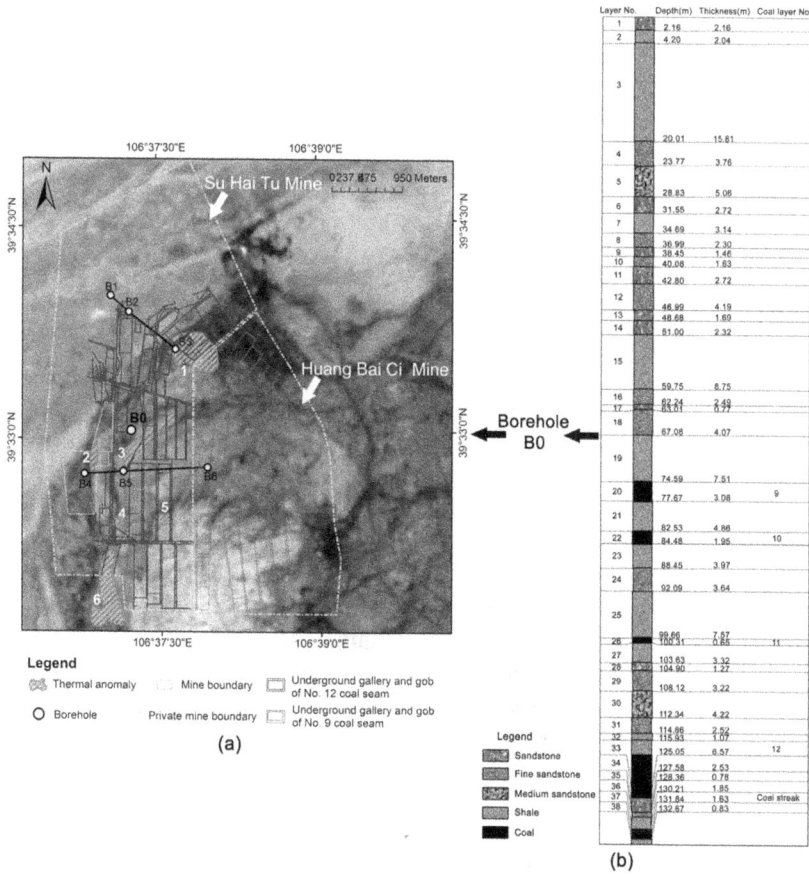

FIGURE 3.4   (a) Horizontal profiles of thermal anomalies merged with underground mining galleries and goafs (b) B0 Bore hole data of Su Hai Tu Coal mine.

be due to it being a deeper coal seam, as well as it being below a layer of stiff overlying rock. Herein, case studies on Su Hai Tu Coal Mine and Wu Hu Shan Coal Mine are presented, but any on Huang Bai Ci Coal Mine have been excluded.

3.4.1 INTERPRETATION OF HYBRID DATA AND IDENTIFICATION OF HAZARDS

Figure 3.4 (a) shows a horizontal profile of surface thermal anomalies merged with subsurface mining galleries and goafs. There are six thermal anomaly zones (No. 1–No. 6 Fire Zones). The extracted coal in the Su Hai Tu Coal Mine originated from the 9th, 10th, and 12th coal seams. UCFs happened at the 9th and 10th coal seams. Note that, due to incomplete data for the 9th and 10th coals seams, the subsurface map of mining galleries and goafs shown in Figure 3.4 (a) is the 12th coal seam. However, the subsurface map still provides useful reference information for underground mining galleries and goafs at the 9th and 10th coal seams.

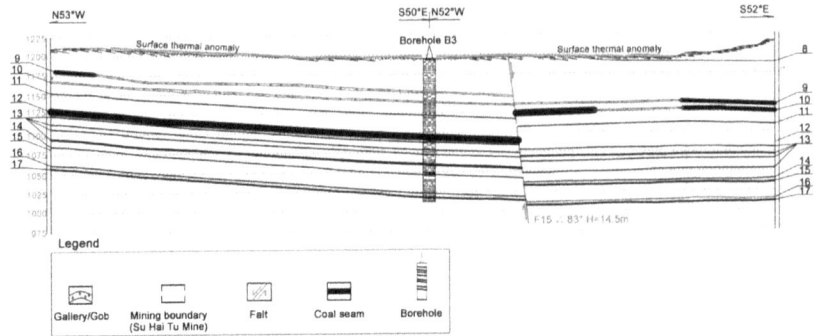

FIGURE 3.5 Vertical profile of thermal anomalies merged with underground mining galleries and goafs as well as B3 borehole data for Su Hai Tu Coal Mine.

Seven boreholes (B0–B6) were drilled during the geological surveys. Herein, B0 borehole data is presented in Figure 3.4 (b). In addition, B3–B6 borehole data was included in the construction of the vertical profiles shown in Figures 3.5 and 3.6.

Figure 3.4 indicates that only one UCF was caused by private mining activities, and most UCFs were triggered by long-wall mining activities in the Su Hai Tu Coal Mine. The long-wall mining technique has been widely employed in China. The extracted coal seam forms a caved zone, i.e. goaf, and overlying rocks collapse, leading to subsidence. Ventilation pathways, such as cracks and fissures, occur if the depth of the extracted coal seam is shallow enough. Particularly, Figure 3.4 (b) shows that the main overlying rock layers are comprised of shale, which is prone to mechanical failure that creates fissures and cracks. Field work in the Wuda coalfield confirmed that ventilation pathways existed in the UCF zone. Figure 3.4 (a) and Figure 3.5 show that No. 1 fire zone overlapped with goafs at the 9th and 10th coal seams. We inferred the fire depth of No. 1 UCF to be around −50 m. Furthermore, the surface thermal anomaly near borehole B3 was spatially consistent with the underground goafs, as shown in Figure 3.5. It indicates that No. 1 UCF was caused by underground long-wall mining at the 9th and/or the 10th coal seam.

Figure 3.6 presents a vertical profile of thermal anomalies merged with underground mining galleries and goafs for the Su Hai Tu Coal Mine. Three boreholes

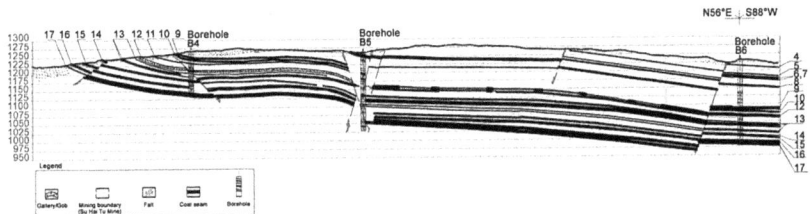

FIGURE 3.6 Vertical profile of thermal anomalies merged with underground mining galleries and goafs as well as B4, B5, and B6 borehole data for Su Hai Tu Coal Mine.

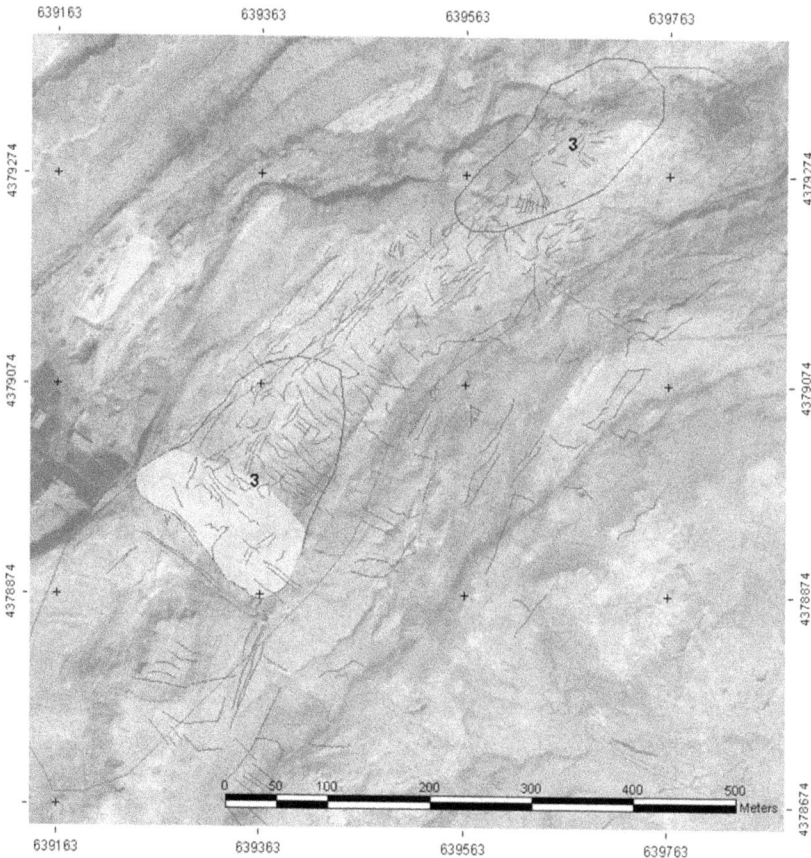

FIGURE 3.7 Field surveys on cracks occurred in No. 3 UCF of the Wuda coalfield. (Kuenzer et al., 2008.)

(B4–B6) were drilled along the profile. Figure 3.4 shows that No. 3 UCF occurred around borehole B5 and B0. Borehole data shown in Figures 3.4 and 3.6 suggests that UCF originated from the 9th coal seam, which is approximately 25 m below the surface (see Figure 3.6). Additionally, Figure 3.6 illustrates that several faults exist around the No. 3 UCF. The shallow depth along with the faults make the generation of ventilation pathways more likely if the coal seam is extracted. Field surveys confirmed that many cracks were formed in the No. 3 UCF zone, as shown in Figure 3.7 (Kuenzer et al., 2008), fueling the UCF. Figure 3.4 indicates that the direction of the No. 3 UCF propagation was in accordance with the extracted caves (goafs). It demonstrates that the coal fire in zone 3 was triggered by the extraction of underground coal seams.

Along with the fire in zone 3, most UCFs—for example, No. 4–6 (see Figure 3.4), No. 7, and No. 10 (see Figure 3.8)—in the Wuda coalfield exhibit the same propagation behavior. They propagated along the outlines of goafs shown in the merged

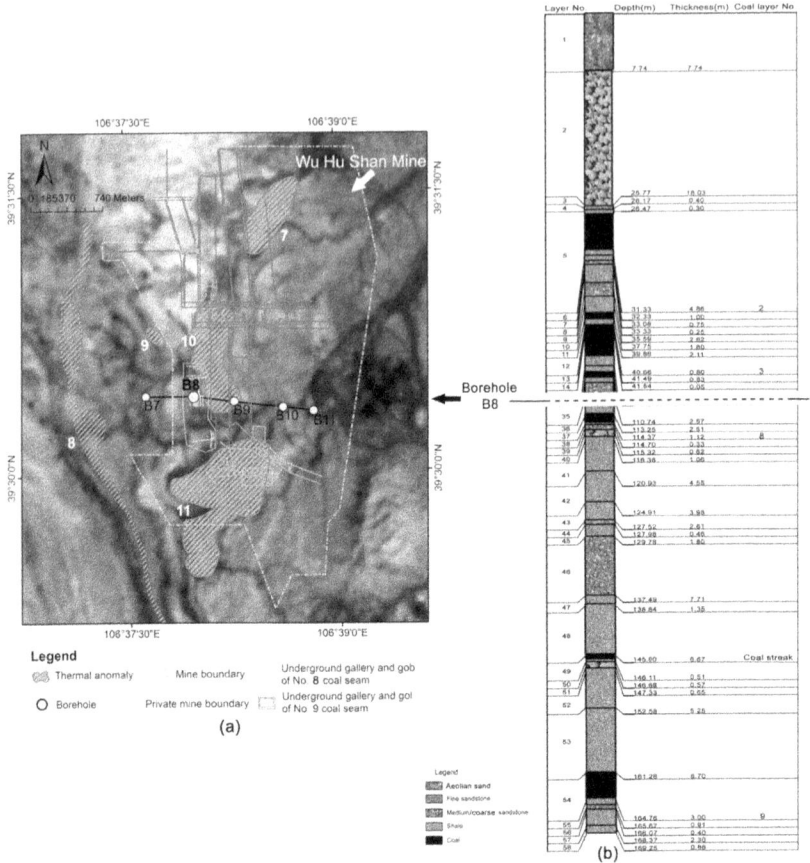

FIGURE 3.8 (a) Horizontal profiles of thermal anomalies merged with underground mining galleries and goafs (b) B8 Bore hole data of Wu Hu Shan Coal mine.

maps. Note that No. 8 fire zone shown in Figure 3.8 was mainly associated with outcrops of the coal seam, while No. 11 fire zone was caused by private mining activities. Hence, their propagation behavior is not relevant to the goafs created by the long-wall mining activities. Also note that the UCFs that occurred in the Wu Hu Shan Coal Mine (see Figure 3.8) happened in the 3rd coal seam, but the coal data for that seam was missing. The underground maps of galleries and goafs for the 8th and 9th coal seams might also provide useful information for the interpretation. Figure 3.8 (b) indicates that the fire depths of UCFs in the Wu Hu Shan Coal Mine are around 45 m.

3.4.2 BOW-TIE ANALYSIS

A bow-tie diagram model for UCF is shown in Figure 3.9. A fault tree (FT) model is presented at the left-hand side of the diagram, and an event tree (ET) model

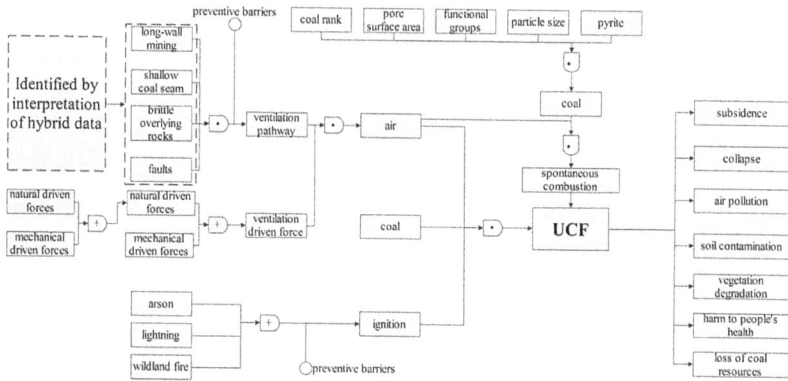

FIGURE 3.9   Bow-tie diagram model for UCF.

is on the right-hand side. Figure 3.9 indicates that UCFs start through compli-cated causes and lead to severe consequences, which, in theory, explains why the prevention and mitigation of UCFs prove to be so challenging. The ET model includes multiple consequences, including subsidence, collapses, air pollution, soil contamination, vegetation degradation, harm to human health, and loss of coal resources. These consequences could occur simultaneously or successively.

In the bow tie analysis, two FT models lead to the top event—an underground coal fire. UCFs can be ignited by arson, lightning, and fires in the surface eco-system, as can be seen at the bottom of the bow-tie diagram in Figure 3.9. UCFs can also be ignited by the spontaneous combustion of coal, as is presented at the top of the bow-tie diagram. Experimental research reported that coal is prone to spontaneous combustion only if the physical (pore surface area and particle size) and chemical properties (coal rank, functional groups, and pyrite) of coal meet certain criteria. Further details about this topic can be found in the relevant literature. Whether through forced ignition or spontaneous combustion, air avail-ability is the common and controlling factor for UCFs. Air availability depends on ventilation pathways and driving forces. It has been acknowledged that the driving forces of UCF ventilation consists of mechanical forces, such as mechani-cal fans used in coal mines, and natural forces, such as topographic effects and atmospheric fluctuations.

However, factors that create the ventilation pathways are not well understood, due to the complicated causal relationship and tempo-spatial patterns among UCFs, mining activities, and geological setups. This chapter proposed an approach to address this challenge. The factors triggering the ventilation pathways were identified via the interpretation of hybrid data from field surveys, underground mining maps, and stratigraphy. It was found that long-wall mining, depth of coal seam, overlying rock properties, and geological faults were all significant factors leading to the generation of ventilation pathways. Actually, save for geological faults, these factors are highly associated with some geohazards, such as subsid-ence and collapses.

Both the ignition vectors and the ventilation pathways are potential avenues for prevention and mitigation of UCFs, as shown in Figure 3.9. The most important barrier is to avoid the generation of ventilation pathways. Figure 3.9 indicates that the mining technical procedures and selection of geological sites must be carefully considered. If the depth of a coal seam is shallow, e.g., <40 m, and the overlying rock is brittle, that site is at a high risk for UCFs if the long-wall mining technique is used. For this scenario, room and pillar mining could be a safer alternative. Figure 3.9 also suggests that blocking ventilation pathways, through the injection of grouts or gels widely utilized in fire-fighting engineering projects, are beneficial in the mitigation of UCFs and their consequent hazards.

3.5 CONCLUSIONS

UCFs are an environmental catastrophe as well as an important safety concern for the coal mining industry. It is of extreme importance to discover the causes and consequences of these fires in order to prevent and mitigate them. The bow-tie diagram provides a valuable technical framework for analyzing the causes and effects of UCFs. However, the task is made more challenging by the complicated causal relationship and tempo-spatial patterns among UCFs, mining activities, and geological setups. In this chapter, we proposed a novel technical solution for this challenge. Hybrid data including field surveys, underground mining maps and stratigraphy was merged so that a tempo-spatial pattern emerged and the causal relationships were revealed. Then, for the first time, a bow-tie diagram model for UCFs was constructed along with the identified factors triggering ventilation pathways, as interpreted from the hybrid data. The bow-tie diagram illustrates the complicated causes and severe consequences of UCFs that lead to the practical challenges of preventing and mitigating UCFs. The most significant avenue of prevention and mitigation is to avoid the generation of ventilation pathways via appropriate mining procedures and the reasonable selection of geological sites. A quantitative analysis for the bow-tie model will be conducted in the future.

ACKNOWLEDGMENTS

This work was funded by National Natural Science Foundation of China (No. 51804168), as well as the China and Postdoctoral Science Foundation (No. 2018T110492 and No. 2017M620209).

REFERENCES

Dai, S., Ren, D., Chou, C.-L., Finkelman, R.B., Seredin, V.V., Zhou, Y., 2012. Geochemistry of trace elements in Chinese coals: A review of abundances, genetic types, impacts on human health, and industrial utilization. Int J Coal Geol 94, 3–21.

Hower, J.C., Henke, K., O'Keefe, J.M.K., Engle, M.A., Blake, D.R., Stracher, G.B., 2009. The Tiptop coal-mine fire, Kentucky: Preliminary investigation of the measurement of mercury and other hazardous gases from coal-fire gas vents. Int J Coal Geol 80, 63–67.

Hower, J.C., O'Keefe, J.M.K., Henke, K., Bagherieh, A., 2011. Time series analysis of CO concentrations from an Eastern Kentucky coal fire. Int J Coal Geol 88, 227–331.

Hower, J.C., O'Keefe, J.M.K., Henke, K.R., Wagner, N.J., Copley, G., Blake, D.R., Garrison, T., Oliveira, M.L.S., Kautzmann, R.M., Silva, L.F.O., 2013. Gaseous emissions and sublimates from the Truman Shepherd coal fire, Floyd County, Kentucky: A re-investigation following attempted mitigation of the fire. Int J Coal Geol 116–117, 63–74.

Kuenzer, C., 2005. Demarcating coal fire risk areas based on spectral test sequences and partial unmixing using multi sensor remote sensing data. Technical University Vienna, Vienna.

Kuenzer, C., Stracher, G.B., 2012. Geomorphology of coal seam fires. Geomorphology 138, 209–222.

Kuenzer, C., Zhang, J., Hirner, A., Bo, Y., Jia, Y., Sun, Y., 2008. Multitemporal in-situ mapping of the Wuda coal fires from 2000 to 2005—assessing coal fire dynamics. UNESCO, Beijing.

Kuenzer, C., Zhang, J., Sun, Y., Jia, Y., Dech, S., 2012. Coal fires revisited: The Wuda coal field in the aftermath of extensive coal fire research and accelerating extinguishing activities. Int J Coal Geol 102, 75–86.

Kuenzer, C., Zhang, J., Tetzlaff, A., van Dijk, P., Voigt, S., Mehl, H., Wagner, W., 2007. Uncontrolled coal fires and their environmental impacts: Investigating two arid mining regions in north-central China. Appl Geogr 27, 42–62.

Li, C., Liang, H., Chen, Y., Bai, J., Cui, Y., 2018a. Distribution of surface soil mercury of Wuda old mining area, Inner Mongolia, China. Hum Ecol Risk Assess 24, 1421–1439.

Li, C., Liang, H., Liang, M., Chen, Y., Zhou, Y., 2018b. Mercury emissions flux from various land uses in old mining area, Inner Mongolia. China J Geochem Explor 192, 132–141.

Liang, Y., Zhu, S., Liang, H., 2018. Mercury enrichment in coal fire sponge in Wuda coalfield, Inner Mongolia of China. Int J Coal Geol 192, 51–55.

O'Keefe, J.M., Henke, K.R., Hower, J.C., Engle, M.A., Stracher, G.B., Stucker, J.D., Drew, J.W., Staggs, W.D., Murray, T.M., Hammond, M.L., Adkins, K.D., Mullins, B.J., Lemley, E.W., 2010. CO2, CO, and Hg emissions from the Truman Shepherd and Ruth Mullins coal fires, eastern Kentucky, USA. Sci Total Environ 408, 1628–1633.

Prakash, A., Gens, R., Prasad, S., Raju, A., Gupta, R.P., 2013. Coal Fires in the Jharia Coalfield, India, in: Stracher, G.B., Prakash, A., Sokol, E.V. (Eds.), Coal and peat fires: A global perspective. Elsevier, Oxford, UK, pp. 154–177.

Song, Z., Kuenzer, C., 2014. Coal fires in China over the last decade: A comprehensive review. Int J Coal Geol 133, 72–99.

Song, Z., Kuenzer, C., 2017. Spectral reflectance (400–2500 nm) properties of coals, adjacent sediments, metamorphic and pyrometamorphic rocks in coal-fire areas: A case study of Wuda coalfield and its surrounding areas, northern China. Int J Coal Geol 171, 142–152.

Song, Z., Kuenzer, C., Zhu, H., Zhang, Z., Jia, Y., Sun, Y., Zhang, J., 2015. Analysis of coal fire dynamics in the Wuda syncline impacted by fire-fighting activities based on in-situ observations and Landsat-8 remote sensing data. Int J Coal Geol 141–142, 91–102.

Stracher, G.B., 2004. Coal fires burning around the world: A global catastrophe. Int J Coal Geol 59, 1–6.

Stracher, G.B., Finkelman, R.B., McCormack, J.K., Schroeder, P.A., Pone, D., Annegarn, H., Blake, D.R., 2013. Coalfield Fires of South Africa, in: Glenn, B., Stracher, A.P., Ellina V. Sokol (Eds.), Coal and peat fires: A global perspective. Elsevier, Oxford, UK, pp. 408–426.

Wessling, S., Kuenzer, C., Kessels, W., Wuttke, M.W., 2008. Numerical modeling for analyzing thermal surface anomalies induced by underground coal fires. Int J Coal Geol 74, 175–184.

Zhang, J., Kuenzer, C., 2007. Thermal surface characteristics of coal fires 1: Results of in-situ measurements. J Appl Geophys 63, 117–134.

Zhang, J., Kuenzer, C., Tetzlaff, A., Oertel, D., Zhukov, B., Wagner, W., 2007. Thermal characteristics of coal fires 2: Results of measurements on simulated coal fires. J Appl Geophys 63, 135–147.

4 Technologies to Inhibit Coal Fires

Jun Deng, Yang Xiao, and Qing-Wei Li
School of Safety Science and Engineering, and
Shaanxi Key Laboratory of Prevention and Control
of Coal Fire, Xi'an University of Science and
Technology, Xi'an, People's Republic of China

Hao Zhang
School of Safety Science and Engineering,
Xi'an University of Science and Technology,
Xi'an, People's Republic of China

CONTENTS

DOI: 10.1201/9781003140382-4

4.1 INTRODUCTION

In the coming decades [1–4], coal energy will remain an important fossil fuel. However, coal fires are a serious hazard that can occur during coal mining, storage, or transportation processes. Coal spontaneous combustion (CSC) is a worldwide phenomenon that has the potential to cause significant social and ecological problems [5–7]. The distribution of major coal fires in Figure 4.1 clearly indicates that these fires happen globally [8].

The ongoing coal fire disaster in China is listed as one of the "world's five major ecological disasters" by the American media—the more serious among these fires are the ones caused by CSC [9]. According to some data, more than 90% of the coal seams mined in China are flammable. CSC threatens the continuation of thick coal seam mining because coal fires lead to significant economic losses and hidden safety risks [10, 11]. In addition to destroying coal resources, the gases produced by coal fires cause serious environmental problems, such as haze and the greenhouse effect. These fires also endanger human health by causing underground accidents, as well as by increasing the frequency of respiratory diseases. China is rich in coal resources; there are 14 large coal bases covering diverse geological conditions. To date, 130 large- and medium-sized mining areas have spontaneously ignited. Any reputable coal mine safety production plan emphasizes that "development must not be at the expense of safety". Therefore, safety must be a prerequisite to every step in the coal production process, and the prevention and control of CSC is a primary factor in creating that safety.

Figure 4.2 describes the process and the main characteristics of CSC [12]. CSC is a complex process. Certain conditions—broken coal pillars, coal walls, concentrated floating coal, and a certain amount of air and oxygen—trigger physical and chemical changes in the coal itself: oxygen absorption and oxidation [13, 14].

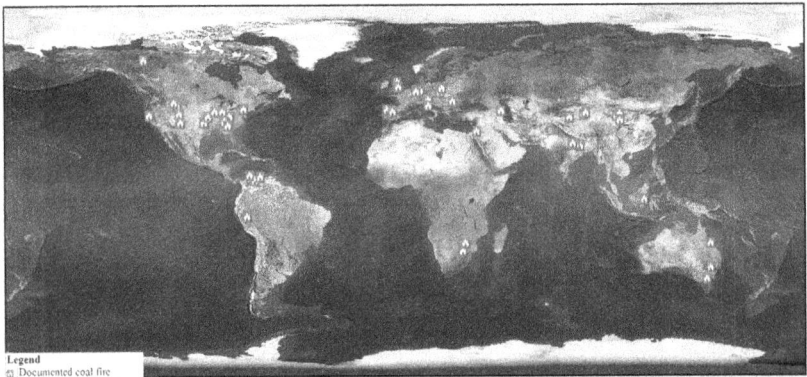

Legend
⬚ Documented coal fire

FIGURE 4.1 Distribution of major coal fires around the globe [8].

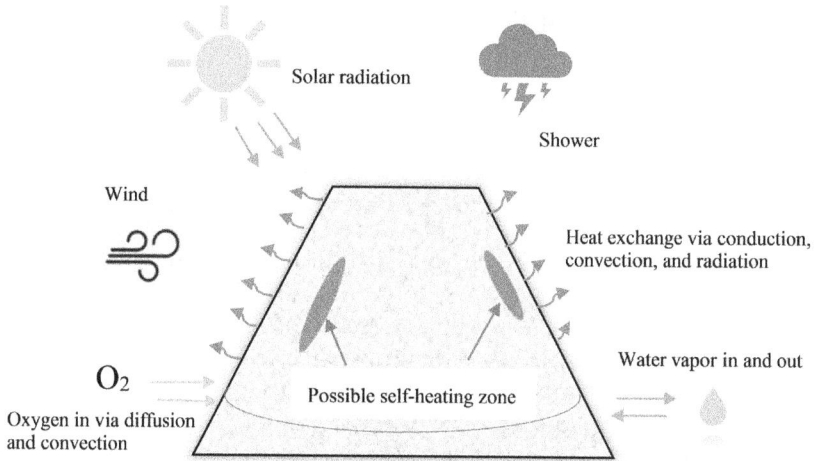

FIGURE 4.2 Process schematic and major characteristics of CSC [12].

The process involves a chemical reaction between coal and oxygen, which leads to a gradual accumulation of heat, which creates favorable conditions for combustion [15–17].

Currently, the most common CSC fire-fighting technologies are grouting, chemical inhibitors, pressure equalization, and the addition of inert gases. Among these technologies, chemical inhibitors are widely used to prevent and suppress the process of CSC. In recent years, scientific researchers have conducted in-depth research on the inhibitory effect of different chemicals, as well as the variety of mechanisms through which they inhibit CSC. Significant progress has been made. The prevention and suppression of CSC are based on the disruption of the basic conditions that cause CSC to occur. When any condition is broken, it will delay or stop the CSC process.

The focus of this chapter is primarily to classify different inhibitors according to their mechanism of inhibiting CSC. In addition, the advantages and disadvantages of various kinds of CSC inhibitors are analyzed and compared according to their practical application in the field.

4.2 CSC AND THE INHIBITION MECHANISM

Coal is a mixture of complex organic compounds and minerals. Due to the different time frames in which coal is formed, its physical and chemical properties vary greatly. Therefore, CSC is an extremely complicated physical and chemical phenomenon. The study of CSC began in the 17th century, but people have still not yet reached a consensus understanding of the CSC mechanism.

4.2.1   MECHANISM OF THE COAL SPONTANEOUS COMBUSTION REACTION

The CSC mechanism reveals the chemical reactions involved in that process, which provides a theoretical basis for a comprehensive understanding of its combustion characteristics. Among many reaction mechanisms, coal-oxygen reaction theory is accepted as the most probable by experts and scholars. The focus is on coal and oxygen. The adsorption and oxidation interaction between coal and oxygen can be proven through experimental research.

From a macro point of view, CSC can be divided into four stages according to the change of temperature, characterization of phenomenon, and gas release: the incubation, the heat storage, the combustion, and the extinction stages. The incubation stage refers to the time when the coal seam is in contact with the air. The normal temperature of the coal begins to increase. During this period, the physical adsorption of oxygen gas is the primary characteristic, with traces of chemical adsorption and chemical reaction. Gas products, such as CO appear, and the heat exchange is weak. In addition, the length of the incubation period depends on the nature of the coal as well as external conditions. The heat storage stage marks the time when the temperature of the coal seam rises to the ignition point. Coal seam adsorption of air changes from physical adsorption to chemical adsorption. Chemical reactions—producing a large amount of CO, CO_2, CH_4, C_2H_4, C_2H_2, H_2, and other gases—increase which facilitates a large amount of heat exchange and transfer. This stage is also the focus of a common technology: the CSC gas early warning system. During the combustion stage, the temperature of the coal reaches the ignition point. When sufficient oxygen is present, combustion will occur. An open flame will appear that is accompanied by a large amount of gas release and heat exchange. The last stage is the extinction of the process. When the coal resources are fully combusted or the fire is controlled by human factors, the coal stops burning and the temperature drops.

Numerous scientific researchers have conducted in-depth studies on CSC from a macro perspective, using various thermal analysis methods. For example, adiabatic oxidation method [18], CPT method [19], C80 microcalorimeter [20, 21], program temperature [22], thermogravimetric analysis, differential thermal analysis, and differential scanning calorimetry [23]. From a microcosmic perspective, coal is a complex macromolecular structure. In the study of the surface molecular structure of CSC, different methods are used to study the microstructure of coal molecules. Among them, in situ Fourier transform infrared (FTIR) analysis [24], X-ray computed tomography [25, 26], FTIR spectroscopy [27], electron spin resonance (ESR) [28, 29], scanning electron microscopy [30], and nuclear magnetic resonance [31] have been applied widely to explore the microcosmic reactions involved in CSC.

Based on the summary of scholarly research, the active group on the surface of coal is the primary driver of oxidative CSC [32–34]. The reaction happens in three steps, and heat is generated in each of them. Researchers, while accounting for the variety in chemical structures of different types of coal, simulated the oxidation process between coal and oxygen molecules. The conditions leading

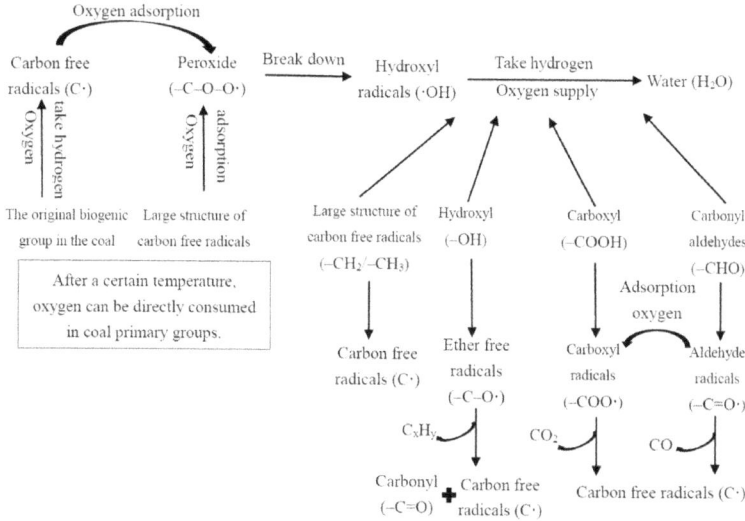

FIGURE 4.3 CSC in the chemical reaction diagram [35].

to the release of H_2O, CO, CO_2, and other gases were studied. This revealed the key chemical kinetic mechanism in the oxidation process. The reactions of active groups are illustrated in Figure 4.3 [35].

During the chemical reaction, there is a reduction in the aliphatic C–H key of the reactive group and an increase in the oxygen-containing carbonyl and carboxyl functional groups. The aliphatic substituents are the first to be oxidized. The product in the form of peroxide adsorbs into the surface of the coal molecule because each functional group has a different capacity to be oxidized. Therefore, the energy needs for the oxidation process are not the same across functional groups.

This is also closely related to the degree of coal metamorphism. The higher the degree of coal mineralization, the more likely the ether bond is attacked and broken, while the number and types of substituted benzenes on the surface decrease. In FTIR, the ether-bond peak absorbance of anthracite is relatively weak. Combined with the research made by experts and scholars, it has been found that, with the increase of the degree of coal, the changes in the structure of associating hydroxyl groups, $-CH_2$, $-CH_3$, C=O, C–O, minerals, and substituted benzenes in coal molecules cause changes to the surface-active groups of coal molecules. At the same time, the number of aromatic C=C structures and graphite carbon microcrystalline structures increase. From the microscopic point of view, the variation in oxidation activity of each coal sample is caused by the difference of functional groups in the molecular structure.

4.2.2 MECHANISM OF INHIBITION

The technique of inhibition is to add substances that inhibit CSC and to suppress or reduce the exothermic heat of CSC by spraying or pressing the coal. Based on

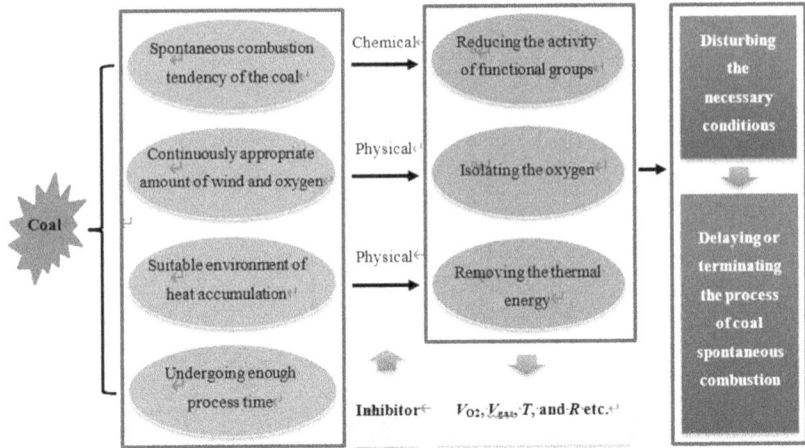

FIGURE 4.4 Inhibiting mechanism for CSC and the characterization parameters [36]. (Note: V_{O_2} and V_{gas} represent the rates of oxygen consumption and gas production, respectively; T stands for the characteristic temperature; R is the inhibiting efficiency).

the mechanism of CSC, the inhibitors can be divided into two categories: physical inhibitors and chemical inhibitors. Physical inhibitors generally focus on the influence of temperature and oxygen concentration on the coal oxidation process as well as external factors—such as endothermic cooling and oxygen isolation— to reduce the temperature of the coal which slows down or stops the oxidation process. Chemical inhibitors usually change the type of active sites on the coal surface and reduce the number of active groups and free radicals by adding the appropriate chemicals that interfere with the microscopic mechanism of the oxidation reaction. The CSC inhibition mechanism and the characterization of macro parameters are presented in Figure 4.4 [36].

4.3 TYPES AND FUNCTIONS OF CSC INHIBITORS

An inhibitor is a material that prevents coal from combining with oxygen, reduces coal oxidation activity, or is used to inhibit oxidation and prevent CSC. In this section, various inhibitors are classified and introduced, starting with those that inhibit CSC.

4.3.1 PHYSICAL INHIBITORS

Physical inhibitors are a class of agents that prevent CSC by changing the environment surrounding the coal or the physical conditions of the coal itself. They mainly act on the physical adsorption stage of CSC. There are currently many studies investigating these materials. The success of physical inhibitors is achieved through two mechanisms: one is to form a protective layer of liquid film on the surface of the coal by covering the surface of the active center. This prevents the coal and oxygen from making contact. The other aspect is to add a large amount of water to the coal body. As the water evaporates, heat is transferred away from the coal body which reduces the rate of oxidation.

4.3.2 WATER-ABSORBING SALT INHIBITORS

The most commonly used water-absorbing salt inhibitors are chlorine salts, such as $MgCl_2$, KCl, $CaCl_2$, and NaCl [37]. By creating a barrier between the coal and oxygen and by absorbing water, these inhibitors can decelerate the oxidation of coal. These compounds are typically active for only a relatively short time, and they have a low efficiency. As a consequence, high concentrations (>20%) of these salt inhibitors are required to achieve effective inhibition [38, 39]. Zheng [40] investigated the effects of $MgCl_2$, $NH_4H_2PO_4$, and NH_4HCO_3 on the inhibition of CSC and discovered that the inhibiting effect was considerable only when the concentration of the inhibitor exceeded 20%.

Yu et al. [41] reported that $MgCl_2$ performed favorably in the early stages of CSC, but its inhibiting effects were weakened in later stages so much that they became catalytic [42]. Its success in the initial stages was because of the moisture-absorbing and water-retaining properties of $MgCl_2$, which enabled it to form a liquid film that isolated air and reduced the temperature of the coal. However, as the moisture evaporated, the inhibiting effect ceased Slovák and Taraba [43] compared the inhibiting effects of $CaCl_2$ and urea by calculating the apparent activation energy (E_a) and concluded that the presence of urea in coal led to a reduction in the E_a of the oxidation process, at temperatures in excess of 200°C. Urea is a catalyst for coal oxidation, but at temperatures of up to 300°C, it can be used as an inhibitor.

To solve these issues, Cui et al. [44] proposed using thermo-responsive inhibitors (TRIs) to inhibit CSC. The team discovered that the most stable inhibiting effect was achieved with $CaCl_2 \cdot 6H_2O$; during heating, its rate of inhibition rose steadily to reach a maximum of 79.9% at 200°C. This information is valuable for the future development of inhibitors.

4.3.3 AMMONIUM SALT INHIBITORS

Ammonium salts are well-known inhibitors that undergo endothermic reactions during the pyrolysis process. These salts effectively absorb the heat produced during the coal oxidation reaction. In addition, ammonium salts decompose into NH_3 and CO_2 which can dilute the oxygen concentration, thereby reducing the coal oxidation reaction rate [45]. The pyrolysis of ammonium salts enhances the capture of free radicals formed in the coal oxidation reaction, thereby terminating the chain reactions [46]. However, due to the poor stability of some ammonium salts—decomposition at 30°C—as well as the creation of NH_3 from that decomposition reaction, they pose a health hazard for underground mining personnel.

4.3.4   ALKALINE INHIBITORS

Because most coal is rich in sulfur-containing pyrite, CSC is prone to hydrolysis, which leads to an oxidation reaction and the release of thermal energy [47]. The oxidation process causes the pyrite's volume to expand, thereby increasing the contact area between the coal and oxygen, which speeds up the coal-oxygen reaction. $Ca(OH)_2$ is a common alkaline inhibitor that is typically used in a solution of water at concentrations of 10–15%. It breaks the self-oxidation cycle of highly sulfurous coal. The chemical action of $Ca(OH)_2$ on highly sulfurous coal is as follows [48]:

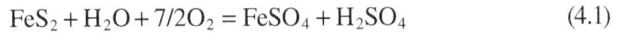

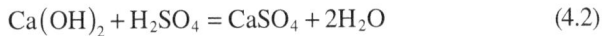

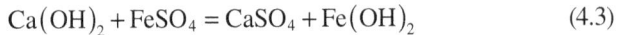

$$FeS_2 + H_2O + 7/2O_2 = FeSO_4 + H_2SO_4 \tag{4.1}$$

$$Ca(OH)_2 + H_2SO_4 = CaSO_4 + 2H_2O \tag{4.2}$$

$$Ca(OH)_2 + FeSO_4 = CaSO_4 + Fe(OH)_2 \tag{4.3}$$

$$3Ca(OH)_2 + Fe_2(SO_4)_3 = 3CaSO_4 + 2Fe(OH)_3 \tag{4.4}$$

$Ca(OH)_2$ inhibits the combustion of high-sulfur coal in three ways—through physical, chemical, and negative catalysis. $CaSO_4$ has low solubility in water, and $Fe(OH)_2$ and $Fe(OH)_3$ are colloidal substances; these characteristics enable a favorable coverage and a filling effect to be achieved. These chemicals react with unreacted $Ca(OH)_2$ on the surface of pyrite. This reaction forms a hydrophilic membrane that covers the active surface center of the coal, thereby decreasing the number of effective collisions between reactant molecules and effectively inhibiting the oxidation reaction.

4.3.5   COLLOIDAL INHIBITORS

Colloidal inhibitors comprise an inorganic mineral gel and composite colloids of fly ash, loess, and sand. These inhibitors act as a barrier to limit the air supply to the coal-oxygen reaction [49–53]. Experiments conducted by Zhong [54] revealed that colloidal materials have an inhibiting effect on the heating, CO production, and oxygen consumption rates of coal. In terms of chemical composition and water-saving capacity, the preferential order of colloidal inhibitor use is polymer colloids > water glass > fly ash > composite colloids. Because of the exploitation of coal in Northwest China and water shortages, the cost of conventional, loess-containing grouting is high. In addition, if fly ash or sand is simply injected into water, the ratio of water to sand/fly ash must reach 7:1–15:1. Xu et al. [55] found that the compound appending a material comprising a sand-suspended colloid was composed of inorganic mineral gel—which formed the basis of the colloid—organic polymers, and a dispersant. They also confirmed that these materials have an inhibiting effect on CSC. For the prevention of CSC, a new type of gel—corn straw-co-AMPS-co-AA hydrogel—was obtained by Cheng et al. [56] through the graft co-polymerization of corn straw, 2-acrylamide-2-methylpropanesulfonic acid (AMPS), and acrylic acid (AA). The gel was employed to forestall re-ignition. Therefore, it is considered to be superior to the aforementioned gels.

4.3.6   INERT GAS AND FOAM INHIBITORS

N_2 and CO_2 are the main inert gas inhibitors, and when pumped into an enclosed area, these inhibitors dilute the concentration of oxygen in the space, thereby reducing the coal oxidation rate [57, 58]. The main foam inhibitors are inert foam, high-magnification micro-capsule foam, three-phase foam [59], and paste foam [60].

Smith [61] introduced CO_2 into the bottom of heat-treated coal through a liquid shunt to study the mechanism and application of CO_2 for the heat treatment of coal piles. Smith confirmed that this reduced the rate of oxidation and inhibited CSC. Qin [62] revealed that three-phase foam comprising fly ash, nitrogen, and water exhibited excellent coal fire extinguishing characteristics. Therefore, it has considerable potential to mitigate the coal fire problem. However, its application is limited by its inability to produce a substantial amount of gaseous CO_2 in underground spaces. Liu et al. [63] addressed this concern by independently developing a new apparatus that yielded more than 205 times as much CO_2 as dry ice under normal conditions by efficiently sublimating dry ice. Lu [60] revealed that extensive water slurry loss must be avoided after the rupture of foam cells. This research also invented the paste foam inhibitor that blocks the fracture network in coal along the radial direction. With the temperature of the coal particles at approximately 670°K, a more favorable bubble configuration was maintained, and this had the advantageous effect of cooling the coal's surface. In addition, due to the strong stacking ability of the foam material, it effectively covered the surface of the coal, thus isolating the coal from oxygen. At the same time, the water in the foam absorbed heat, after the absorption of heat foam dehydration burst. In this process, the inert gas was released and diluted the concentration of oxygen, so as to inhibit CSC. However, it was difficult to maintain the adhesion of the liquid film on the surface of the coal after the foam was broken, which was caused by the relatively low stability of the foam. Therefore, in order to improve its ability to inhibit CSC, it is necessary to improve the foam's stability.

4.3.7   ORGANIC FOAM INHIBITORS

Rock Off, Marie San, and Agloni are examples of organic foam inhibitors. Polyurethane foams [64] containing isocyanate and polyether polyols in the cross-linking agent, foaming agent, surfactant, and other compounds are applied to the coal and rock reinforcement, gas sealing, and filling and plugging air. Phenolic foam [65] is a phenolic resin that contains a foam agent, a curing agent, a surfactant, and other additives and is polymerized into a closed-cell rigid foam. It is suitable for use in fireproof, closed walls; filling high-risk fire areas; and gas drilling sites in abandoned spaces. Urea-formaldehyde foam [66] is composed of urea-formaldehyde resin, surfactants, hardeners, foaming agents, and other additives. It is suggested to use mechanically-stirred or chemically prepared polymer foam to fill the surface area.

4.3.8   OTHER INHIBITORS

There are some other inhibitors that can be used to inhibit CSC, such as polymer inhibitors, aerosol inhibitors, and light paste inhibitors.

Polymer inhibitors are polymers containing surfactants and additives. These inhibitors adsorb onto the surface of the coal, and the heat energy of the coal is transferred to the air via evaporation, thereby cooling the coal. In addition, the polymer emulsion covers the coal surface which isolates the coal from oxygen. Xiao and Du [67] used polymer molecules, water glass, calcium oxide, and surfactants to prepare a polymer inhibitor with a significant inhibitory effect; at about 100°C, a polymer concentration of 0.12% achieved an inhibition rate of 90%. Wang [68] studied the polyvinyl acetate emulsion inhibitor and found that it had a significant cooling effect as well as effectively acting as an oxygen barrier.

Aerosol inhibitors are used as a novel fire prevention technology in which chemical inhibitors and other synergistic mechanisms achieve fire prevention through physical cooling. Deng et al. [69] found that a supersonic atomizing nozzle (MAL1130B1) resulted in particles of the optimal fire-preventing size when a liquid pressure of 1.2 MPa and air pressure of 0.6 MPa were employed. A cold fog was formed that favorably extinguished fires [70].

Light paste inhibitors are used primarily because of their favorable properties, such as a certain compressive strength, their capacity to expand and fill the caving area, as well as plugging air leakage channels. Xiao et al. [71] researched a lightweight paste material comprising, by mass, 20% PB-type activator, 2% compound enhancer, 1% compound bulking agent, and 77% material additives. After filling in a goaf, it expanded to prevent gas from flowing to a main roadway, and inhibiting CSC.

4.3.9   CHEMICAL INHIBITORS

The chemical inhibitor first reacts with the active center groups on the coal surface to form a relatively stable chain ring, which destroys or reduces the amount of active groups in the coal in advance, inhibits the chemical adsorption of the coal, and makes the coal sample difficult at low temperature. The inhibitor reacts chemically with oxygen to slow down the oxidation reaction rate, or it captures free radicals through chemical reactions to generate a relatively stable intermediary product, which stops the chain reaction process of coal oxidation and combustion, thereby preventing coal from combusting spontaneously. In general, the inhibition efficiency is relatively high.

4.3.10   ANTIOXIDANT INHIBITORS

Antioxidant inhibitors are key chemical inhibitors that include antioxidant A, urea, diphenylamine, and polyethylene glycol. The inhibition mechanism involves the generation of active radicals at low temperatures, followed by a chemical reaction with the inhibitor. These inhibitors interrupt the free radical chain reaction in coal, which prevents the coal oxidation reaction from proceeding.

For instance, diphenylamine is a strong oxygen inhalator that consumes excess O_2 and terminates free radical chains in coal. Details of the chemical degradation are presented as follows [72]:

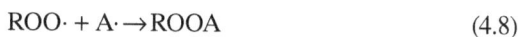

$$R\cdot + AH \rightarrow RH + A\cdot \qquad (4.5)$$

$$ROO\cdot + AH\cdot \rightarrow ROOH + A \qquad (4.6)$$

$$A\cdot + A\cdot \rightarrow A\text{–}A \qquad (4.7)$$

$$ROO\cdot + A\cdot \rightarrow ROOA \qquad (4.8)$$

During the free radical reaction in CSC, the chain-growth reaction is terminated. If there is no R· and ROO·—representing a coal molecule, such as an alkyl, alkenyl, or complex macromolecular radical—this demonstrates that the CSC is a free radical reaction. CSC can be effectively controlled through the addition of free radical inhibitors.

Wang et al. [73] employed the chemical inhibitors poly (ethylene glycol) 200 (99% mass purity) and catechin [74] to prevent CSC; they found that certain active functional groups (–OH) were eliminated and the oxidation of methyl and methylene groups prevented.

Urea only increases the E_a at temperatures below 200°C. Hence, it can be employed to inhibit the coal oxidation process. However, above 200°C, urea has a catalytic [43] effect.

Vitamin C is an essential antioxidant; it works primarily through controlling the chain reaction of free radicals and hydrogen peroxide in order to inhibit the oxidation of coal. However, the use of vitamin C as an inhibitor of coal oxidation has been limited because tests suggest that it reacts with oxygen at approximately 200°C [75].

A free radical scavenger, 2,2,6,6-tetramethyl-1-piperidinyloxy (TEMPO), readily combines with free radicals to generate inactive substances and is regarded to be a superior inhibitor to vitamin C. During the coal oxidation process, TEMPO combines with alkyl free radicals and forms stable substances that regulate the reaction rate of active free radicals and, thus, inhibits CSC [76].

4.3.11 Ionic Liquid (IL) Inhibitors

ILs are a type of salt comprising anions and cations and are liquids at or near room temperature. ILs are a new generation of green solvents. They are widely used in coal chemistry because of coal's solubility in them and their ability to change the morphological characteristics of coal which reduces the number of active functional groups (Figure 4.5) [77].

Wang et al. [78] and Zhang et al. [79] found that an IL could break the hydrogen bond in –OH during coal oxidation. This dissolved the coal, but the IL also reacted with the coal through the complex production of carboxyl groups. Studies

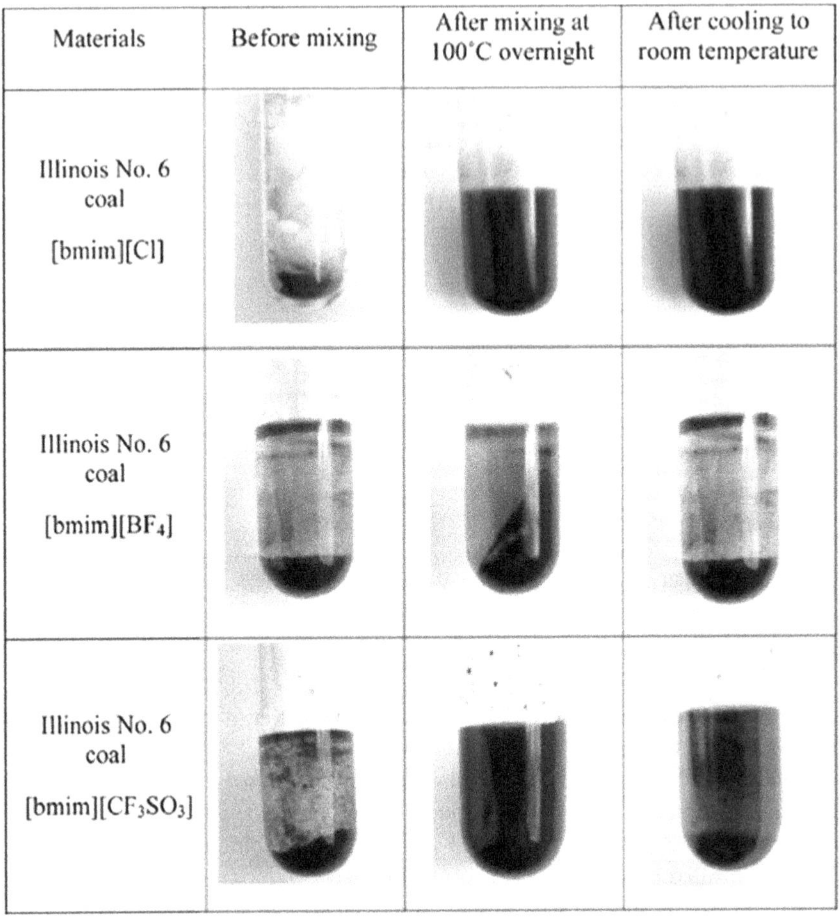

Materials	Before mixing	After mixing at 100°C overnight	After cooling to room temperature
Illinois No. 6 coal [bmim][Cl]			
Illinois No. 6 coal [bmim][BF₄]			
Illinois No. 6 coal [bmim][CF₃SO₃]			

FIGURE 4.5   Appearance of coal after mixed with ILs [77].

show that IL anions containing halides can only be stable at 150°C, while IL and other anions are stable up to 400°C [77]. Cui et al. [80] confirmed that [MMIM] [I] was the functional group most capable of reducing −OH groups in lignite and inhibiting its spontaneous combustion. Cummings et al. [81, 82] concluded that ILs do not drastically change the thermal characteristics of coal and found that some ILs break the macrostructure of coal, resulting in an increase in short-chain aliphatic hydrocarbons. To obtain a low-cost IL, To et al. [83] used the simple acid-base reactions between two naturally occurring chemicals—choline and amino acids—as a basis for manufacturing low-cost choline-amino acid-based ILs, which were discovered to be optimal pre-treatment agents for lignite coal and thermally bituminous coal.

ILs have been gradually applied in the field of coal chemistry and have attracted attention for their effective physical and chemical properties. They are non-flammable, have low melting points, high thermal stability, solubility, and recyclability. It is also possible to configure different ILs by changing their cation and anion combination. This only increases the already high application potential of ILs. Current studies concluded that ILs dissolve and destroy active functional groups in coal which reduces their oxidation activity. To date, imidazole-based ILs dissolve and destroy coal the most effectively. The damaged structures mainly include hydrogen bonds, aromatic structures, and aliphatic chains. Hydrocarbons, hydrocarbon groups, carbonyl groups, and other oxygen-containing functional groups to achieve the purpose of inhibiting CSC. However, due to the high process cost of the IL and the still maturing technology and equipment, ILs have not yet entered the stage of industrial application. The cost or timeline to reach that stage is not easy to estimate. Nonetheless, the unique properties of ILs are worthy of further research and the pursuit of new breakthroughs to harness their ability to inhibit CSC.

4.3.12 COMPOSITE INHIBITORS

The main driver of CSC is coal reacting with oxygen. This reaction generates a large amount of heat which heats the coal until it spontaneously ignites. In the process, a series of physical and chemical reactions occur, including physical adsorption, chemical adsorption, oxidation, and the synthesis and decomposition of products. The spontaneous combustion process is a chain reaction, and the coal-oxygen composite reaction process is divided into three stages, which are closely linked. The blocking action of physical inhibitors is mainly targeted at the first stage of the reaction. This blocking mechanism has two primary aspects. First, after contact between the inhibitor and the coal, a protective film is formed on the surface of the coal, which isolates it from the oxygen and reduces the chance of interaction between the coal and the oxygen, thus achieving the blocking effect. In addition, after the water attached to the inhibitor enters the coal body, the evaporation of that water will transfer away some of that heat, thus reducing the temperature of the coal body, destroying the heat storage environment, and reducing the self-heating rate. If the physical inhibitor can achieve the above effect, the CSC will be inhibited through these physical mechanisms.

The chemical inhibition of composite inhibitors works on the second and third stages of the spontaneous combustion process. The primary mechanism is to reduce the amount of active molecules on the coal surface of the coal. Also, the chemical inhibition reduces the creation of a variety of products which results in a low enough temperature so that the reaction rate is slowed considerably or stopped altogether. This reduction of the accumulation of heat energy interrupts the spontaneous combustion process which results in the inhibition of CSC. According to the analysis above, composite inhibitors work at all three of the stages of the reaction. This diversity of mechanisms, at both the physical and chemical levels, means that composite inhibitors have the potential to be much more effective inhibitors which reduce the risk of CSC.

Some scholars, by employing existing inhibitor compounds, have developed physical and chemical composite inhibitors. Tang [84] used $MgCl_2$ and phosphate to synergistically inhibit CSC. In the early stage, inorganic chlorides merely inhibited CSC; by contrast, at high temperatures, phosphorous-containing inhibitors also exerted considerable flame-retardant effects. Chen et al. [85] revealed that the presence of $MgCl_2$ significantly improved the surface activity of sodium dodecyl sulfate (SDS) in addition to increasing foam stability. Moreover, ammonium inhibitors substantially enhanced the foam characteristics of SDS, thus eliciting synergetic effects in inhibiting coal oxidation. Wang et al. [86] developed a novel material—foamed gel—through the combination of high-water absorbency gel and a compound foaming agent. This new material effectively controlled the spontaneous combustion process. Xi and Li [87] proposed applying a blend of thermoplastic coal powder (TCP) as a foam carrier. They revealed that the optimal coal self-heating inhibitor was a froth solution composed of approximately 30% TCP, by mass.

Qin et al. [15] introduced composites of ascorbic acid inhibitors and superabsorbent hydrogels to inhibit CSC. Moreover, a novel type of composite inhibitor was proposed that realized the cage-wrapping and sustained-release functions of polymers and thus acted as a high-activity chemical inhibitor. A composite inhibitor was prepared containing poly (AA)/sodium alginate super absorbent and (+)-catechin (PAA/SA-CC) [88]. Pandey et al. [89] reported that a halogen inhibitor had a synergistic effect, which improved inhibition. Taraba et al. [90] reported the effects of organic and inorganic solutions, the action of additives on coal, and the thermal effect on coal during immersion in the additive solution. The additive with the most favorable inhibiting effect—up to 70%—at a low temperature was urea. Zhou et al. [91] used coal/Zn-Al-layered double hydroxide (Zn-Al-LDH) composites that were synthesized using Shenfu coal (Shenfu, Shaanxi Province, China) as a raw material and co-precipitation. The results indicated that coal provided acidic functional groups to the interlayer anion of composites. The coal char on the surface of the composites formed an insulation layer, causing Zn-Al-LDHs to have a synergistic fire-retardant effect which inhibited CSC.

Yang et al. [92] explored the synergetic effects of Zn/Mg/Al–CO_3 LDHs and observed that LDHs were highly compatible with coal, forming a crystalline structure. This structure interrupted the diffusion of oxygen for combustion and consequently inhibited CSC. Furthermore, LDHs generated considerable amounts of CO_2 and H_2O during the endothermic reaction, which effectively diluted the oxygen concentration in the environment and reduced the surface temperature of the coal. Xu et al. [93] investigated a new compound material—a sand-suspended colloid—consisting of an organic polymer and a mineral inorganic gel. This material avoided excess water utilization and enhanced the quality of sand injection while grouting. Raymond et al. [94] revealed that surfactants, sulfonate salts, and inorganic phosphate form stable crosslinking carbon-oxygen complexes, thereby increasing both the thermal stability of coal and the thermal runaway temperature.

In addition, some studies have changed or deactivated the active groups in coal, leading to compound-based chemical inhibitors. A similar observation was made by Qi et al. [95], who manufactured a controlled-release inhibitor based on the available halogen inhibitors and added catechin and a co-polymer. They observed that the effect of the controlled-release inhibitor was maintained for long periods, thereby inhibiting the process of coal self-heating more effectively than pre-existing halogen inhibitors. Xi et al. [96] used a polymorphic foam (PF) synthesized from a sol foam/polycaprolactone solution and organic acid to inhibit CSC. This foam also inhibited the formation of new $-CH_2/CH_3$ and $-OH$ functional groups. The results revealed that the PF delayed and controlled CSC. Yang et al. [97] discovered that the extraction of soluble organic matter increased the permeability of coal; as a result, inhibitors readily penetrated the coal and prevented active groups from being oxidized.

4.4 APPLICATIONS

Ventilation technologies were previously used in coal mines to minimize the leakage of air in areas at risk of fires [98–102]. However, in recent years, emerging materials [103–105], such as inhibitors, inert gases, grout, foams, and gels, have been injected to prevent coal fires and their disastrous consequences. Traditional methods for preventing the extinguishing substance from reaching the fire location, such as grouting and nitrogen injection, have been limited due to their poor coverage [106–109]. The working section of a mine can become polluted by the use of grouting material, which tends to flow into the working surface. By contrast, liquid N_2, or liquid CO_2, [110–114] is a potentially more efficient heat transfer medium. The liquid is vaporized to absorb a considerable amount of heat and leads to an 800-fold increase in the volume of the cryogenic material. This expansion leads to an isotropic distribution of the heat-absorbing gas, thereby forcing out the hot combustion gases and displacing oxygen completely [115–117].

Recently, researchers have attempted to produce a novel phase-transition aerosol using ultrasonic waves [118]. Because of the reaction between aerosol particles of acrylamide and H_3BO_3, this aerosol was eventually deposited and adsorbed on the coal surface. The results revealed that residual coal formed dense protective layers with thin sheets.

4.5 CONCLUSIONS

The mechanism, effectiveness, cost, and suitable application situation for various inhibitors of CSC have been reviewed and compared in this study. The advantages and disadvantages of these materials are listed in Table 4.1.

Most conventional inhibition of CSC involves a single mechanism, such as endothermic cooling or the creation of an oxygen barrier. In contrast, some chemical inhibitors, such as antioxidant inhibitors, acid inhibitors, and ILs, work by retarding the chemical reactions involved in the coal oxidation process. In general,

TABLE 4.1

Advantages and Disadvantages of Different Materials for CSC Prevention [36]

Classification	Materials	Advantages	Disadvantages
Physical-based materials	Chlorine salts (e.g. $MgCl_2$, $CaCl_2$, and $NaCl$, etc.)	Low cost, nontoxic, and considerable inhibition effect on the early CSC.	Short inhibition life
	Ammonium salts (e.g. $NH_4H_2PO_4$, NH_4Cl, and NH_3HCO_3, etc.)	Action of thermal decomposition strengthens the action of heat adsorption; Gaseous products dilute the oxygen concentration.	Poor thermal stability makes it difficult to maintain the inhibitory effect; Generated NH_3 threats the human health.
	Alkalis (e.g. $Ca(OH)_2$)	Considerable inhibition effect on high sulfur coal	Strong causticity makes it easy to corrode equipment and endanger human health.
	Inert gases (e.g. N_2 and CO_2)	Significant dilution of the oxygen concentration; Gasification of liquid inert gases adsorbs a large amount of heat.	Strong diffusibility makes it difficult to isolate oxygen for a long time and present poor control effect on the areas with poor seal.
	Foams (e.g. three-phase foam and paste foam)	Easy pileup leads to better coverage of coal, especially for the high-level areas; Release of encased inert gases further isolates the oxygen; Considerable control effect in the short term.	Ineffective heat transfer due to low specific heat; Poor foam stability causes that the foam is fragile and the liquid film adheres with difficulty to the surface of coal.
	Colloids (e.g. polymer gel, fly ash complex gel, sand-suspended gel etc.)	High water content and strong water retention remove a large amount of thermal energy during evaporation. Good permeability makes it effectively cover the coal and seal the cracks after gelling; Strong thermal stability and high inhibition life.	High cost; Complex application process.
	Polymers (e.g. polyvinyl acetate emulsion)	A stable solid-phase film is formed to isolate the oxygen.	Failure at high temperature due to thermal decomposition; The generated flammable gases increase the fire risk.
	Aerosols	Easy to adhere to the coal evenly.	Fine particles are easily inhaled into the body; Low conversion rate when preparing.

(Continued)

TABLE 4.1 (Continued)
Advantages and Disadvantages of Different Materials for CSC Prevention [36]

Classification	Materials	Advantages	Disadvantages
	LHDs (e.g. Zn-Al-LDHs and ZnMgAl-CO$_3$-LDHs)	Multi-step thermal decomposition absorbs heat at different stages.	High cost; Difficulty in large-scale application; Immature application process
Chemical-based materials	Antioxidants (e.g. diphenylamine, polyethylene glycol, and urea etc.)	Inhibit the active sites and capture the free radicals to destroy the chain reaction during CSC, essentially reducing the spontaneous combustion tendency.	The oxidation or thermal decomposition may aggravate the process of CSC.
	ILs (e.g. [BMIM][BF$_4$], and [BMIM][Cl] etc.)	High thermal stability; Recyclability.	High cost; Immature application technology and equipment.
Composite materials		Synergize the advantages of different materials to improve the inhibition effect.	The mechanism of synergistic suppression for CSC and the selection, proportion, and optimization of ingredients need to be studied further.

traditional inhibitors, such as water-absorbing salts, alkaline inhibitors, and colloid inhibitors, have the advantage of a lower cost. However, the newer generation of inhibitors such as antioxidants, acid resistance agents, IL inhibitors, and composite inhibitors have the potential to be more effective. Some of these inhibitors, such as ammonium salt inhibitors and aerosol inhibitors, are less commonly used than others due to their limits in effectiveness or their higher cost. The composite inhibitors have yet to see applications in the field. There is still active research in developing efficient, easy-to-use, and low-cost inhibitors. In the meantime, more studies to analyze the effectiveness and adaptability of existing inhibiting technologies are needed.

REFERENCES

1. Nature Editorial (2016) News: Energy hit. *Nature* 534:152.
2. Gianfrancesco AD (2017) Worldwide overview and trend for clean and efficient use of coal. *Mater Ultra Supercrit Adv Ultra supercrit Power Plants* :643–87.
3. Musa SD, Tang Z, Ibrahim AO, Habib M (2018) China's energy status: A critical look at fossils and renewable options. *Renew Sustain Energy Rev* 81:2281–90.
4. Mao JD, Schimmelmann A, Mastalerz M, Hatcher PG, Li Y (2010) Structural features of a bituminous coal and their changes during low-temperature oxidation and loss of volatiles investigated by advanced solid-state NMR spectroscopy. *Energy Fuel* 24(4):2536–44.
5. Deng J, Xiao Y, Lu J, Wen H, Jin Y (2015) Application of composite fly ash gel to extinguish outcrop coal fires in China. *Nat Hazards* 79(2):881–98.
6. Zeng Q, Tiyip T, Wuttke MW, Guan W (2015). Modeling of the equivalent permeability for an underground coal fire zone, Xinjiang region, China. *Nat Hazards* 78(2):957–71.
7. Lee SS, Wilcox J (2017) Behavior of mercury emitted from the combustion of coal and dried sewage sludge: The effect of unburned carbon, Cl, Cu and Fe. *Fuel* 203:749–56.
8. Melody SM, Johnston FH (2015). Coal mine fires and human health: What do we know? *Int J Coal Geol* 152:1–14.
9. Deng J, Xiao Y, Li Q, Lu J, Wen H (2015) Experimental studies of spontaneous combustion and anaerobic cooling of coal. *Fuel* 157:261–69.
10. Wessling S, Kuenzer C, Kessels W, et al (2008) Numerical modeling for analyzing thermal surface anomalies induced by underground coal fires. *Int J Coal Geol* 74:175–84.
11. Liu L, Zhou FB (2010) A comprehensive hazard evaluation system for spontaneous combustion of coal in underground mining. *Int J Coal Geol* 82:27–36.
12. Hao W, Glarborg P, Frandsen FJ, Dam-Johansen K, Jensen PA, Sander B (2013). Trace elements in co-combustion of solid recovered fuel and coal. *Fuel Process Technol* 105:212–21.
13. Zhang Y, Wu J, Chang L, et al (2013) Changes in the reaction regime during low-temperature oxidation of coal in confined spaces. *J Loss Prev Process Ind* 26:1221–9.
14. Ran VKS (2013) Spontaneous heating and fire in coal mines. *Procedia Eng* 62:78–90.
15. Qin B, Dou G, Wang Y, Xin H, Ma L, Wang D (2017) A superabsorbent hydrogel–ascorbic acid composite inhibitor for the suppression of coal oxidation. *Fuel* 190:129–35.

16. Smith M A, Glasser D (2005) Spontaneous combustion of carbonaceous stockpiles. Part I: the relative importance of various intrinsic coal properties and properties of the reaction system. *Fuel* 84(9):1151–1160.
17. Wang H, Dlugogorski BZ, Kennedy EM (2003) Coal oxidation at low temperatures: Oxygen consumption, oxidation products, reaction mechanism and kinetic modelling. *Prog Energy Combust Sci* 29(6):487–513.
18. Jones RE, Townend DTA (1945) Mechanism of the oxidation of coal. *Nature* 155:424–5.
19. Zubíček V, Adamus A (2013) Susceptibility of coal to spontaneous combustion verified by modified adiabatic method under conditions of Ostrava–Karvina Coalfield, Czech Republic. *Fuel Process Technol* 113:63–6.
20. Xu YL, Wang LY, Tian N, Zhang JP, Yu MG, Delichatsios MA (2017) Spontaneous combustion coal parameters for the Crossing–Point Temperature (CPT) method in a Temperature–Programmed System (TPS). *Fire Safety J* 91:147–54.
21. Zhao X, Wang Q, Xiao H, Mao Z, Chen P, Sun J (2013) Prediction of coal stockpile autoignition delay time using micro-calorimeter technique. *Fuel Process Technol* 110:86–93.
22. Qi G, Wang D, Chen Y, Xin H, Qi X (2014) The application of kinetics based simulation method in thermal risk prediction of coal. *J Loss Prev Process Ind* 29:22–9.
23. Qi X, Wang D, Zhong X, Gu J, Xu T (2010) Characteristics of oxygen consumption of coal at programmed temperatures. *Min Sci Technol (China)* 20(3):372–7.
24. Slovák V, Taraba B (2010) Effect of experimental conditions on parameters derived from TG-DSC measurements of low-temperature oxidation of coal. *J Therm Anal Calorim* 101(2):641–6.
25. Qi X, Wang D, Xin H, Qi G (2013) In Situ FTIR study of real-time changes of active groups during oxygen-free reaction of coal. *Energy Fuel* 27(6):3130–6.
26. Mathews JP, Campbell QP, Xu H, Halleck P (2017) A review of the application of X-ray computed tomography to the study of coal. *Fuel* 209:10–24.
27. Xiao Y, Lu JH, Wang CP, Deng J (2016) Experimental study of high-temperature fracture propagation in anthracite and destruction of mudstone from coalfield using high-resolution microfocus X-ray computed tomography. *Rock Mech Rock Eng* 49(9):3723–34.
28. Wang Y, Wu J, Xue S, Wang J, Zhang Y (2017) Experimental study on the molecular hydrogen release mechanism during low-temperature oxidation of coal. *Energy Fuel* 31(5):5498–506.
29. Bandara TS, Kannangara GSK, Wilson MA, Boreham CJ, Fisher K (2005) The study of Australian coal maturity: Relationship between solid-state NMR aromaticities and organic free-radical count. *Energy Fuel* 19(3):954–9.
30. Zhong XX, Wang DM, Xu YL, Xin HH (2010) The variation characteristics of free radicals in coal oxidation. *J China Coal Soc* 35(6):960–3. (in Chinese)
31. Li Z, Liu D, Cai Y, Ranjith PG, Yao Y (2017) Multi-scale quantitative characterization of 3-D pore-fracture networks in bituminous and anthracite coals using FIB-SEM tomography and X-ray μ-CT. *Fuel* 209:43–53.
32. Li S, Tang D, Xu H, Yang Z (2012) Advanced characterization of physical properties of coals with different coal structures by nuclear magnetic resonance and X-ray computed tomography. *Comput Geosci* 48:220–7.
33. Xu JC, Zhang XH, Wen H, et al (2000) Procedure of reaction between coal and oxygen at low temperature and calculation of its heat emitting intensity. *J China Univ Min Technol* 3:31–35.
34. Luo YF, Li WH (2004) X-ray diffraction analysis on the different macerals of several low-to-medium metamorpic grade coals. *J China Coal Soc* 29(3):338–41.

35. Wei A, Li Z, Pan S, Yang Y (2007). Experimental study on free radical reaction of coal initiated by ultraviolet light. *J China U Min Technol* 36(5):582–5. (in Chinese)

36. Li QW, Xiao Y, Zhong KQ, et al (2020) Overview of commonly used materials for coal spontaneous combustion prevention. *Fuel* 275:117981.

37. Singh AK, Sahay N, Ahmad I, Mondal S (2002) Role of inorganic compounds as inhibitor in diminishing self-heating phenomena of Indian coal. *J Mines Metals Fuels* 50(9):356–9.

38. Sujanti W, Zhang DK (2000) Investigation into the role of inherent inorganic matter and additives in low-temperature oxidation of a Victorian brown coal. *Combust Sci Technol* 152(1):99–114.

39. Watanabe WS, Zhang DK (2001) The effect of inherent and added inorganic matter on low-temperature oxidation reaction of coal. *Fuel Process Technol* 74(3):145–60.

40. Zheng LF (2010) Test and analysis on salty retardants performance to restra in coal oxidized spontaneous combustion. *Coal Sci Technol* 38:70–2. (in Chinese)

41. Yu SJ, Xie FC, Jia BY, Zhang PF (2012) Influence study of organic and inorganic additive to coal combustion characteristic. *Procedia Environ Sci* 12:459–67.

42. Le Manquais K, Snape C, Barker J, McRobbie L (2012) TGA and drop tube furnace investigation of alkali and alkaline earth metal compounds as coal combustion additives. *Energy Fuel* 26(3):1531–9.

43. Slovák V, Taraba B (2012) Urea and $CaCl_2$ as inhibitors of coal low-temperature oxidation. *J Therm Anal Calorim* 110(1):363–7.

44. Cui C, Jiang S, Shao H, Zhang W, Wang K, et al (2018) Experimental study on thermo-responsive inhibitors inhibiting coal spontaneous combustion. *Fuel Process Technol* 175:113–22.

45. Wu K, Wang Z, Hu Y (2008) Microencapsulated ammonium polyphosphate with urea–melamine–formaldehyde shell: Preparation, characterization, and its flame retardance in polypropylene. *Polym Advan Technol* 19(8):1118–25.

46. Liodakis S, Bakirtzis D, Lois E, Gakis D (2002) The effect of $(NH_4)_2HPO_4$ and $(NH_4)_2SO_4$ on the spontaneous ignition properties of pinus halepensis pine needles. *Fire Safety J* 37(5):481–94.

47. Wen H, Zhang F, Jin Y, Liu W (2011) Experimental research on effect of sulfur on characteristic parameters of coal spontaneous combustion. *Safety Coal Mines* 42:5–7. (in Chinese)

48. Yang S (1996) Experiment study and mechanism analysis of $Ca(OH)_2$ as the retarder of high-sulfur coal. *J China U Min Technol* 04:68–72. (in Chinese)

49. Deng J, Zhang X, Jing J, Ma L, Zhang Y, Wen H (2010) Gel as one potential extinguishing method for controlling coal fires, ICCFR2, Berlin, Germany.

50. Zhai X, Deng J, Wen H, Zhang X (2010) Composite gel as a barrier between mine workings and coal fires. ICCFR2, 19–21 May, Berlin, Germany.

51. Donatello S, Kuenzel C, Palomo A, Fernández-Jiménez A (2014) High temperature resistance of a very high volume fly ash cement paste. *Cement Concrete Comp* 45:234–42.

52. Qin B, Lu Y, Li F, Jia Y, Zhu C, Shi Q (2015) Preparation and stability of inorganic solidified foam for preventing coal fires. *Adv Mater Sci Eng* 2014:1–10.

53. Twardowska I, Stefaniak S (2006) Fly Ash as a Sealing Material for Spontaneous Combustion and Acid Rock Drainage Prevention and Control. In: Sajwan K.S., Twardowska I., Punshon T., Alva A.K. (eds) Coal Combustion Byproducts and Environmental Issues. Springer, New York, NY.

54. Zhong X (2008) Research on oxidation kinetics testing method for the propensity of coal to spontaneous combustion. Doctoral Dissertation, Xuzhou, PR China: China U Min Technol, China. (in Chinese)

55. Xu YL, Wang DM, Wang LY, Zhong XX, Chu TX (2012) Experimental research on inhibition performances of the sand-suspended colloid for coal spontaneous combustion. *Saf Sci* 50(4):822–7.
56. Cheng W, Hu X, Xie J, Zhao Y (2017) An intelligent gel designed to control the spontaneous combustion of coal: Fire prevention and extinguishing properties. *Fuel* 210:826–35.
57. Ren WX, Kang ZH, Wang DM (2011) Causes of spontaneous combustion of coal and its prevention technology in the tunnel fall of ground of extra-thick coal seam. *Procedia Eng* 26:717–24.
58. Shi B, Zhou F (2014) Impact of heat and mass transfer during the transport of nitrogen in coal porous media on coal mine fires. *Sci World J* 2014:293142.
59. Zhou FB, Ren WX, Wang DM, Song TL, Li X, Zhang YL (2006) Application of three-phase foam to fight an extraordinarily serious coal mine fire. *Int J Coal Geol* 67(1–2):95–100.
60. Lu Y (2017) Laboratory study on the rising temperature of spontaneous combustion in coal stockpiles and a paste foam suppression technique. *Energy Fuel* 31(7):7290–8.
61. Smith RH (1980) Inhibiting spontaneous combustion of coal. U.S. Patent, 41993,25.
62. Qin B, Lu Y, Li Y, Wang D (2014) Aqueous three-phase foam supported by fly ash for coal spontaneous combustion prevention and control. *Adv Powder Technol* 25(5):1527–33.
63. Liu W, Qin Y, Yang X, Wang W, Chen Y (2018) Early extinguishment of spontaneous combustion of coal underground by using dry-ice's rapid sublimation: A case study of application. *Fuel* 217:544–52.
64. Serrano A, Borreguero AM, Garrido I, Rodríguez JF, Carmona M (2017) The role of microstructure on the mechanical properties of polyurethane foams containing thermoregulating microcapsules. *Polym Test* 60:274–82.
65. Yun MS, Lee WI (2008) Analysis of bubble nucleation and growth in the pultrusion process of phenolic foam composites. *Compos Sci Technol* 68(1):202–8.
66. Ma Y, Zhang W, Wang C, Xu Y, Li S (2013) Preparation and characterization of melamine modified urea-formaldehyde foam. *Int Polym Proc* 28(2):188–98.
67. Xiao H, Du C (2006) Experiments on a new kind polymer-inhibitor for suppressing coal spontaneous ignition. *J Saf Environ* 6:46–8. (in Chinese)
68. Wang L (2007) Synthesis of PVAC for preventing coal spontaneous combustion and study on behaviors. Master Thesis, Qingdao, PR China: Shandong U Sci Technol. (in Chinese)
69. Deng J, Hu A, Ma L, Dong G, Zhang Y (2015) Test on aerosol preparation for coal spontaneous combustion prevention and control. *Safety Coal Mines* 4:30–3. (in Chinese)
70. Deng J, Wu H, Ma L, Song X (2012) Experimental research on atomizing performance of coal spontaneous combustion inhibitors. *Safety Coal Mines* 5:15–8. (in Chinese)
71. Xiao Y, Deng J, Li SG, Liu X, Wu HP (2011) Study and application on fire extinguishing and preventing with light-paste material for coalmine. *J Coal Sci Eng* 17(3):340–4.
72. Li Z, Kong B, Wei A, Yang Y, Zhou Y (2016) Free radical reaction characteristics of coal low-temperature oxidation and its inhibition method. *Environ Sci Pollut Res* 23(23):593–605.
73. Wang D, Dou G, Zhong X, Xin H, Qin B (2014) An experimental approach to selecting chemical inhibitors to retard the spontaneous combustion of coal. *Fuel* 117:218–23.

74. Dou G, Wang D, Zhong X, Qin B (2014) Effectiveness of catechin and poly (ethylene glycol) at inhibiting the spontaneous combustion of coal. *Fuel Process Technol* 120:123–7.
75. Qin B, Dou G, Wang D (2016) Thermal analysis of vitamin C affecting low-temperature oxidation of coal. *J Wuhan U Technol* 31(3):519–22.
76. Li J, Li Z, Yang Y, Kong B, Wang C (2018) Laboratory study on the inhibiting effect of free radical scavenger on coal spontaneous combustion. *Fuel Process Technol* 171:350–60.
77. Pulati N, Sobkowiak M, Mathews JP, Painter P (2012). Low-temperature treatment of Illinois No. 6 coal in ionic liquids. *Energy Fuels* 26:3548–52.
78. Wang LY, Xu YL, Jiang SG, Yu MG, Chu TX, Zhang WQ, et al (2012) Imidazolium based ionic liquids affecting functional groups and oxidation properties of bituminous coal. *Saf Sci* 50:1528–34.
79. Zhang W, Jiang S, Wang K, Wang L, Wu Z, Kou L, et al (2011) Study on coal spontaneous combustion characteristic structures affected by ionic liquids. *Procedia Eng* 26:480–5.
80. Cui FS, Laiwang B, Shu CM, Jiang JC (2018) Inhibiting effect of imidazolium-based ionic liquids on the spontaneous combustion characteristics of lignite. *Fuel* 217:508–14.
81. Cummings J, Shah K, Atkin R, Moghtaderi B (2015) Physicochemical interactions of ionic liquids with coal; The viability of ionic liquids for pre-treatments in coal liquefaction. *Fuel* 143:244–52.
82. Cummings J, Tremain P, Shah K, Heldt E, Moghtaderi B, Atkin R, et al (2017) Modification of lignites via low temperature ionic liquid treatment. *Fuel Process Technol* 155:51–8.
83. To TQ, Shah K, Tremain P, Simmons BA, Moghtaderi B, Atkin R (2017) Treatment of lignite and thermal coal with low cost amino acid based ionic liquid-water mixtures. *Fuel* 202:296–306.
84. Tang Y (2018) Experimental investigation of applying $MgCl_2$ and phosphates to synergistically inhibit the spontaneous combustion of coal. *J Energy Inst* 91(5):639–45.
85. Chen P, Huang F, Fu Y (2016) Performance of water-based foams affected by chemical inhibitors to retard spontaneous combustion of coal. *Int J Min Sci Technol* 26(3):443–8.
86. Wang G, Yan G, Zhang X, Du W, Huang Q (2016) Research and development of foamed gel for controlling the spontaneous combustion of coal in coal mine. *J Loss Prev Process Ind* 44:474–86.
87. Xi Z, Li A (2016) Characteristics of thermoplastic powder in an aqueous foam carrier for inhibiting spontaneous coal combustion. *Process Saf Environ Prot* 104:268–76.
88. Ma L, Wang D, Wang Y, Dou G, Xin H (2017) Synchronous thermal analyses and kinetic studies on a caged-wrapping and sustained-release type of composite inhibitor retarding the spontaneous combustion of low-rank coal. *Fuel Process Technol* 157:65–75.
89. Pandey J, Mohalik NK, Mishra RK, Khalkho A, Kumar D (2015) Investigation of the role of fire retardants in preventing spontaneous heating of coal and controlling coal mine fires. *Fire Technol* 51(2):227–45.
90. Taraba B, Peter R, Slovák V (2011) Calorimetric investigation of chemical additives affecting oxidation of coal at low temperatures. *Fuel Process Technol* 92(3):712–5.
91. Zhou A, Liu Bo, Xu W (2012) Effect of coal on the structure and properties of coal/ LDHs composites. *Adv Mater Res* 399–401:1075–8.

92. Yang Y, Tsai YT, Zhang Y, Shu CM, Deng J (2017) Inhibition of spontaneous combustion for different metamorphic degrees of coal using Zn/Mg/Al–CO$_3$ layered double hydroxides. *Process Saf Environ Prot* 113:401–12.
93. Xu Y, Wang L, Chu T, Liang D (2014) Suspension mechanism and application of sand-suspended slurry for coalmine fire prevention. *Int J Min Sci Technol* 24(5):649–56.
94. Raymond CJ, Farmer J, Dockery CR (2016) Thermogravimetric analysis of target inhibitors for the spontaneous self-heating of coal. *Combust Sci Technol* 68(8):379–410.
95. Qi X, Wei C, Li Q, Zhang L (2016) Controlled-release inhibitor for preventing the spontaneous combustion of coal. *Nat Hazards* 82(2):1–11.
96. Xi Z, Li D, Feng Z (2017) Characteristics of polymorphic foam for inhibiting spontaneous coal combustion. *Fuel* 206:334–41.
97. Yang Y, Li Z, Si L, Li J, Qin B, Li Z (2017) SOM's effect on coal spontaneous combustion and its inhibition efficiency. *Combust Sci Technol* 189(12):2266–83.
98. Xia T, Zhou F, Wang X, Zhang Y, Li Y, Kang J, et al (2016) Controlling factors of symbiotic disaster between coal gas and spontaneous combustion in longwall mining gobs. *Fuel* 182:886–96.
99. Lu W, Cao YJ, Tien JC (2017) Method for prevention and control of spontaneous combustion of coal seam and its application in mining field. *Int J Min Sci Technol* 27(5):839–46.
100. Qi X, Wang D, Xin H, Zhong X (2013) Environmental hazards of coal fire and their prevention in China. *Environ Eng Manage J* 12(10):1915–9.
101. Deng J, Xiao Y, Luo ZM, Zhang YN, Zheng XZ (2017) Technologies of monitoring, alerting and controlling for coal fires in China. Geophysical Research Abstracts, Vol. 19, EGU2017-PREVIEW, Vienna, Austria, 23–28 April.
102. Evseev V (1985) New methods for the prevention of spontaneous fires in underground coal mines. Paper in Proceedings of 21st International Conference of Safety in Mines Research Institutes Sydney, Sydney, Australia 481–3.
103. Zhang L, Qin B, Shi B, Wu Q, Wang J (2016) The fire extinguishing performances of foamed gel in coal mine. *Nat Hazards* 81(3):1957–69.
104. Shao Z, Wang D, Wang Y, Zhong X, Tang X, Hu X (2015) Controlling coal fires using the three-phase foam and water mist techniques in the Anjialing Open Pit Mine, China. *Nat Hazards* 75(2):1833–52.
105. Tripathi DD (2008) New approaches for increasing the incubation period of spontaneous combustion of coal in an underground mine panel. *Fire Technol* 44(2):185–98.
106. Colaizzi GJ (2004) Prevention, control and/or extinguishment of coal seam fires using cellular grout. *Int J Coal Geol* 59(1–2):75–81.
107. Ray SK, Singh RP (2007) Recent developments and practices to control fire in underground coal mines. *Fire Technol* 43(4):285–300.
108. Singh RVK, Singh VK (2004) Mechanised spraying device–A novel technology for spraying fire protective coating material in the benches of opencast coal mines for preventing spontaneous combustion. *Fire Technol* 40(4):355–65.
109. Smith AC, Miron Y, Lazzara CP (1988) Inhibition of spontaneous combustion of coal. Report of Investigations, US Department of the Interior, Bureau of Mines, USA.
110. Zhou F, Shi B, Cheng JW, Ma LJ (2015) A new approach to control a serious mine fire with using liquid nitrogen as extinguishing media. *Fire Technol* 51(2):325–34.
111. Zhang X, Zhang D, Wen H, Li S (2014) Parameter of liquid nitrogen injection against fires in gobs. Sun City, Northwest Province, South Africa. IMVC.

112. Shu YB, Li WJ, Li ZX (2012) The technology of liquid CO_2 used for fire prevention and the related device. *Adv Mater Res* 347–353:1642–6.
113. Zhang X, Wen H, Cheng F, Ma L, Bu R, Sun J (2010) Extinguishing coal spontaneous combustion with liquid carbon dioxide, ICCFR2, Berlin, Germany.
114. Ma L, Xiao Y, Zhang X, Zhao Y (2010) Liquid CO_2 for controlling spontaneous combustion in large-scale coal bunkers, ICCFR2, Berlin, Germany.
115. Adamus A (2001) Review of nitrogen as an inert gas in underground mines. *J Mine Vent Soc S Afr* 54(3):60–1.
116. Morris R (1987) A review of experiences on the use of inert gases in mine fires. *Min Sci Technol* 6(1):37–69.
117. Kim AG (2004) Cryogenic injection to control a coal waste bank fire. *Int J Coal Geol* 59(1–2):63–73.
118. Tang Y (2016) Inhibition of low-temperature oxidation of bituminous coal using a novel phase-transition aerosol. *Energy Fuel* 30(11):9303–9.

5 Micro-Scale Piloting Test for Coal and Gas Outburst Prevention by Liquid Carbon Dioxide Injection and In-Situ Experiment

Zhenbao Li
School of Petrochemical Engineering, Lanzhou University
of Technology, Lanzhou, People's Republic of China

CONTENTS

DOI: 10.1201/9781003140382-5

5.1   INTRODUCTION

As an environmentally friendly and efficient form of energy, coalbed methane (CBM) has generally been used to generate power in residential and industrial sectors. However, high CBM pressure in coal seams is one of main induced factors causing coal and gas outburst (CGO) disasters [1, 2]. In China, over 70% of coal mines with a high methane content and low permeability belong to high gas and CGO mines, which seriously threatens the safety of miners [3]. As the depth and scale of mining has expanded, so have more and more coal mines encountered low permeability with a high methane content in coal seams, conditions challenging to ECBM [4, 5]. Extensive research relating to hydraulic fracturing, inert gas injection (CO_2, N_2, and hot flue gas), and blasting technology have been employed for ECBM over the past 10 decades [6–9].

Considering its low-temperature damage effect and predominant adsorption capacity, liquid carbon dioxide (LCO_2) has been used to prevent and control CGO disasters over recent years [10–12]. New pores and cracks, developed by temperature gradient in coal seams, are generated in the process of LCO_2-ECBM, which mainly results from thermal stress. The CBM extraction efficiency can be evaluated by the features of the pores and cracks during the process of LCO_2-ECBM. Zou et al. [13] described the interactions between permeability and porosity, and classified the 15 coal samples into three types: pore dominated samples, cleat dominated samples, and combined effect samples. Cai et al. [14] tested the effect of liquid nitrogen on the thermal damage in rock and reported an increase in pore size, a drop in pore volume, and an expansion in the cracks. Wen et al. [15] investigated the changes in pore microstructure and crack network in coal before and after LCO_2 freezing-thawing and revealed that both the total pore area and the volume increased by 3.2%–18.6%. Xu et al. [16] focused on the crack variation caused by the temperature effect of cyclic LCO_2 injected into coal seams and deduced that the crack generation was primarily caused by the freezing of pre-existing water within the pores which led to a 9% volume increase. Vishal [17] evaluated the permeability transformation in CO_2-treated coal at different phases (0–20 h, 20–40 h, and 40–60 h) and determined that the permeability reduction for LCO_2 in porous coal was 25%, 13.5%, and 1% in the three saturation periods, which was a smaller reduction in coal permeability compared with supercritical CO_2 over the three periods.

Owing to the heat exchange between LCO_2 and the coal matrix, the low-temperature LCO_2 transforms into a gaseous state and CO_2–CH_4 competitive adsorption occurs during the LCO_2-ECBM process. The adsorbed CBM accounts for about 85% of the total quantity in the coal seams, which seriously increases the residual methane content and the gas extraction efficiency during the LCO_2-ECBM process. Inclusive experiments on the characteristics of CO_2/CH_4 competitive adsorption in different coal ranks have been conducted [18–20]. The quantity of CO_2 adsorbed in the coal surface is approximately 2–4 times that of CH_4 at the same pressure and temperature conditions [21, 22]. Experimental and theoretical research on the competitive adsorption behaviors of multi-component gases in coal has revealed that CO_2 is preferentially adsorbed in the pore surface of coal samples compared to CH_4 [23–25]. The CO_2 adsorption leads to coal matrix swelling, which decreases the spaces of cleat channels at reservoir conditions and thus reduces the permeability [26, 27]. Meanwhile, the desorption of CH_4 results in coal matrix shrinkage and permeability improvement [28, 29]. Therefore, the process of LCO_2-ECBM is variable and complex.

Previous studies have shown that the type of adsorbed gas, pressure, temperature, and coal rank, among other factors, all affect the results of CO_2 displacement [30–33]. However, the associated variations in pores and cracks produced by thermal damage and displacement efficiency over time during the LCO_2-ECBM process for preventing and controlling CGO need further investigation. Micropilot tests and in-situ experiments of LCO_2-ECBM were quantitatively explored. The key purpose of this study is as follows: (1) to experimentally reveal the microstructure variation of LCO_2 saturated coal, (2) to simulate the evolution of CO_2–CH_4 displacing efficiency, and (3) to explore the practical procedure and parameters by field test. This study provides the theoretical and practical basis to prevent and control CGO accidents by LCO_2-ECBM technology.

5.2 FEATURES OF PHASE CHANGE DURING LCO_2-ECBM INJECTION

CO_2 has gas, liquid, solid, and supercritical states which are mainly determined by the parameters of temperature and pressure. At present, more than 90% of coal seam mining in China occurs where the original ground temperature is below 30°C. Heat is exchanged between coal and gas- or liquid-CO_2 during the LCO_2 injection process. The temperature distribution of the coal seam and the characteristic parameters of the carbon dioxide determine the phase change characteristics during the LCO_2-ECBM injection and are exhibited in Figure 5.1. The pressure versus temperature curve that occurs during the LCO_2-ECBM process can be divided into five stages.

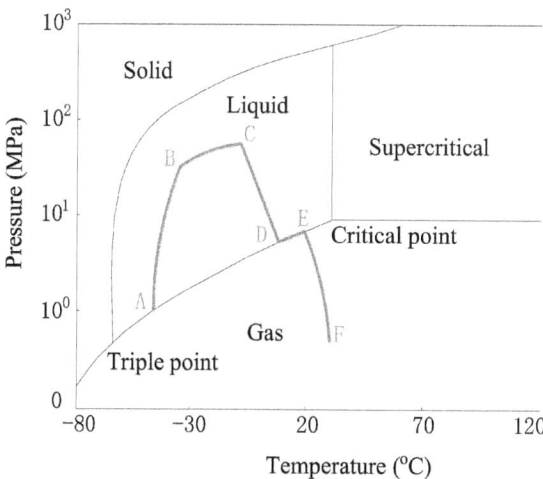

FIGURE 5.1   Phase change curve during LCO_2-ECBM injection.

1. Pressure increasing stage (stage I, A→B)

    LCO$_2$ is stored in the tank at a temperature of −50°C to −30°C and at a pressure of 1.5–2.8 MPa and is transported to an underground coal mine which consists of both gas and liquid phases (point A in Figure 5.1). Then, LCO$_2$ is pumped into a coal seam borehole causing a rapid rise in pressure and a slight temperature rise (points A→B in Figure 5.1).

2. Heat change stage (stage II, B→C)

    After LCO$_2$ is injected into the coal seam borehole, heat is exchanged between LCO$_2$ and the borehole wall and the temperature increases sharply, whereas the temperature decreases slightly (points B→C in Figure 5.1).

3. LCO$_2$ seepage stage (stage III, C→D)

    LCO$_2$ seepage along the cracks of the coal matrix occurs when it is injected into the coal seam. On the one hand, heat exchange convectively occurs between the coal surface and LCO$_2$, and the fluid temperature rises. On the other hand, the fluid pressure continues to decrease as the seepage radius increases. The parameters and phase characteristics of this process are shown as the C–D segment in Figure 5.1.

4. Phase change stage (stage IV, D→E)

    As the temperature rises and the pressure decreases, LCO$_2$ in the fissure of the coal seam gradually transforms into a gas phase. Due to the effects of two-phase transformation and phase change pressurization, there is an obvious increase in the coal pore pressure with LCO$_2$ phase change at this stage. The characteristics of the parameters are shown as the D–E segment in Figure 5.1.

5. Seepage and diffusion stage (stage V, E→F)

    When LCO$_2$ completely transforms into a gaseous state, CO$_2$ enters the pores and cracks of the coal. The temperature and pressure of the CO$_2$ gas tend to be consistent with the corresponding parameters of the coal seam with regard to the seepage and diffusion process. Finally, CO$_2$ is absorbed into the coal pores, or it is extracted from the borehole in a gaseous phase. The characteristic parameters are shown as the E–F segment in Figure 5.1.

    Based on the above analysis, the effects of LCO$_2$-ECBM include improving the permeability of the LCO$_2$ with low-temperature damage and efficient CO$_2$–CH$_4$ displacement caused by LCO$_2$ phase transformation which can significantly enhance drainage efficiency. Therefore, the pore variation in LCO$_2$ saturated coal was tested, and the evolution of gas CO$_2$–CH$_4$ displacement was explored below using experimental methods.

5.3 LCO$_2$ DAMAGE CHARACTERISTICS OF COAL PORES

5.3.1 EXPERIMENTAL METHOD OF PORE STRUCTURE

In accordance with the Hodot classification of coal pores, four kinds of coal pores were classified in this study: Macropores (> 1000 nm), mecropores (100–1000 nm), transition pores (10–100 nm), and micropores (< 10 nm) [34, 35]. Transition pores and micropores were both attributed to "adsorbed pores" (ADP), whereas the residual pores were termed "seepage pores" (SEP), which respectively served as the diffusion interspaces and seepage channels for ECBM [36, 37]. Extensive experimental methods, such as nuclear magnetic resonance, nitrogen adsorption-desorption (LPN), and mercury intrusion porosimetry (MIP), have been used to achieve the parameters of different coal pores. LPN is relatively adopted for ADP, and MIP is perfectly suitable for SEP. Therefore, the testing methods of MIP and LPN were applied to analyze the distribution of coal pores during LCO_2-ECBM.

The specific surface area and diameter of ADP in coal were examined by the Micromeritics Instrument ASAP 2020. The particle size of coal was 0.30–0.178 mm in LNP. The pore diameter parameters and specific surface area were respectively acquired by Barrett–Joyner–Halenda (BJH) [38] and Brunauer–Emmett–Teller (BET) formulas [39]. MIP was implemented by the Quantachrome Auto Pore 9500. The coal samples used for MIP were crushed into 2–4 cm^3 cubic blocks. The range of mercury filling pressure was 0.0099–423.15 MPa, and the pore size was obtained using the Washburn formula [40]. The experimental setup was as follows:

1. The MIP and LNP coal samples were saturated in a sealed cylinder with a length of 30 cm and a diameter of 20 cm (filled with liquid CO_2). A Dewar bottle was used to provide liquid CO_2 at a low temperature of $-40°C$ and a pressure of 1.8 MPa for 2 h.
2. To remove the mixed gases of CO_2/CH_4 and the adsorbed water in the coal pores, the coal samples were dried and desorbed in a vacuum chamber at a temperature of 60°C for 12 h.
3. MIP and LNP were implemented to obtain the ADP and SEP distributions of the raw and LCO_2 saturated coal samples.

5.3.2 EVOLUTION OF COAL PORE STRUCTURE

5.3.2.1 LNP Isotherms and ADP Analyses

The adsorption and desorption curves of LNP on the raw and treated coal are displayed in Figure 5.2. The critical values of relative pressure (P/P_0) were 0.5 and 0.95, which divided the curve into three stages. Monolayer adsorption, multilayer adsorption on the ADP surface, and capillary condensation in the SEP space occurred in corresponding stages [41, 42]. The turning points occurred distinctly at the desorption curves of $P/P_0 = 0.5$, which was caused by the narrow slit and ink-bottle-shaped pores [43]. It can be concluded that the pore structure of the coal seam is suitable for CBM storage but comparatively inferior for CBM flow.

The adsorption-desorption curves did not perfectly overlap at the stage of monolayer adsorption, illustrating that extensive micropores existed in the sample that restricted gas desorption. Moreover, the widths of the hysteresis loops for the LCO_2 treated coal samples (Figure 5.2b) were larger than those of the raw samples (Figure 5.2a), implying that the ADP was generated after LCO_2 treatment. The specific surface area in the coal sample relates to ADP [44]. To precisely determine the damage effect, the ADP volume, surface area, and diameters were obtained and are shown in Table 5.1.

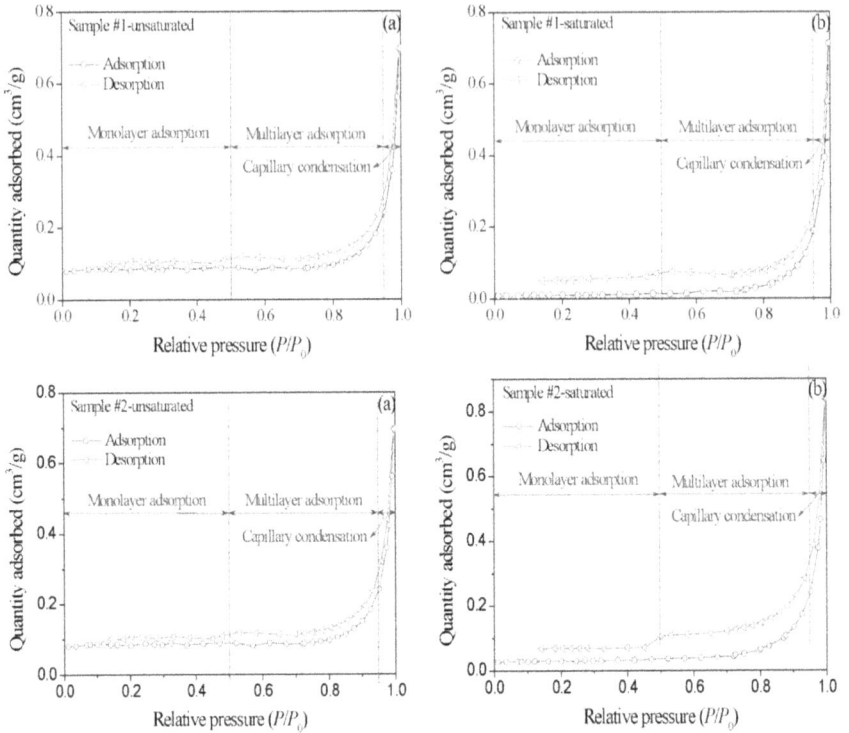

FIGURE 5.2 LNP curves for (a) raw samples and (b) LCO$_2$ saturated samples.

TABLE 5.1
Pore Structure Parameters Obtained by LNP

Sample	$S_{BET, Ad}$ (cm^2/g)	S_{mi+tra} (cm^2/g)		$V_{BJH, Ad}$ (cm^3/g)	V_{mi+tra} (cm^3/g)		A_{PD-1} (nm)
		S_{mi}	S_{tra}		V_{mi}	V_{tra}	
Sample #1-unsaturated	1450.40	565.14	885.26	0.000876	0.000221	0.000655	4.41
Sample #1-saturated	2577.78	1425.32	1152.46	0.001069	0.000304	0.000765	7.93
Sample #2-unsaturated	1489.38	466.23	1023.15	0.000799	0.000255	0.000544	4.22
Sample #2-saturated	2720.44	1254.63	1465.81	0.000929	0.000233	0.000696	8.03

Note: $S_{BET, Ad}$, specific surface area with the diameter of adsorbed pores; S_{mi+tra}, specific surface area of micropores and transition pores; S_{mi}, specific surface area of micropores; S_{tra}, specific surface area of transition pores; $V_{BJH, Ad}$, volume with the diameter of adsorbed pores per unit mass; V_{mi}, volume of micropores per unit mass; V_{tra}, volume of transition pores per unit mass; A_{PD-1}, average pore diameter.

In terms of the volume, average increases of 19.1% and 81.2% in the specific areas of the treated coal (samples #1 and #2) were observed and compared to the original samples. Furthermore, the increments of the average pore diameters of ADP in the two treated coal samples were respectively 3.52 and 3.81 nm than in the corresponding raw coal. A distinct variation of ADP in the treated coal samples was discovered. Low temperature conditions during LCO_2 injection resulted in thermal stress in the coal matrix [45], and a certain amount of micropores and transition pores in the coal were formed, thus resulting in apparent rises in the specific surface area and volume of ADP.

5.3.2.2 MIP Curves and SEP Analyses

MIP curves of the raw and saturated coal are exhibited in Figure 5.3. The mercury volumes that remained in the raw coal pores during the desorption process were 0.0181 and 0.0103 cm³/g (Figure 5.3a), while in the homologous treated coal pores they were 0.0246 and 0.0192 cm³/g (Figure 5.3b), which suggests that the ADP existed in both the original and saturated samples (capillary pressure generated in micropores to prevent mercury extrusion) [46]. The volumes of residual mercury in the saturated coal were larger than those in the untreated coal, indicating that a greater amount of ADP was generated in the treated samples, which was similar to the evolution of LNP.

As seen in Figure 5.3, three filling types (interpore filling, intrapore filling, and matrix compressibility) were displayed in the samples during mercury injection. In addition, two key points (turning and yield points) obtained by the slope of the mercury intrusion curve divided the curves into three phases [47, 48]. Based on the Washburn equation [49], the pressure of the turning point at phase-1 was 0.015 MPa (corresponding pore size: 0.2 mm) for the original and treated coal. The turning points of the original and treated samples were 47 and 13 MPa, respectively, which corresponded to the pore size values of 100 and 30 nm. Therefore, part of the transition pores (distributing in the 30–100nm) and SEP were effectively tested without coal matrix compression. The pore parameters

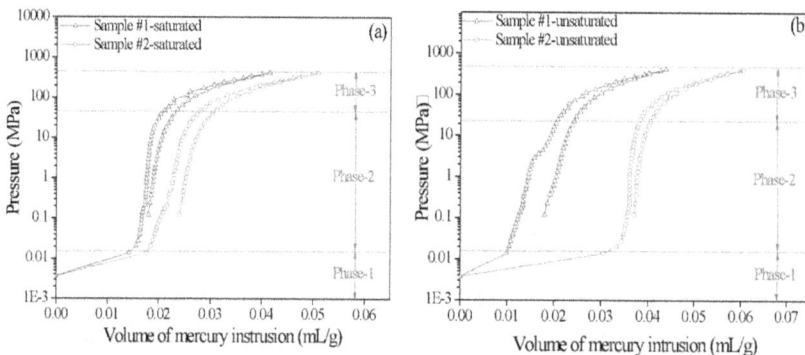

FIGURE 5.3 MIP curves for (a) unsaturated samples and (b) LCO_2 treated samples.

TABLE 5.2
Pore Parameters Obtained by MIP

Samples No.	ϕ (%)	K (mD)	A_{PD-2} (nm)	S_t (cm²/g)	S_{me+ma} (cm²/g)		V_t (cm³/g)	V_{me+ma} (cm³/g)	
					S_{me}	S_{ma}		V_{me}	V_{ma}
Sample #1-unsaturated	5.62	16.35	11.31	163207	553	256	0.0325	0.0037	0.0212
Sample #1-saturated	6.11	25.61	13.15	170455	689	281	0.0497	0.0055	0.0276
Sample #2-unsaturated	5.78	24.57	16.32	157892	376	168	0.0468	0.0014	0.0285
Sample #2-saturated	7.11	27.22	20.05	159776	446	187	0.0515	0.0023	0.0314

Note: ϕ, porosity; ρ_b, bulk density; ρ_s, skeletal density; K, permeability; A_{PD-2}, average pore diameter; S_t, total special surface area; S_{me+ma}, special surface area sum of macropores and mecropores; S_{me}, special surface area of mecropores; S_{ma}, special surface area of macropores; V_t, total pore volume; V_{me+ma}, pore volume sum of macropores and mecropores; V_{me}, pore volume of mecropores; V_{ma}, pore volume of macropores.

of the two coal samples obtained by MIP are shown in Table 5.2. After LCO_2 saturation, the coal porosity increased by 8.7% and 23% for samples #1 and #2. The surface areas of the saturated coal were 7248 and 1184 cm²/g higher than those of the raw samples, and the pore volumes of the treated coal were 0.0172 and 0.0047 cm³/g higher than those of the original coal. Furthermore, new pores evidently formed during LCO_2 saturation. These conclusions were similar to the ADP evolution obtained by LNP experiment. The distribution of the SEP had great influence on the gas seepage rate in the coal seams. The permeability of the LCO_2 treated samples grew an average of 33.7% more than those of the original samples.

5.4 CO_2–CH_4 DISPLACING EFFECT

5.4.1 Testing Method of CO_2–CH_4 Displacement

As shown in Figure 5.4, the experimental displacement system consists of a binary gas supply unit (CO_2/CH_4 with 99.99% purity), a temperature controlling chamber, a large-capacity sample tank (with a diameter of 30 cm and length of 250 cm), a vacuum pump, and a gas detecting unit (flow meter and gas chromatograph).

According to GB/T 19560-2008, the original coal sample #1 was screened into a fine grain with particle size of 0.18–0.25 mm and the inner water in the coal pores was dried using a drying oven at 60°C for 12 h. Then, 430 g of the sample was placed in the sample tank. The adsorbed pressures of CH_4 in the coal pores were designed to be 0.8, 1.2, and 1.6 MPa, respectively, which corresponded to the methane pressures during a different LCO_2-ECBM period. The displacement flows at the tank outlet were set to 6, 10, and 15 mL/min, respectively. During the CO_2–CH_4 displacement process, the output pressure at the CO_2 gas cylinders

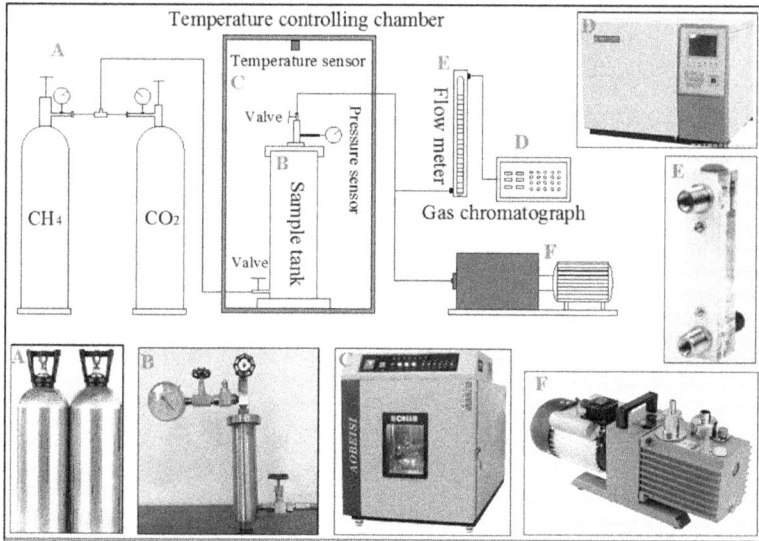

FIGURE 5.4 CO_2-CH_4 displacement system: (A) CO_2/CH_4 gas cylinders, (B) large-capacity sample tank, (C) temperature controlling chamber, (D) gas chromatograph, (E) flow meter, and (F) vacuum pump.

was equal to the adsorbed pressures of the CH_4. The displacement procedure was as follows: (1) Before starting the testing, the experimental system was checked to ensure that there were no leaks. The residual gases were adsorbed in the coal pores and stored in the free space of the sample tank and were wholly replaced by the vacuum pump at the absolute pressure of 100 Pa and a temperature of 20°C for 12 h; (2) the adsorption process for each adsorbed pressure and displacement flow was carried out for 6 h; and (3) meanwhile, the released gases from the sample tank were constantly measured using the gas chromatograph until there was no more than a 0.05% concentration of methane.

CH_4 in the coal pores included adsorption and dissociation phases. Two primary reactions occurred in the coal pores during ECBM: CO_2-CH_4 replacement in the ADP pores and mitigation in the SEP pores. During the dissociation phase, CH_4 is assumed to be consistent with an ideal gas. The calculation method of CH_4 volume under different temperatures and pressures can be shown as [50, 51]:

$$V_{SC}^{CH_4} = \frac{P_i V_{P_i}^{CH_4} T_0}{TP_0} \tag{5.1}$$

where $V_{SC}^{CH_4}$ and $V_{P_i}^{CH_4}$ are the CH_4 volumes under standard conditions (the pressure value of P_0 is 0.101325 MPa and the temperature value of T_0 is 273.15 K) and experimental conditions, respectively; mL; P_i is the experimental pressure, MPa, $i = 1, 2$, and 3, with a corresponding value of $P_i = 0.8, 1.2$, and 1.6 MPa; T is the experimental temperature of 293.15 K.

CH_4 volume under experimental conditions can be shown as:

$$V_{P_i}^{CH4} = \int_0^{t_c} q_j \psi_t dt \tag{5.2}$$

where q_j is the displacement flow rate, mL/min, $j = 1, 2$, and 3, corresponding to the value of $q_j = 6$, 10, and 15 mL/min; ψ_t is the methane concentration at the tank outlet and acquired by the displacement testing, %; t_c is the experimental time, min.

Substituting Eq. (5.2) into Eq. (5.1), CH_4 volume under standard conditions in the displacement process is defined as:

$$V_{SC}^{CH4} = \frac{P_i V T_0 \int_0^{t_c} q_j \psi_t dt}{T P_0} \tag{5.3}$$

5.4.2 VARIATION OF CO_2–CH_4 DISPLACEMENT

5.4.2.1 Gas Concentration of Displacement

The variation of the CH_4 concentration under displacement flow rates and pressures is exhibited in Figure 5.5. As the displacement time increased, the CH_4 concentration at the tank outlet plummeted. Based on the slope of the CH_4 concentration over time, one critical point was identified that divided the curves into two stages: rapid falling stage (stage I) and stable decreasing stage (stage II). For better analysis of the displacement effect, the critical point T_C was defined as the testing time when the range of CH_4 concentration went from 100% to 5%. T_C under different displacement flow and pressure is revealed in Table 5.3.

T_C was influenced by displacement pressures and flow rates. A positive correlation was displayed between T_C and displacement pressure at stage I, which indicated an easier displacement in the coal seam with a lower CBM pressure than of that with a larger CBM pressure. The T_C and displacement flow rate at this stage showed a negative relation, indicating a large displacement efficiency

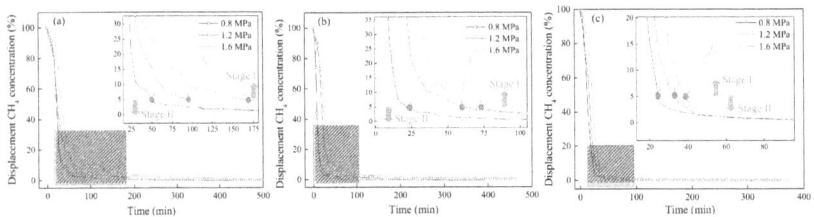

FIGURE 5.5 Displacement of CH_4 concentration with different displacement flow rates. (a) The displacement flow rate of 6 mL/min, (b) 10 mL/min, and (c) 15 mL/min.

TABLE 5.3

Critical Times of the Displacement Experiment

	Critical Time T_C (min)		
Flow Rate (mL/min)	0.8 MPa	1.2 MPa	1.6 MPa
6	50	95	160
10	25	60	75
15	20	35	40

due to the high flow rate during the CO_2-ECBM for preventing and controlling CGO. Conversely, the curves of the CH_4 concentration at the tank outlet under different flow rates and displacement pressures showed an overlapping trend at stage II. The curves of the CH_4 concentration at different pressures entirely overlapped at the displacement time of 340, 230, and 50 min, which corresponded to the displacement flow rates of 6, 10, and 15 mL/min, respectively. It intimated a similar displacement efficiency under different flow rates and pressures when the displacement time reached one certain value.

5.4.2.2 CH_4 Volume of Displacement

The CH_4 volume is shown in Figure 5.6. Negative and positive relations of displacement pressure versus total CH_4 volume were respectively exhibited at stage I and II. Taking the displacement pressure of 1.6 MPa as an example, CH_4 volumes at T_C with displacement flow rates of 6, 10, 15 mL/min were 188.6, 269.3, and 290.8 mL, respectively. The CH_4 volume increased with an increase in the flow rate. As the time increased, the two stages were displayed in the curves of the CH_4 volume, which was consistent with the trend of the CH_4 concentration. Furthermore, the curves of CH_4 volume presented parallel lines beyond the T_C under different testing conditions.

According to previous studies, CH_4 existing in coal pores includes dissociation and adsorption phases [52, 53]. In the dissociation phase, CH_4 was removed at stage I, and the amount of CH_4 at the tank outlet mainly related to the gas content. With the gas pressure rising, more dissociated CH_4 in a compressed state existed

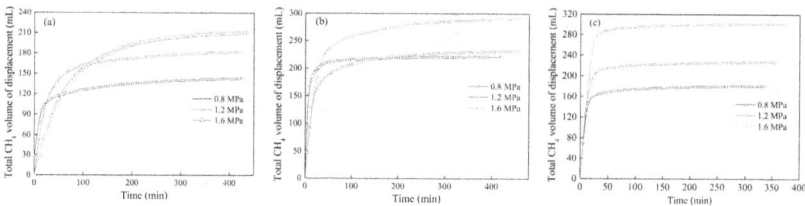

FIGURE 5.6 Displacement of CH_4 volume at different flow rates. (a) The displacement flow rate of 6 mL/min, (b) 10 mL/min, and (c) 15 mL/min.

in the SEP, and more CH_4 was driven out in the displacement process. Moreover, competitive adsorption acted as a dominant effect in stage II, and the CH_4 volume increments sharply decreased, which was mainly dependent on the desorption rate of CH_4 in the adsorption phase of the ADP. This indicates that different displacement flow rates should be suitable for reducing the usage of CO_2 during the displacement process.

5.5   MECHANISMS OF LCO$_2$-ECBM RECOVERY

According to the comprehensive analysis above, the process of LCO_2-ECBM includes two main mechanisms: permeability improvement and CO_2-CH_4 displacement (competitive adsorption and driving force). A heat exchange occurred when low-temperature LCO_2 (below $-30°C$) was injected into the coal seams and seeped into the pore surface of the coal matrix (approximately 25–30°C). The temperature gradient generated thermal stress in the coal. ADP and SEP formed when the thermal stress reached the mechanical threshold of the coal body (Figure 5.7a), and coal diffusivity and permeability improved.

Moreover, LCO_2 gradually changed into a gas phase as heat was exchanged, and the volume and pressure of the gas CO_2 rose after phase transition in the

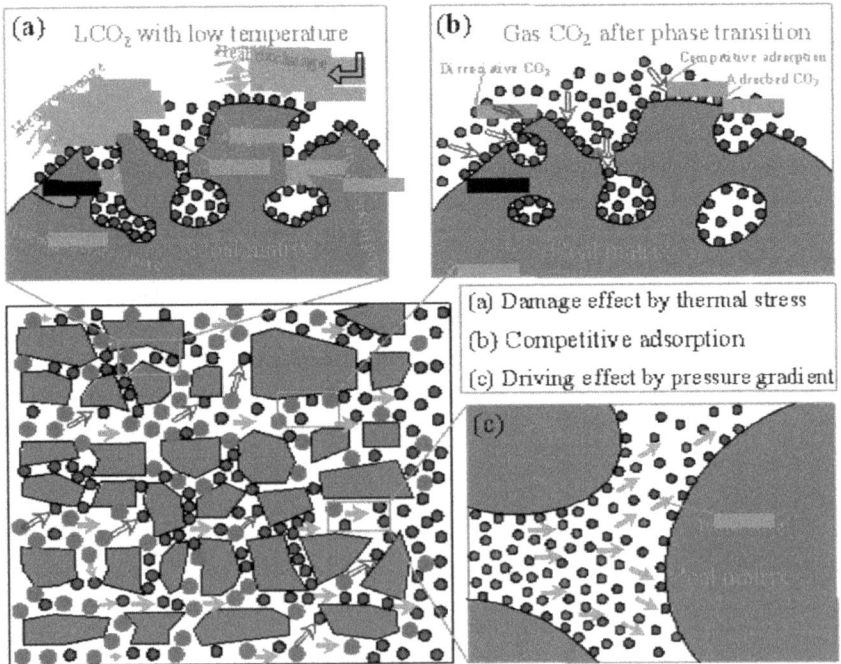

FIGURE 5.7   Mechanisms of LCO$_2$-ECBM recovery.

coal pores and cracks. The driving force, caused by the pressure gradient in the SEP, effectively counteracted the flow resistance during the gas drainage process (Figure 5.7c). Furthermore, competitive adsorption occurred between the adsorbed CH_4 and the dissociative CO_2 on the ADP surface, and part of the adsorbed CH_4 transformed into a free phase (Figure 5.7b), which was favorable for CH_4 extraction. However, the matrix swelled when free CO_2 was adsorbed at the coal pore surface, and this reduced the permeability and was disadvantageous to CH_4 desorption. The compound effects of both the enhancing and inhibiting mechanisms are complicated and need further study.

5.6 IN-SITU EXPERIMENT AND DRAINAGE EFFECT

5.6.1 GEOLOGICAL CHARACTERIZATION AND SAMPLE COLLECTION

Zhangji mine (shown in Figure 5.8), a typical CGO coal mine in the Huainan mining area of Anhui Province, China, covers an area of 70.5 km^2 and has 1.78 billion tons of measured coal reserves and an annual productive capacity of 9.6 million tons [54].

Currently, one of the major mining seams in Zhangji mine, No. 6 coal seam, is buried at a depth of 517.5–628.3 m, and its average thickness is 5.2 m with a dig angle of 2°–5°. Based on the geological data, the firmness coefficient of the coal sample is 0.21–0.37, the initial gas release velocity is 16–17 mL/s, and the permeability coefficient of this seam is 0.62–1.24 mD. The maximum value of the CH_4 content in No. 6 coal seam is 12.38 m^3/t, and the corresponding methane pressure is 0.78–1.47 MPa which obviously exceeds the critical indicators (gas pressure of 0.74 MPa and CH_4 content of 8 m^3/t) and can be identified as a CGO coal seam [1].

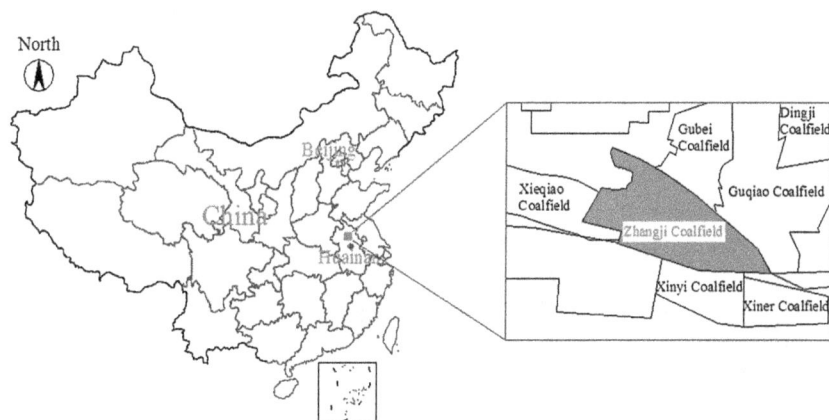

FIGURE 5.8 Location of Huainan mining area and the sampling site.

5.6.2 APPARATUS AND BOREHOLE LAYOUT

A new mining roadway (17246 working face) in the No. 6 coal seam of Zhangji mine was chosen to be the in-situ testing site for LCO_2-ECBM recovery, which was first implemented in the Huainan coal mining area to prevent and control CGO. As seen in Figure 5.9, the in-situ apparatus consisted of four parts: a LCO_2 supply system, a monitoring system, an injection system, and a pipeline system. The supply system of LCO_2 involved three LCO_2 tanks with a volume of 2 m³. The dynamic injection part was a $BPCO_2$-1000/60 plunger pump with a working pressure of 0–6 MPa and a maximum flow of 1.5 m³/h. The monitoring system included a Turbine flowmeter, a Venturi flow meter, a pressure gauge, a data collecting instrument, and a relief valve group. The data detecting instrument ensured real-time recording of the flow and pressure parameters during LCO_2 injection. The pipelines included high-pressure gas drainage pipes and LCO_2 injection hoses.

The distribution of the boreholes in the No. 6 seam is displayed in Figure 5.9. Two injection boreholes (named S1 and S2) and 18 drainage boreholes (Nos. #1 to #18) were completed. All boreholes were 140 m in length, 113 mm in diameter, and 5 m in interval. The outer 30 m of the two injection boreholes were sealed to inhibit the leakage of injected CO_2 into the roadway. In order to compare the CH_4 drainage effect of the injection area versus the non-injection area, 18 identical extraction boreholes with the same geological parameters (named the contrasting area) were drilled. Two sets of Turbine flowmeters were respectively installed in the testing area and contrasting area to obtain gas drainage parameters.

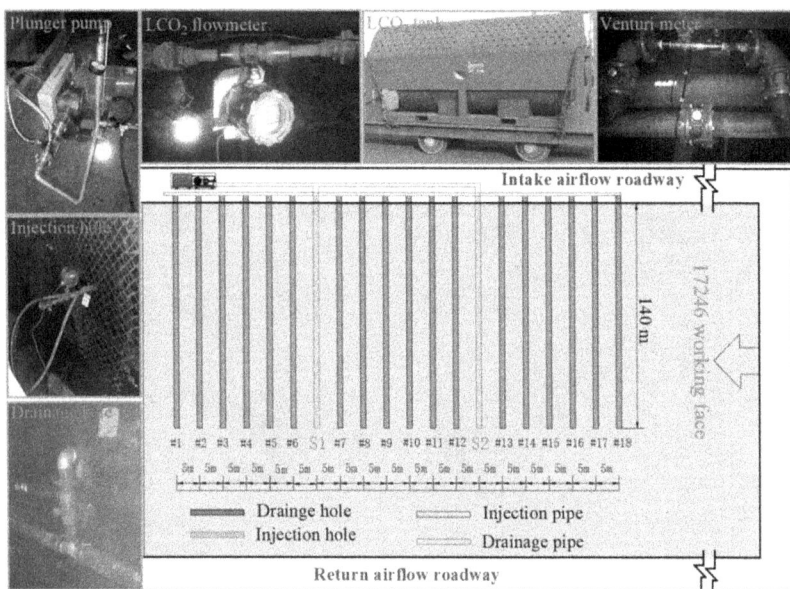

FIGURE 5.9 In-situ testing apparatus and borehole layout in 17246 working face.

5.6.3 Technological Procedure

The LCO_2-ECBM procedure consisted of three main steps: apparent connection, LCO_2 injection, and gas monitoring/drainage. The detailed operative steps were reported in a previous study by Wen et al. [10].

1. Apparent connection. Three tanks filled with LCO_2 were connected in parallel by high-pressure hoses and a four-way valve with a plunger pump inlet. Then, the Turbine flow meter and injection borehole were connected sequentially. The extraction boreholes (Nos. #1–#18) were connected to the gas extraction pipeline and the Venturi flow meter was centrally installed. Date cables were used to join the Turbine flow meter and pressure gauge to the data collecting instrument.
2. LCO_2 injection. The 18 extraction holes were all closed before the LCO_2 injection. A plunger pump and three LCO_2 tanks were opened and then checked for any gas leakage. Moreover, a data collecting instrument was used to obtain the pressure and flow parameters. The two borehole injections were performed separately.
3. Gas drainage and monitoring. All borehole extractions occurred steadily at a pressure of –30 kPa after LCO_2 injection. The parameters of CH_4 concentration, CO_2 concentration, and flow in the drainage boreholes and pipeline were consequently detected.

5.6.4 Displacement Effect of LCO_2-ECBM

5.6.4.1 Injection Parameters

The accumulated LCO_2 volumes and pump pressures of the injection boreholes during the LCO_2-ECBM process are displayed in Figure 5.10. The maximum values of the pump pressures for the S1 and S2 boreholes were 3.83 and 3.77 MPa, respectively, and the corresponding accumulated volumes of injected LCO_2 were 5.39 and 5.41 m^3, respectively.

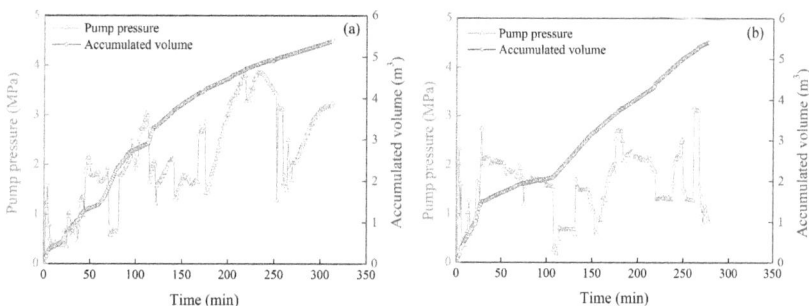

FIGURE 5.10 Pump pressures and volumes of LCO_2 injecting into (a) S1 and (b) S2 boreholes.

During the injection process, the LCO_2 flow rates of the injection boreholes fluctuated from 0.5 to 2 m³/h with a change in pump pressure. It showed a negative relation of flow versus pump pressure. LCO_2 flow reached a minimum, while the pump pressure reached a certain maximum. This illustrates micro-crack (especially SEP) generation and propagation in the coal seam during the LCO_2 injection process [55, 56]. These results were in accordance with the pore structure variation feature obtained by the LNP and MIP experiments.

5.6.4.2 Effective Displacement Radius

After LCO_2 injection, the drainage borehole valves of both the testing and the contrasting areas were synchronously turned on. Gas samples were collected daily from the drainage boreholes (Nos. #1–#18) in the testing and contrasting areas and detection was done using a gas chromatograph. According to the geological information, the original CO_2 concentration in the No. 6 coal seam was no more than 0.3%. Hence, CO_2 can be regarded as an index for determining the effective displacement radius of LCO_2-ECBM. CO_2 concentrations in drainage boreholes during the gas extraction process are displayed in Figure 5.11. The CO_2 concentrations of the #1 and #18 drainage boreholes were 1.34% and 1.32%, which suggested that the effective displacement radius was no less than 30 m.

Furthermore, a fluctuating characteristic in the three stages was observed due to the variation of CO_2 concentrations during the extraction process: a descending trend (first 10 days), an upward trend (10–20 days), and a reducing trend (after the 20th day). It can be deduced that the adsorbed and desorbed CO_2 co-existed in

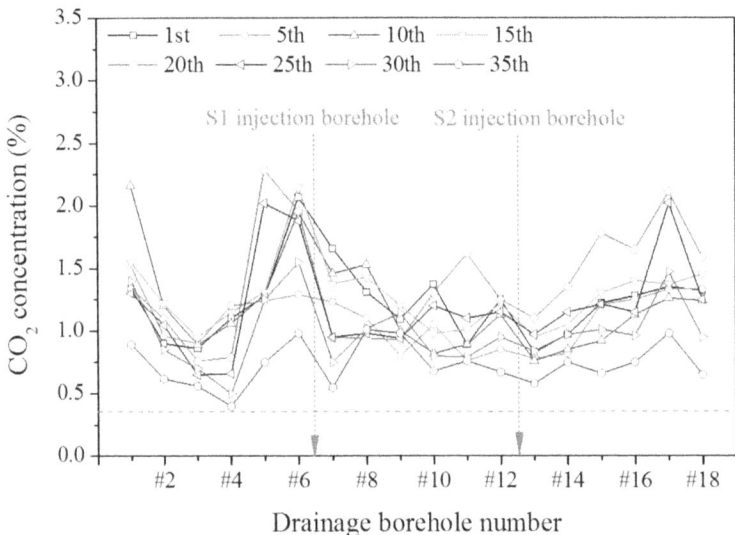

FIGURE 5.11 CO_2 distribution in extraction boreholes during drainage process.

the coal pores and cracks after LCO_2 injection. The desorbed CO_2 concentration in the coal seam decreased with the gas extraction in the first stage. Meanwhile, the adsorbed CO_2 was gradually released as the temperature rose in the coal seam due to heat exchange [24, 57], and the gas concentration reached its maximum value on the 20th day. A falling tendency occurred at the third stage when the CO_2 loss rate was greater than the release rate. This suggests that an optimized interval time is needed to study repeated LCO_2-ECBM.

5.6.5 CH_4 DRAINAGE EFFICIENCY

The most efficacious indicators for judging the LCO_2-ECBM effect are CH_4 concentration and pure CH_4 flow in the drainage pipes. As seen in Figure 5.12, the average CH_4 concentration in the drainage pipe of the testing area was 53.2% for about one month, whereas the corresponding contrasting area was 27.2%, which shows an increase of 1.96 times after LCO_2-ECBM.

Similarly, the pure CH_4 flow in the testing area was 2.28 times larger compared to the contrasting area. The effect of LCO_2-ECBM was remarkable in preventing and controlling CGO in the No. 6 coal seam of the Zhangji mine. Furthermore, the variation of both the CH_4 concentration and the pure CH_4 flow in the testing area exhibited a periodic feature: a descending trend (first ten days), an upward trend (10–20 days), and a reducing trend (after the 20th day) which is fundamentally consistent with a change in CO_2 distribution. However, the mixed gas flow measured using the Venturi meter in the drainage pipe of the testing area was close to that of the contrasting area. It showed a greater impact on CH_4 concentration than that of the CH_4 flow during LCO_2 injection into the No. 6 coal seam with the in-situ experimental parameters in this study. Combined with the results in Figure 5.11, CO_2–CH_4 competitive adsorption had a dominant effect during the LCO_2-ECBM process. Further studies should be carried out to explore the displacement effect under other in-situ experimental parameters (such as injection pressure, flow rate, and total LCO_2 quantity).

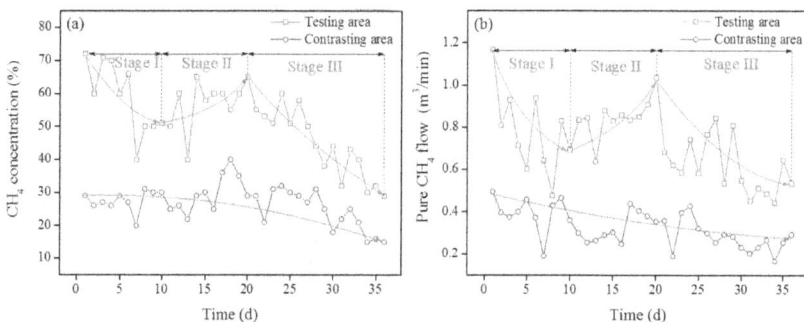

FIGURE 5.12   (a) CH_4 concentration and (b) CH_4 flow rate of drainage pipes in testing and contrasting areas.

5.7   CONCLUSIONS

Enhancing the efficiency of CBM extraction by injecting LCO_2 into a coal seam is a novel technique for preventing and controlling CGO disasters. The mechanisms during the LCO_2-ECBM process, including the damaging effects on coal microstructure due to low temperature conditions and displacement by CO_2–CH_4 competitive adsorption, were quantitatively analyzed in this paper using LNP, MIP, and displacement tests. Meanwhile, an in-situ experiment in the No. 6 coal seam of the Zhangji mine was implemented to explore LCO_2-ECBM feasibility and CH_4 drainage efficiency.

1. The variation of ADP and SEP in the coal samples before and after LCO_2 saturation was revealed. Compared to the raw samples, the ADP specific area and the SEP volume in the saturated coal samples were an average of 81.2% and 0.011 cm^3/g larger. The permeability of the saturated samples had an average increase of 33.7% compared to the raw samples. The evolution of the microstructure in the saturated coal featured new pore formation and the transformation of primary pores into larger ones.

2. The experimental apparatus for displacement was self-developed and the CO_2–CH_4 displacement efficiency under different pressures and flow rates was analyzed. The CH_4 concentration in the displacement flow showed periodic features before and after the critical time. It showed a negative correlation between displacement efficiency and flow rate at the rapid falling stage, and an irrelevant correlation at the stable decreasing stage, which mainly related to the conversion rate between the adsorbed and dissociative CH_4 in the coal pores at different displacement stages.

3. An in-situ experiment was carried out and CH_4 drainage efficiency was further analyzed. The effective displacement radius was more than 30 m when the highest pump pressure was 3.83 MPa and the total injected LCO_2 volume was approximately 5.4 m^3. Both pure CH_4 flow and concentration were increased 2.28 and 1.96 times. Thus, LCO_2-ECBM technology can be effective in preventing and controlling CGO hazards.

REFERENCES

1. Jun Tang, Chaojie Wang, Yujia Chen, Xiaowei Li, Dingding Yang, Jie Liu. Determination of critical value of an outburst risk prediction index of working face in a coal roadway based on initial gas emission from a borehole and its application: A case study. Fuel 2020;267:117229.
2. Gui Fu, Xuecai Xie, Qingsong Jia, Wenqing Tong, Ying Ge. Accidents analysis and prevention of coal and gas outburst: Understanding human errors in accidents. Process Safety and Environmental Protection 2020;134:1–23.
3. Yi Xue, Feng Gao, Yanan Gao, Hongmei Cheng, Yingke Liu, Peng Hou, Teng. Quantitative evaluation of stress-relief and permeability-increasing effects of overlying coal seams for coal mine methane drainage in Wulan coal mine. Journal of Natural Gas Science and Engineering 2016;32:122–137.

4. MCM Nasvi, PG Ranjith, J Sanjayan. Effect of different mix compositions on apparent carbon dioxide (CO_2) permeability of geopolymer: Suitability as well cement for CO_2 sequestration wells. Applied Energy 2014;114:939–48.
5. Jinxin Huang, Guang Xu, Yunpei Liang, Guozhong Hu, Ping Chang. Improving coal permeability using microwave heating technology—A review. Fuel 2020; 266:117022.
6. GH Ni, K Dong, S Li, Q Sun. Gas desorption characteristics effected by the pulsating hydraulic fracturing in coal. Fuel 2019;236:190–200.
7. C Zhai, M Li, C Sun, JG Zhang, W Yang, Q Li. Guiding-controlling technology of coal seam hydraulic fracturing fractures extension. International Journal of Mining Science and Technology 2012;22:831–836.
8. Congmeng Hao, Yuanping Cheng, Liang Wang, Hongyong Liu, Zheng Shang. A novel technology for enhancing coalbed methane extraction: Hydraulic cavitating assisted fracturing. Journal of Natural Gas Science and Engineering 2019;72:103–110.
9. Nie Baisheng, Li Xiangchun. Mechanism research on coal and gas outburst during vibration blasting. Safety Science 2012;50:741–744.
10. Hu Wen, Xiaojiao Cheng, Jian Chen, Chunru Zhang, Zhijin Yu, Zhenbao Li, Shixing Fan, Gaoming Wei, Bangkai Cheng. Micro-pilot test for optimized pre-extraction boreholes and enhanced coalbed methane recovery by injection of liquid carbon dioxide in the Sangshuping coal mine. Process Safety and Environmental Protection 2020;136:39–48.
11. Guozhong Hu, Wenrui He, Miao Sun. Enhancing coal seam gas using liquid CO_2 phase-transition blasting with cross-measure borehole. Journal of Natural Gas Science and Engineering 2018;60:164–173.
12. Vikram Vishal. In-situ disposal of CO_2: Liquid and supercritical CO_2 permeability in coal at multiple down-hole stress conditions. Journal of CO_2 Utilization 2017; 17:235–242.
13. Mingjun Zou, Chongtao Wei, Zhiquan Huang, Simin Wei. Porosity type analysis and permeability model for micro-trans-pores, meso-macro-pores and cleats of coal samples. Journal of Natural Gas Science and Engineering 2015;27:776–784.
14. Chengzheng Cai, Gensheng Li, Zhongwei Huang, Zhonghou Shen, Shouceng Tian, Jiangwei Wei. Experimental study of the effect of liquid nitrogen cooling on rock pore structure. Journal of Natural Gas Science and Engineering 2014;21: 507–517.
15. Hu Wen, Gaoming Wei, Li Ma, Zhenbao Li, Changkui Lei, Jianchi Hao. Damage characteristics of coal microstructure with liquid CO_2 freezing-thawing. Fuel 2019; 249:169–177.
16. Jizhao Xu, Cheng Zhai, Shimin Liu, Lei Qin, Ruowei Dong. Investigation of temperature effects from LCO_2 with different cycle parameters on the coal pore variation based on infrared thermal imagery and low-field nuclear magnetic resonance. Fuel 2018;215:528–540.
17. Vikram Vishal. Saturation time dependency of liquid and supercritical CO_2 permeability of bituminous coals: Implications for carbon storage. Fuel 2017;192:201–207.
18. Anna Pajdak, Mateusz Kudasik, Norbert Skoczylas, Mirosław Wierzbicki, Leticia Teixeira Palla Braga. Studies on the competitive sorption of CO_2 and CH_4 on hard coal. International Journal of Greenhouse Gas Control 2019;90:102789.
19. Song Yu, Jiang Bo, Lan Fengjuan. Competitive adsorption of $CO_2/N_2/CH_4$ onto coal vitrinite macromolecular: Effects of electrostatic interactions and oxygen functionalities. Fuel 2019;235:23–38.
20. Siyuan Wu, Cunbao Deng, Xuefeng Wang. Molecular simulation of flue gas and CH4 competitive adsorption in dry and wet coal. Journal of Natural Gas Science and Engineering 2019;71:102980.

21. Basanta Kumar Prusty. Sorption of methane and CO_2 for enhanced coalbed methane recovery and carbon dioxide sequestration. Journal of Natural Gas Chemistry 2008;17(1):29–38.
22. Sujoy Chattaraj, Debadutta Mohanty, Tarkeshwar Kumar, Gopinath Halder. Thermodynamics, kinetics and modeling of sorption behaviour of coalbed methane–A review. Journal of Unconventional Oil and Gas Resources 2016;16:14–33.
23. Gregory N. Okolo, Raymond C. Everson, Hein W.J.P. Neomagus, Richard Sakurovs, Mihaela Grigore, John R. Bunt. The carbon dioxide, methane and nitrogen high-pressure sorption properties of South African bituminous coals. International Journal of Coal Geology 2019;209:40–53.
24. Cheng Guan, Shimin Liu, Chengwu Li, Yi Wang, Yixin Zhao The temperature effect on the methane and CO_2 adsorption capacities of Illinois coal. Fuel 2018;211:241–250.
25. Amer Syed, Sevket Durucan, Ji-Quan Shi, Anna Korre. Flue gas injection for CO_2 storage and enhanced coalbed methane recovery: Mixed gas sorption and swelling characteristics of Coals. Energy Procedia 2013;37: 6738–6745.
26. Richard Sakurovs. Relationships between CO_2 sorption capacity by coals as measured at low and high pressure and their swelling. International Journal of Coal Geology 2012;90:156–161.
27. Rui Wang, Qizhi Wang, Qinghe Niu, Jienan Pan, Haichao Wang, Zhenzhi Wang. CO_2 adsorption and swelling of coal under constrained conditions and their stage-change relationship. Journal of Natural Gas Science and Engineering 2020;76:103205.
28. Barbara Dutka. CO_2 and CH_4 sorption properties of granular coal briquettes under in situ states. Fuel 2019;247:228–236.
29. Zofia Majewska, Stanisław Majewski, Jerzy Ziętek. Swelling of coal induced by cyclic sorption/desorption of gas: Experimental observations indicating changes in coal structure due to sorption of CO_2 and CH_4. International Journal of Coal Geology 2010;83(41):475–483.
30. Cuijuan Luo, Dengfeng Zhang, Zengmin Lun, Chunpeng Zhao, Haitao Wang, Zhejun Pan, Yanhong Li, Jin Zhang, Shuaiqiu Jia. Displacement behaviors of adsorbed coalbed methane on coals by injection of SO_2/CO_2 binary mixture. Fuel 2019;247:356–367.
31. Xiaogang Zhang, P.G. Ranjith. Experimental investigation of effects of CO_2 injection on enhanced methane recovery in coal seam reservoirs. Journal of CO_2 Utilization 2019;33:394–404.
32. Chuanjie Zhu, Jiamin Wan, Tetsu K. Tokunaga, Na Liu, Baiquan Lin, Hourong Wu. Impact of CO_2 injection on wettability of coal at elevated pressure and temperature. International Journal of Greenhouse Gas Control 2019;91:102840.
33. Jiawen Cai, Shengqiang Yang, Xincheng Hu, Wanxin Song, Yawei Song, Qin Xu, Shuai Zhang. Risk assessment of dynamic disasters induced by gas injection displacement in coal seams. Process Safety and Environmental Protection 2019;128:41–49.
34. K.S.W. Sing. Reporting physisorption data for gas/solid systems with special reference to the determination of surface area and porosity (Recommendations 1984). Pure and Applied Chemistry 1985;57:603–619.
35. Mahbubul Muttakin, Sourav Mitra, Kyaw Thu, Kazuhide Ito, Bidyut Baran Saha. Reporting physisorption data for gas/solid systems with special reference to the determination of surface area and porosity. International Journal of Heat and Mass Transfer 2018;122:795–805.
36. Haijiao Fu, Dazhen Tang, Ting Xu, Hao Xu, Shu Tao, Song Li, ZhenYong Yin, Baoli Chen, Cheng Zhang, Linlin Wang. Characteristics of pore structure and fractal dimension of low-rank coal: A case study of Lower Jurassic Xishanyao coal in the southern Junggar Basin, NW China. Fuel 2017;193:254–264.

37. Yanbin Yao, Dameng Liu, Dazhen Tang, Shuheng Tang, Wenhui Huang, Zhihua Liu, Yao Che. Fractal characterization of seepage-pores of coals from China: An investigation on permeability of coals. Computers Geosciences 2009;35(6):1159–1166.

38. C.R. Clarkson, R.M. Bustin. The effect of pore structure and gas pressure upon the transport properties of coal: A laboratory and modeling study. 1. Isotherms and pore volume distribution. Fuel 1999;78:1333–1344.

39. P.J. Crosdale, B.B. Beamish, M. Valix. Coalbed methane sorption related to coal composition. International Journal of Coal Geology 1998;35:147–158.

40. E.W. Washburn. The dynamics of capillary flow. Physical Review 1921;17:273–283.

41. E.A. Takara, E. Quiroga, D.A. Matoz-Fernandez, N.A. Ochoa, A.J. Ramirez-Pastor. Fractional statistical theory of nite multilayer adsorption. Applied Surface Science 2016;360:14–19.

42. Wu Jianjun, Li Xia, Zhou Guoli, Miao Zhenyong, Li Huirong, Li Guoning, Hu Xuelian. Effect of pore structure on moisture re-adsorption of dewatered lignite. Journal of China University of Mining and Technology 2013; 42: 806–811 (in Chinese).

43. F.B. Scheufele, C.E. Módenes, C. Ribeiro, F.R. Espinoza-Quiñones, R. Bergamasco, N.C. Pereira. Monolayer–multilayer adsorption phenomenological model: Kinetics, equilibrium and thermodynamics. Chemical Engineering Journal 2016;284:1328–1341.

44. J. Liu, J. Zhu, J. Cheng, J. Zhou, K. Cen, Pore structure and fractal analysis of Ximeng lignite under microwave irradiation. Fuel 2015;146:41–50.

45. X. Sun, Z. Wang, B. Sun, W. Wang. Research on hydrate formation rules in the formations for liquid CO_2 fracturing. Fuel 2016;33:1390–1401.

46. Hu Wen, Zhenbao Li, Jun Deng, Chi-Min Shu, Bin Laiwang, Qiuhong Wang, Li Ma. Influence on coal pore structure during liquid CO_2-ECBM process for CO_2 utilization. Journal of CO_2 Utilization 2017;21:543–552.

47. C. Liu, G. Wang, S. Sang, W. Gilani, V. Rudolph. Fractal analysis in pore structure of coal under conditions of CO_2 sequestration process. Fuel 2015;139:125–132.

48. C. Liu, G. Wang, S. Sang, V. Rudolph. Changes in pore structure of anthracite coal associated with CO_2 sequestration process. Fuel 2010;89:2665–2672.

49. Song Yu, Jiang Bo, Shao Pei, Wu Jiahao. Matrix compression and multifractal characterization for tectonically deformed coals by Hg porosimetry. Fuel 2018; 211;661–675.

50. Chaojun Fan, Derek Elsworth, Sheng Li, Lijun Zhou, Zhenhua Yang, Yu Song. Thermo-hydro-mechanical-chemical couplings controlling CH_4 production and CO_2 sequestration in enhanced coalbed methane recovery. Energy 2019;173:1054–1077.

51. Yuannan Zheng, Qingzhao Li, Chuangchuang Yuan, Qinglin Tao, Yang Zhao, Guiyun Zhang, Junfeng Liu. Influence of temperature on adsorption selectivity: Coal-based activated carbon for CH_4 enrichment from coal mine methane. Powder Technology 2019;347:42–49.

52. Zhaofeng Wang, Weiwei Su, Xu Tang, Jiahao Wu. Influence of water invasion on methane adsorption behavior in coal. International Journal of Coal Geology 2018;197:74–83.

53. Guoxi Cheng, Bo Jiang, Ming Li, Jiegang Liu, Fengli Li. Effects of pore structure on methane adsorption behavior of ductile tectonically deformed coals: An inspiration to coalbed methane exploitation in structurally complex area. Journal of Natural Gas Science and Engineering 2020;74:103083.

54. Bin Tang, Hua Cheng, Yongzhi Tang, Zhishu Yao, Chuanxin Rong, Xiaojian Wang, Weipei Xue, Shiwei Guo. Experiences of gripper TBM application in shaft coal mine: A case study in Zhangji coal mine, China. Tunnelling and Underground Space Technology 2018;81:660–668.

55. Ruxin Zhang, Bing Hou, Peng Tan, Yeerfulati Muhadasi, Weineng Fu, Xiaomu Dong, Mian Chen. Hydraulic fracture propagation behavior and diversion characteristic in shale formation by temporary plugging fracturing. Journal of Petroleum Science and Engineering 2020;190:107063.
56. Xinglong Zhao, Bingxiang Huang, Jie Xu. Experimental investigation on the characteristics of fractures initiation and propagation for gas fracturing by using air as fracturing fluid under true triaxial stresses. Fuel 2019;236:1496–1504.
57. John H. Levy, Stuart J. Day, John S. Killingley. Methane capacities of Bowen Basin coals related to coal properties. Fuel 1997;76(9):813–819.

6 Fighting Fires in the Coal Mining Industry

Jun Deng, Yang Xiao and Qing-Wei Li
School of Safety Science and Engineering, and
Xi'an University of Science and Technology,
Shaanxi Key Laboratory of Prevention and Control
of Coal Fire, Xi'an University of Science and
Technology, Xi'an, People's Republic of China

Kun-Hua Liu
School of Safety Science and Engineering,
Xi'an University of Science and Technology,
Xi'an, People's Republic of China

CONTENTS

DOI: 10.1201/9781003140382-6

6.1 COAL FIELD FIRES

Fires caused by spontaneous combustion of coal occur worldwide and seriously threaten the natural environment and human health, resulting in huge loss of resources and environmental pollution. Spontaneous combustion fires in coal seams are common in China, the United States, India, Russia, Australia, Indonesia, Central Asia, and other countries and regions [1]. The environmental hazards caused by coal fires include: (1) excessive CO_2 emission [2, 3], (2) serious air pollution, (3) loss of plant and animal life, (4) safety and health risks for miners and the surrounding population.

The causes of coalfield fire formation are highly complex. Disasters caused by combustion of coal may be triggered by a natural occurrence, such as lightning. In addition, spontaneous combustion of coal may arise as a consequence of weathering of seam outcrops, leading to destruction of or damage to coal mines. Ignition of a coal seam and the formation of coal seam fire could also be the result of human activity, such as mining.

The surface of a coalfield fire area is usually covered with rocks. The combined effects of high temperature, combustion of coal, and gravity lead to the disintegration of the layers of sand and mudstone that cover the area, producing crevices and fissures of various sizes, while also causing the rocks in the vicinity of the fire to metamorphose and even turn into lava slag. The high temperature caused by burning causes natural convection, driving air into the surface of the coal mass through rock fissures, thus intensifying the burning. A subsurface fire requires fuel, oxygen, and heat, and the removal of any one of these components will extinguish the fire. Numerous conventional technologies, such as grouting, three-phase foam, gelatin infusion, inert gas injection, and so on, have been developed and applied to prevent and mitigate the hazards associated with coal fires [4–11]. Furthermore, water, slurries, or liquefied nitrogen can be injected to cool down the coal or cut off the air supply to extinguish the fire. The three-phase foam, which is composed of mud, nitrogen, and water, is used to fight subsurface fires. Surface sealing is an oxygen exclusion method used to control subsurface coal fires. This method is relatively inexpensive and suppresses surface evidence of combustion. Smaller fires can also be handled by removing the burning coal by excavation, but it does not work well for extinguishing larger coal fires.

In this paper, the application of composite fly ash gel to extinguish open-pit coal fires is presented using the Haibaoqing coal fire as an example.

6.2 HAIBAOQING COAL FIRES

The Baoding mining area of Panzhihua City is the largest coal production district in Sichuan Province, China, with an estimated 400 million tons of coal, most of which (300 million tons) is located in the deep parts.

Haibaoqing coal fires are located in the Baoding mining area in Panzhihua City, Sichuan Province (Figure 6.1). The coal fires are 1000 m wide and cover an area

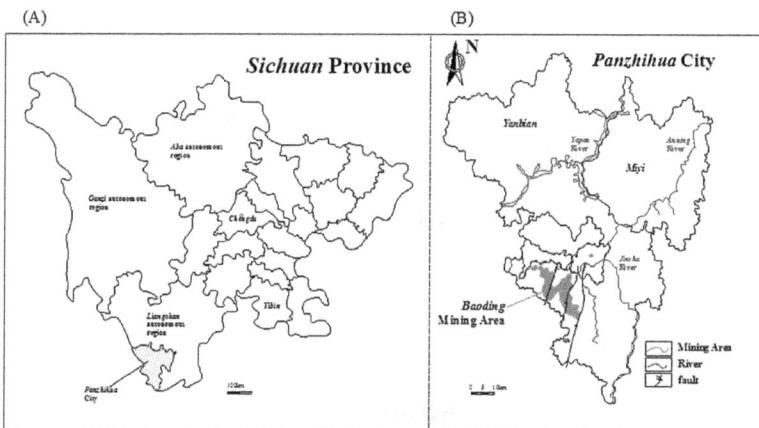

FIGURE 6.1 Location (a) and tectonic map (b) of the Baoding mining area. (b) Provides an enlarged view of the blue area in (a).

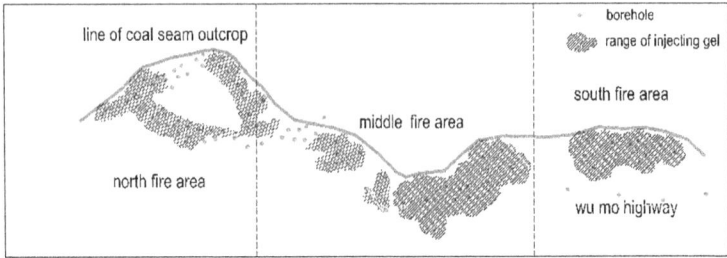

FIGURE 6.2   The coal fire areas of Haibaoqing were investigated. The coal fires stretch across an area 1000 m in width (red line outlines the northern most extent), covering a total area of 60000–70000 m². The borehole and areas of gel injection are shown. The Wumo Highway is in light blue.

measuring 60000–70000 m² (Figure 6.2). Wumo Road runs through this fire area. The fires pose a danger to humans, animals, roads, and buildings within and surrounding the area. The Baoding Haibaoqing No. 4 coal seam combustion area has complex conditions, water shortage, poor altitude, and poor transportation facilities.

Fly ash generated by the Panzhihua Power Station is used in composite gel materials for fire prevention and extinguishing. The advantages of this material are as follows: low cost, high heat absorption to reduce the temperature of the burning coal seam, high sealing capacity to reduce air exchange, and long-term stability in high-temperature environments. The injection system is simple to operate and is suitable for controlling large coal seam outcrop fires. The area has cavities, weak spots, and gas leaks, thus further complicating the geological situation. Based on the environment of the affected area, the integrated control measures mainly include filling and covering the outcrop coal fire with composite fly ash gel, according to the following procedure:

1. Simulate the development of an outcrop fire and accurately detect the extent of the burning area by the radon-test method.
2. Determine the extent, degree, and trend of the burning area, which is divided into three fire areas: north, middle, and south.
3. Position and install holes to inject composite fly ash gel material and monitor carbon monoxide and temperature in the area of ignition.
4. Establish a barrier area in the middle of the burning area by injecting gel to prohibit outward expansion. The fire in the area can be extinguished with continuous gel injection.
5. Deep hole blasting method is used for the high-temperature area, caused by combustion of the coal seam in the mining area. The composite fly ash gel can be effectively covered by settling after sand blasting.
6. Stripping the high-temperature coal mass in the coal seam of the outcrop. To prevent reignition of the outcrop, the excavated area is injected and filled with composite fly ash gel.
7. Seal and fill small hidden pits and cracks with composite fly ash gel separately.

6.3 NUMERICAL MODELING

According to the conservation laws of mass, energy, and diffusion, the two-dimensional mathematical model of spontaneous combustion fire in an outcrop coal seam can be expressed as:

$$
\begin{cases}
\dfrac{\partial}{\partial x}\left(\xi_x \dfrac{\partial H}{\partial x}\right) + \dfrac{\partial}{\partial y}\left(\xi_y \dfrac{\partial H}{\partial y}\right) = 0 \quad \text{(air leakage)} \\[2mm]
\dfrac{\partial C}{\partial \tau} + \bar{Q}_x \cdot \dfrac{dC}{dx} + \bar{Q}_y \cdot \dfrac{dC}{dy} = D_e\left(\dfrac{\partial^2 C}{\partial x^2} + \dfrac{\partial^2 C}{\partial y^2}\right) - V(T) \quad \text{(oxygen concentration)} \\[2mm]
\rho_e \cdot c_e \dfrac{dT}{d\tau} = \lambda_e\left(\dfrac{\partial^2 T}{\partial x^2} + \dfrac{\partial^2 T}{\partial y^2}\right) - \rho_g c_g\left(\bar{Q}_x \cdot \dfrac{dT}{dx} + \bar{Q}_y \cdot \dfrac{dT}{dy}\right) + q(T) \\
\hspace{8cm}\text{(energy conservation)}
\end{cases}
$$

Definite condition of the mathematical model (Table 6.1). Based on the field situation, we obtained the parameters (Table 6.2) of the coal fire area and established a model (Figure 6.3) with a two-dimensional mesh.

TABLE 6.1
Definite Condition of the Mathematical Model

Definite Condition	Flow Distribution of Air Leakage	Oxygen Concentration	Temperature Distribution
Initial Condition		$C\big|_{\tau=0} = C_0$	$T\big|_{\tau=0} = T_0$
first boundary condition	$H\big|_s = H_r$	$C\big|_s = C_t$	
second boundary condition	$\dfrac{dH}{dn}\Big|_s = 0$	$\dfrac{dC}{dn}\Big|_s = 0$	
third boundary condition			$-\lambda_e \dfrac{dT}{dx}\Big|_{\tau=0} = h(T_m - T_g)$

TABLE 6.2
Parameters Used for Modeling Temperature and Seepage Variations

Symbol	Value	Symbol	Value	Symbol	Value
ρ_m	1350	ρ_y	2683	D_0	3.26×10^{-5}
λ_m	1.22	λ_y	2.73	k_1	3.279×10^{-9}
c_m	1610	c_y	1992	k_2	1.386×10^{-10}
ρ_g	1.305	ρ_t	1787	h_m	13
λ_g	0.0213	λ_t	1.28	h_t	10
c_g	995.45	c_t	1196	h_y	11

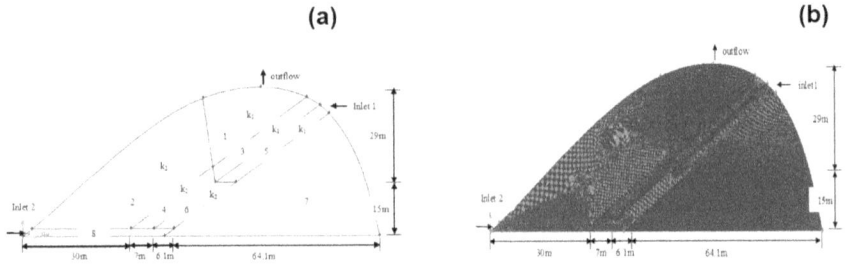

FIGURE 6.3   The simulation model.

In this paper, the temperature distribution at different leak strengths and the seepage field at different temperatures (i.e., the distribution of oxygen concentration) were simulated using Fluent software. The temperature and seepage physical model were built based on a prototype of the No. 4 seam outcrop, Haibaoqing coal fires, Baoding mining area, Panzhihua City, Sichuan Province, China. Using the GAMBIT meshing tool, the physical model had a grid spacing of 0.2 m divided into 1016145 nodes.

There were two air inlets and only one exhaust port in the model, and they were set at the intersection of coal and air. The model size was simulated based on the size of the outcropping fire area, which is a portion of the northern part of the Wumo Highway (Figure 6.2). The outcrop angle of the No. 4 coal seam was 32° and the average depth of the coal seam was 35 m. A typical No. 4 coal seam outcrop fire was representative of fire development throughout the region and could be inferred from simulations. The simulation results (Figures 6.4 and 6.5)

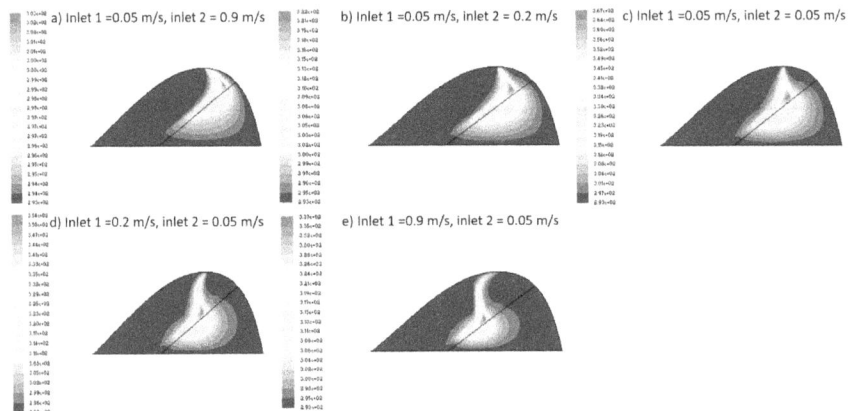

FIGURE 6.4   Temperature field simulation under different leakage intensities. Longitudinal axis represents temperature in Kelvin (K).

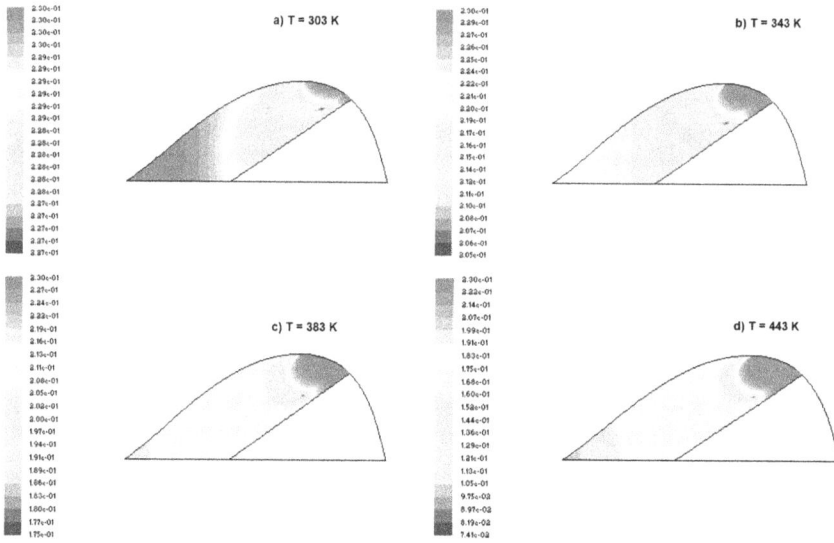

FIGURE 6.5 The seepage flow field (i.e., distribution of the oxygen concentration) simulated under different temperatures. Longitudinal axis represents oxygen concentration in percentage (%).

show that changes in temperature and flow field had important effects on the spontaneous combustion of the coal seam outcrop as well as the amount of gas leakage. As the air leakage increases, oxygen is continuously delivered to the spontaneous combustion area and the coal-oxygen complex reaction intensifies. The temperature continues to rise and expand into deep loose coal, eventually forming an outcrop fire. Therefore, reducing the amount of air leaking into the second boundary is a key method of preventing spontaneous combustion in coal seam outcrop.

6.4 DETECTING AND FIGHTING FIRE

6.4.1 COAL FIRE DETECTION

The burning areas were studied in detail by the radon-test method. As with other nuclear-based techniques, the fundamental principle of the radon-test technique is based on detecting and measuring the energy emitted during the radioactive decay of radon and using the detected and measured information as well as other relevant data to locate extreme subsurface heat sources. A total of 470 detection points were searched for, with a distance of 10–15 m between each detection point.

The detection area was 750 m from north to south and 50 m from east to west, with an altitude range of 1590–1685 m. Analysis based on the location of

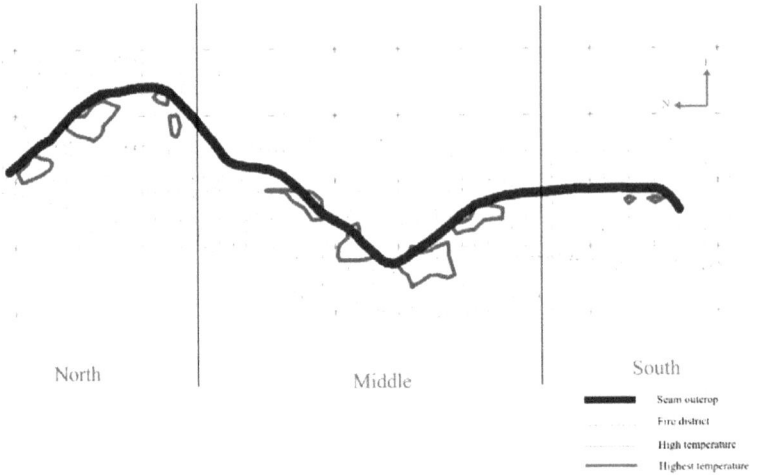

FIGURE 6.6 The three burning areas as detailed by the radon-test method indicate the range of the fire region (dotted red line). The contour of the seam outcrop of the coal is delineated by the blue line. The outline of the high-temperature region is indicated by the bold red line and is the scope of highest temperature of the outcrop coal fire. The light blue outline represents the Wumo Highway.

the burning area and the data obtained from the radon method showed that: (1) the burning area formed a deviated "N" shape; (2) the length was 685 m and the width was 25 m; (3) the whole burning area could be divided into three areas: south, middle, and north, with a total area of 8490 m². The northern burning area was 5110 m², the middle burning area was 2010 m², and the southern burning area was 1370 m² (Figure 6.6).

6.4.2   INJECTING COMPOSITE FLY ASH GEL

6.4.2.1 Drilling

A total of 112 holes were drilled throughout the burning area for gel filling and monitoring. They were 59, 32, and 21 boreholes in the northern, middle, and southern areas, respectively (Figure 6.7). The boreholes were used to monitor the temperature of the coal fire areas and comprised two types: the ck series and the bk series.

    The ck and bk series boreholes were drilled at different times. The bk series boreholes were typically located within the high-temperature area. There were 113 boreholes in the ck series, and 46 boreholes in the bk series. The areas shown by the pink dashed lines correspond to the temperature data of the boreholes, and the areas plotted by brown solid lines are based on the radon data.

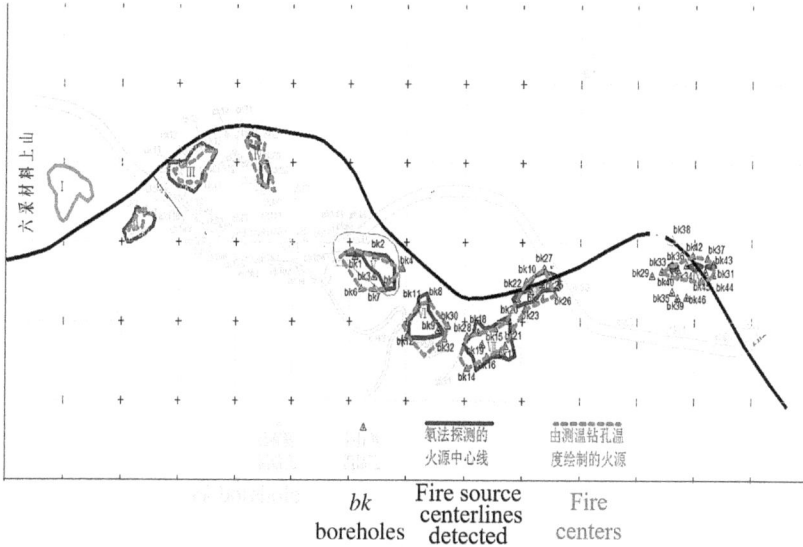

bk boreholes · Fire source centerlines detected · Fire centers

FIGURE 6.7 There were 59, 32, and 21 boreholes for gel filling and monitoring in the northern, middle, and southern areas, respectively. The wide range of boreholes are represented by purple dots and the high-temperature area boreholes as blue triangles. The curves of the fire sources detected by temperature monitoring are shown.

6.4.2.2 Injecting and Covering with Composite Fly Ash Gel

1. The proportion of gel material:

The components and proportions of the composite gel are shown in (Table 6.3). The additive gelatinizes and solidifies the slurry at the designated locations, thereby blocking gaps in the coal and enclosing the high-temperature area. The gelation accumulates to form composite fly ash gel. Due to its water retention and permeability properties, the addition of a suspension agent ensures that the slurry does not segregate during transport, allowing good flow.

TABLE 6.3

Components and Proportions of Composite Fly Ash Gel

Components	Proportion/Percent of Mixture
Water: Fly ash	2:1
Composite gel additive	1‰
Suspending agent	0.5‰

(a)Silicone   (b) Fly ash composite gel   (c) Loess composite colloid   (d) Polymer colloid

(e) Gel injection operation system

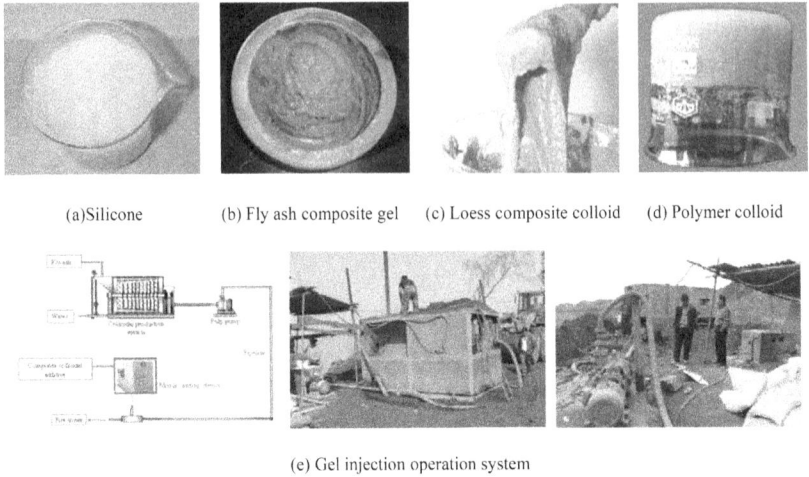

FIGURE 6.8 There are several different materials that can be used to extinguish an outcrop fire including: (a) silicone, (b) fly ash composite gel, (c) loess composite colloid, (d) polymer colloid, and (e) gel injection operation system.

2. Gel injection operation system:

A fly ash injection fire suppression system was established in the southern and northern burning areas. The system operation and equipment are described in (Figure 6.8). The fly ash is transported through the loading equipment to the gel production machine, where a certain concentration of fly ash slurry of specified consistency is produced.

The slurry is transported by pump to the injection location where additives are added to the mixture by a gel injector to form a composite fly ash gel. The composite gel is pressed through the opening into the burning areas and the system works well at a pressure of 1.8 Mpa.

3. Amount of gel injection:

The total amount of gel used for the entire burning area was about 110000 m³. This was composed of 78923 m³ of water, 34575 m³ of fly ash, and 90 tons of additive. The number of boreholes for gel injection was 73, with 28, 30, and 15 boreholes in the northern, middle, and southern burning areas, respectively.

6.4.2.3 Blasting Combustion Space Area and Injecting Composite Fly Ash Gel

Because the burning areas were geologically complicated and highly inclined, the coal seam had been burning for a long period of time. During the burning of a pile or seam, gases are emitted from the borehole and the slurry flows into the pile or space during the injection. Due to the steep inclination, the gel could not stay in the high-temperature area, and thus could not put out the fire. Therefore,

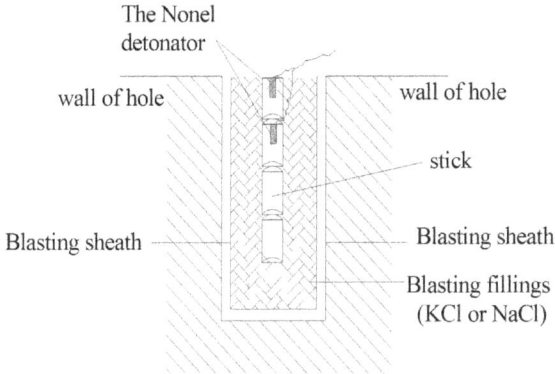

FIGURE 6.9   The structure of blasting borehole.

the areas were sandblasted, then injected and covered with gel. The total blasting area was 454 m², and 41 blasting boreholes were used at depths of 5–7 m. There was obvious subsidence of the subsurface visible at the surface after the blast, and the roofs of the coal mines were near the point of collapse, while heat and smoke burst through leakage points at the surface. A schematic drawing of the blast set-up is shown in (Figure 6.9). (Figure 6.10) depicts the subsidence area, formed after blasting, that was injected with composite fly ash gel.

Excavating the high-temperature coal and injecting composite fly ash gel.

The feasibility of complete excavation depends on the depth of the coal seam and the amount of overburden to be removed. This is a passive control method because the fire will continue to burn until it reaches the obstacle. The feasibility of trench barriers usually depends on the required excavation depth. Four high-temperature areas were identified in the middle burning area based on measured surface data. The overburden was stripped, and the burnt-out rocks were excavated and brought to the surface (Figure 6.11).

The excavated rock was cooled with water followed by injection into the exposed fractures and chambers and was then covered with gel. Sealing of small

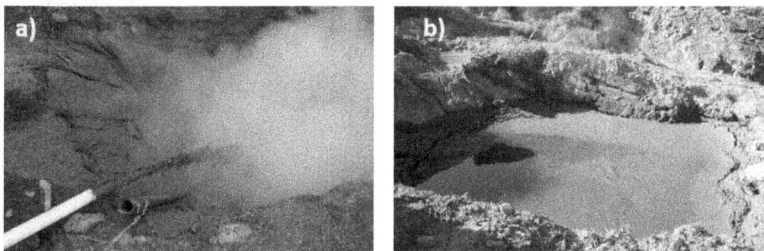

FIGURE 6.10   (a) The outcrop fire situation before applying control measures, (b) The space situation after injecting the gel to extinguish the outcrop fire.

FIGURE 6.11 Typical badlands topography associated with the coal outcrop fires. Excavated burnt-out rocks are brought to the surface. The plume of steam and smoke from this site was visible for miles.

private pits with composite fly ash gel. The widely used traditional fire-controlled surface soil-sealing method was adopted to create an oxygen-deficient environment in the fire area. This technique is most often applied to fires that have developed superficial cracks over a wide area and consists of placing a thick "blanket" of compacted soil material over the affected area, following a complete removal of all overlying vegetation. Soil sealing requires monitoring and regular maintenance as cracks or erosion may occur over time. Soil sealing is sometimes used in combination with other methods. This method is used to check and seal air leaks caused by small private pits near the burning area. These seals are checked regularly to ensure they maintain their capacity. Smaller illegal coal mines were sealed up, cracks were leveled, and sink damage was filled with composite fly ash gel to reduce air leakage (Figure 6.12).

6.4.3 CHECKING AND MONITORING

Monitoring of holes in the burning area was conducted along the exposed coal seam and in the center of the fire (Figure 6.13). A total of 23 monitoring boreholes

FIGURE 6.12 Condition of the hole (a) before, and (b) after being sealed. Emission of thick smoke appeared near the outcrop fire space. After sealing, characteristics of surface combustion disappeared.

FIGURE 6.13 Using the equipment to check and monitor the temperature and gases emitted from the coal fires.

were created, each with a diameter of 108 mm. There are two main monitoring methods to ensure that a fire is extinguished: (1) temperature test; (2) gas analysis. The main monitoring equipment includes: (a) a temperature test thermocouple; (b) gas sampling equipment and container; (c) portable gas testing equipment; (d) gas chromatography in the laboratory.

6.5 EFFECTIVENESS OF CONTROL METHODS IN THE BURNING AREA

6.5.1 ANALYSIS OF COMBUSTION TEMPERATURE OF BURNING AREAS

In the early stages of fighting a coal fire, the surface temperature of the burning area is very high. Temperatures exceeded 500°C in some areas. Depressions and cracks on the surface emitted dense smoke with CO and SO_2 concentrations > 10000 ppm and > 100 ppm, respectively. The rocks in the burned area were badly burned and the high heat capacity of the surrounding rocks caused high temperatures even after the fire had stopped.

After a fire has been extinguished, the surface burning characteristics usually disappear and the surface temperature gradually approaches the ambient temperature. Our analysis found that CO and SO_2 on the surface approached normal atmospheric levels, and there was no smell produced by seam combustion throughout the entire burning area. (Figure 6.14) depicts the fire control effects of injecting composite fly ash gel in the northern area.

6.5.2 ANALYSIS OF BOREHOLE TEMPERATURES IN THE BURNING AREA

Prior to fire suppression, 80–112 boreholes had temperatures exceeding 100°C, with the highest borehole temperature exceeding 600°C. After extinguishing the fires, the temperatures of the 112 boreholes returned to approximately normal temperatures (25°C, surface temperature). (Figure 6.15) shows the temperature changes after injecting gel.

FIGURE 6.14 Surface temperatures of burning areas are extremely high in the early phase of the fire extinguishing process. (a) Thick smoke is emitted from depressions and cracks in the surface. (b) After extinguishing the fires, characteristics of surface combustion generally disappear and surface temperatures approach the normal range.

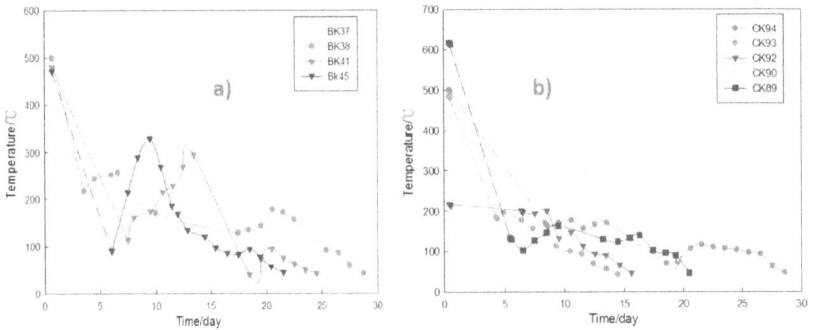

FIGURE 6.15 Temperature changes of the monitoring boreholes' highest temperatures in the (a) south and, (b) the north before and after injection of the gel. The dashed lines indicate the temperature before injection, and solid lines represent the temperature after the fire was extinguished.

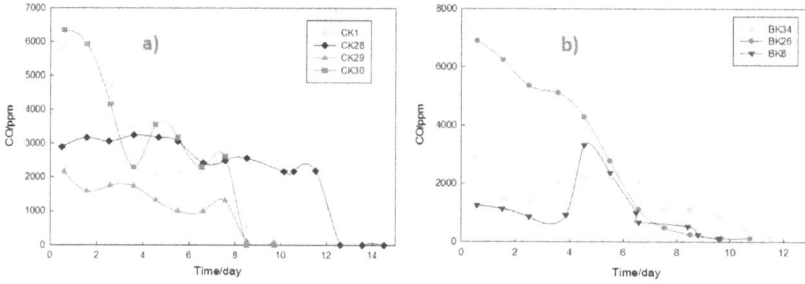

FIGURE 6.16 Changes in concentration of detected CO before and after injection of gel to control the fire in the (a) northern and, (b) southern fire district.

6.5.3 ANALYSIS OF TOXIC GAS OF CO

Before glue injection, a large amount of smoke was being emitted from the borehole as well as surface depressions and cracks, and the concentration of CO in the smoke was > 10000 ppm.

Gas chromatography was used to measure the content of CO in the borehole samples. The gases were taken from the borehole using an air pocket. After injection of the gel to control the coal seam fire, CO could still be detected in a few boreholes, but the concentrations were very low. The surface CO was close to negligible levels and there was no odor of burning seam throughout the burning area. The concentrations of toxic gases were greatly reduced and environmental pollution decreased markedly. (Figure 6.16) shows the concentration changes of CO in the fire areas.

6.6 SUMMARY

After eight months, the outcrop fire areas of the No. 4 seam of the Haibaoqing, Baoding mining area were brought under control and the combustion characteristics on the surface and toxic gases had disappeared. Monitoring after one year confirmed that the burning areas that initially exhibited high temperatures had been extinguished (i.e., temperatures had returned to normal). This signifies that the environment has been improved, coal resources are protected, and the local topography has been stabilized. As a consequence, coal mining can be conducted safely, which in turn provides various social and economic benefits. The results that were achieved using the engineering techniques described above demonstrate that this novel method was an effective approach for fire control in a coal mining area, with broad potential applicability. The method and results can be summarized as follows:

1. Effective detection of the extent of coal seam outcrop fires was accomplished by the radon-test method, and the area was divided into three critical areas based on the severity of burning: north, middle, and south. Three general strategies were initialized based on the conditions in the area: control after detection, control by area, and continuous monitoring.

2. Fly ash produced by the Panzhihua City power station was used to develop a composite fly ash gel material. The advantages of this material in fire prevention and extinguishing are as follows: low cost, high heat absorption to decrease temperature of the burning coal seam, sealing capacity to reduce air exchange, and stability in high-temperature environments. The operation design of the mobile gel-injection system on the ground is easy to use. It is suitable for controlling large areas of coal seam outcrop fires.

3. This study is the first to use the deep hole blasting technique for high-temperature areas of large piles of coal caused by coal seam burning. After sandblasting, a composite gel was injected into the depression to create an effective composite gel cover.

4. A comprehensive technical system of coal seam outcrop fire extinguishing, with gluing and covering as the main methods of control, was implemented. Shallow burning areas were excavated by strip mining. Closure of small mines and high-temperature roof blasting of old piles were employed as supplementary methods. The combination of these methods improved the effectiveness of fire suppression, while reducing costs and preventing reignition of the burned areas.

6.7 UNDERGROUND FIRE

6.7.1 DONGTAN

Most of the mines in the Yanzhou Mining District, China, are deep, especially the Zhao Lou, Dongtan, Jining No. 2, and No. 3 mines. The Baodian mines are more than 500 m deep. The Dongtan mine has a first level depth of 700 m and a second level depth of 800 m. The No. 3 coal seam in the lower part of the Shanxi Group of the Permian system contains moderately degraded gas coal with a tendency to spontaneous combustion. It is a self-combusting level II coal seam, with an ignition period of 3–6 months. The ground temperature at the −660 m level ranges from 28.6°C to 32.3°C, with an average of 29.6°C. The ground temperature is prominent. The mine is located in the central part of the Yanzhou mining area. The mine uses a pair of vertical shafts with two levels, the first level is at −660 m and the second level is at −746 m, and the main mine has three coal seams with an average thickness of 8.76 m and a dip angle of 0–12°, with an average of 6°. The average thickness of the coal seam is 8.76 m. The average dip angle of the seam is 0–12°, with an average of 6°, and the structure is simple. A map showing the location of the Dongtan mine is presented in (Figure 6.17).

The main no. 3 coal seam of the Dongtan coal mine has a short natural ignition period and the coal seam is prone to spontaneous combustion. Since the adoption of integrated release without coal column mining, the natural ignition of the coal seam at the integrated release face has become very serious. From 1996 to 2010, there were 36 ignition accidents, which included 12 occurrences (33%) in the roof of a roadway, and 24 occurrences (67%) in the adjacent open area.

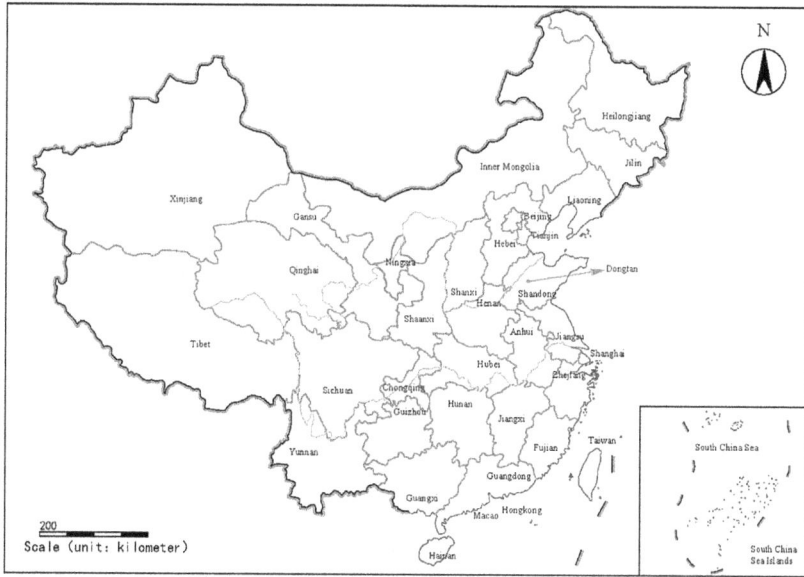

FIGURE 6.17 Location map of Dongtan mine site.

6.7.2 1304 COMPREHENSIVE RELEASE SURFACE NATURAL IGNITION HAZARD ANALYSIS

6.7.2.1 Geological Situation

The 1304 mine is located in the eastern section of a mining area, the north side of which is the 1305 comprehensive release face (not mined), and the south side comprises the 1303 comprehensive release face mining void area, the 1304 intake lane, and the 1303 backwind lane between a 4-m wide small coal column. The workings start from the open cut (10 m from the FS18 fault on the transport side and 6m from the F19 fault on the track side) in the east to the design stop line (10 m from the F2 fault on the transport side and 30 m from the track side to the west). Due to the influence of the Dazhongtong village to protect the coal column, the working face is arranged in the "knife handle" style with large and small cuttings. The working face has a strike length of 2456.43 m, of which the combined face part consists of a strike length of 1906.43 m and an inclination width of 241.50 m. The "knife handle" part has a strike length of 550.00 m. The "knife handle" part is 550.00 m long and 151.40 m wide.

A total of 39 faults are associated with the working face, and 17 faults were revealed during the process of transporting the smooth excavation: F2, EF26, EF26-1, EF61, EF62, EF63, EF18, EF79, EF32, EF81, EF82, DF15, EF83, FS8-1, FS8, EF86, FS18; a total of 17 faults were revealed during the process of rail smooth excavation. F2, EF26, EF52, EF18, EF31, EF56, EF32, EF57, EF81, EF58, EF59, EF71, EF87, EF83, EF85, No. 1 shaft east, F19; one fault was revealed

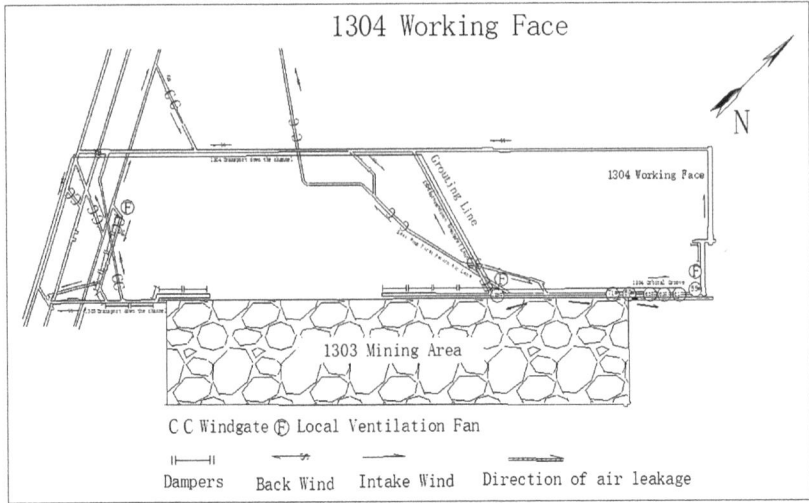

FIGURE 6.18 Schematic diagram of the 1304 working face.

during the hole boring: EF84; four 3D physical faults were detected in the working face: DF1, DF11, DF13, F14. All of the above faults were positive faults.

The coal seam recoverability index of the working face was 1, and the coefficient of variation was 12.36%. The coal seam dip angle was 0–12°, with an average of 6°. According to the result of the gas grade appraisal in 2009, the mine was a low gas mine with a relative gas outflow of 0.05 m³/t. The coal dust has a risk of explosion with an explosion index of 37.42%. The propensity of coal seam to spontaneous combust was grade II, and it was thus deemed to be a spontaneous combustion coal seam with a firing period of 3–6 months. At −520 to −660 m level, low temperature was in the range of 28.6–32.3°C, with an average of 29.6°C, and the ground temperature was more prominent. A schematic diagram of the 1304 working face is shown in (Figure 6.18).

6.7.2.2 1304 Comprehensive Release Face Natural Fire Hazard

Due to the special geological conditions of the 1304 comprehensive release face, the complex mining environmental conditions, the presence of numerous faults, part of the fault drop, loose coal body, coupled with factors along the empty stay lane, risk of impact ground pressure, open leaky wind channel, there is a greatly increased potential risk of natural fire. The main hidden danger arises from this face mining area, the 1303 old eyelet, the 1303 backwind lane old empty area, and over the fault broken belt, which are most affected by the 1303 old eyelet and the 1303 backwind lane old empty area.

The 1304 track is arranged along the 1303 transport, which is along the empty roadway, and readily forms a wind leakage channel, so the prevention of wind leakage is the key to avoiding spontaneous combustion and fire.

6.7.2.3 Factors Affecting 1304 Comprehensive Release Face Natural Fire

6.7.2.3.1 Coal Spontaneous Combustion Period

The natural ignition period of coal is not only related to the oxidative exothermic properties of coal itself (internal factors) but is also related to the oxygen supply conditions and heat storage conditions of the loose coal body (external factors).

1. Experimental shortest natural ignition period:

   Under the best oxygen supply and heat storage conditions provided by the laboratory, the time required for the loose coal body to ignite by spontaneous heating (subject to green smoke) was investigated. The shortest fire period for the experimental 3-layer coal in the Dongtan coal mine was 32 days.

2. The shortest natural ignition period of coal seam:

   Under the actual conditions of the site, the time from exposure of the coal body to air to green smoke was deemed to be the shortest natural ignition period of the coal seam when there was a combination of maximum residual coal thickness, minimum residual coal particle size distribution, highest original surrounding rock temperature, and best air leakage strength. The original temperature of the Dongtan coal seam reached 28–33°C, and the starting temperature of residual coal in the adjacent mining area along the air side reached 35–60°C. The natural firing period was short at the location where the size of residual coal in the fault structure areas and stress concentration areas was less than 3 mm, and the thickness of top coal could reach more than 3.0 m. The shortest firing period of the Dongtan coal mine was 18–26 days.

3. Actual firing period of coal seam:

   The actual firing period was defined as the natural firing period corresponding to a specific location and specific conditions underground. The actual firing period of 3 seam coal in the Dongtan coal mine was 45–75 days.

The natural fire characteristics of the mining area of the heaving surface are as follows:

1. Dynamic movement of air occurs in three areas within the extraction area: dispersion area, oxidation spontaneous combustion area, and asphyxiation area. The range of the three areas is affected by the comprehensive dynamics of the recovery rate, air supply, and advancement of the comprehensive release workface. The range of the three areas in the mining area moves dynamically with the continuous advancement of the workface. The three areas of the Dongtan coal mine can be divided as follows.

   Choking area ($CO < 50$ ppm$\cap O_2 < 8\%$): low oxygen concentration, resulting in insufficient oxygen supply for coal oxidation and suppressing the development of spontaneous combustion; $CO < 50$ ppm reflects the fact that the broken coal body in the mining area is not affected by the ambient temperature, thickness of broken coal, air leakage, or other conditions, and the oxidation and warming process of coal can continue to develop into spontaneous combustion under these low oxygen concentration conditions.

Dispersion area (CO < 10 ppm∩O$_2$ > 16–18%): the intensity of air leakage is high, the oxygen concentration is high, and the heat of oxidation of crushed coal does not readily accumulate, so it will not spontaneously combust; CO < 10 ppm indicates fast heat dissipation caused by large air leakage, thus preventing the development of coal oxidation in a higher oxygen concentration environment.

Area prone to spontaneous combustion (10 ppm < CO∩8% < O$_2$ < 16–18%): this range of air leakage is moderate and oxygen concentration is more adequate. 10 ppm < CO indicates that the oxidation and warming of broken coal body occurs readily, and thus is prone to development of spontaneous combustion.

2. The high-temperature range of spontaneous combustion in the mining area is large. There is a large amount of residual coal left in the mining area of the heaving face, and the heat of coal oxidation gradually accumulates. Once spontaneous combustion occurs, a large amount of heat energy is stored in the mining area, which causes the temperature of the surrounding coal rock body to rise, and therefore, the high-temperature range of the mining area of the heaving face is large.

The thickness of residual coal in the mining area was analyzed and the results showed that when the thickness of residual coal in the mining area was 0.6 m, coal temperature was 65°C, and the lower limit oxygen concentration value exceeded 21%, then under the actual conditions, the residual coal would certainly not spontaneously combust. That is to say, if the thickness of residual coal in the Dongtan coal mine 3 layer coal was less than 0.6 m, the coal temperature would not exceed its critical temperature and spontaneous combustion could occur. As the residual coal thickness increases, the upper limit of wind leakage intensity rises. This means that under the actual conditions, the greater the thickness of residual coal, the more heat is generated, the less heat is dissipated through the top and bottom plates, and the greater the upper limit of wind leakage intensity. When the thickness of residual coal is less than 0.6 m, the upper limit of wind leakage intensity will be negative, and at this time, all of the heat generated by coal oxidation is dissipated through heat conduction, and the residual coal will not warm up naturally, which also indicates that the thickness of residual coal is less than 0.6 m. It also demonstrates that when the thickness of residual coal is less than 0.6 m, the coal temperature will not exceed its critical temperature and spontaneous combustion could occur.

3. It is difficult to extinguish a fire caused by spontaneous combustion in a mining area. It is difficult to achieve effective management of fire in mining areas using currently available technologies for fire prevention

and extinguishing fires due to open air leakage in a mining area of the heaving surface, the large scope of the fire area, the hidden source of the fire, and the impact of the working space at the surface.

The greater the intensity of the air leakage, the more heat was carried away by the wind flow, the more residual coal was required (i.e. the greater the residual coal thickness), and the greater the limit of residual coal thickness. The limit of residual coal thickness is extremely large at about 65°C, and as long as the residual coal thickness is less than the limit of residual coal thickness at this temperature, the residual coal will not spontaneously combust. The air leakage strength of the mining area and coal column is 0.1–0.24 m^3/(min·m^2), and can readily ignite naturally. It is generally believed that the air leakage velocity in the choking area is V<0.02 m/min, while the measured air leakage strength of the stopping line in the Dongtan mining area is about 0.007 m^3/min·m^2, and the loose coal belt may still be at risk of spontaneous combustion.

6.8   FIRE PREVENTION TECHNIQUES

6.8.1   GROUTING

The centralized slurry-making station in the north wind shaft or the slurry-making station in the mine was used to provide hydrostatic slurry injection through pipelines as part of fire prevention measures involving both rubber injection and slurry injection at the working face of 1304 where there was a potential risk of spontaneous combustion, and the pre-buried pipeline slurry injection was used for the stopping line.

1. Rail smooth grouting pipe:

   North wind shaft grouting station (Φ159×6) → North wind shaft (Φ159×6) → East wing south main return airway (Φ159×6) → One mining area return air uphill (Φ108×5) → One mining area return air uphill 5# link (Φ108×5) → 1304 trackway (Φ108×5).

   In-mine grouting station (Φ159×6) → bottom of shaft grouting platform (Φ159×6) → west wing belt stone gate (Φ159×6) → A and B belt liaison lane (Φ159×6) → east wing belt stone gate (Φ159×6) → first quarry return wind uphill (Φ159×6) → first quarry return wind uphill 5# link (Φ108×5) → 1304 intake lane (Φ108×5).

2. The grouting route of backwind lane:

   North wind shaft grouting station (Φ159×6) → North wind shaft (Φ159×6) → East wing south main return airway (Φ159×6) → First mining area return air uphill (Φ108×5) → First mining area return air uphill 6# link (Φ108×5) → 1304 backwind lane (Φ108×5). In-mine grouting station (Φ159×6) → bottom grouting platform (Φ159×6) → west wing belt stone gate (Φ159×6) → A and B belt contact road (Φ159×6) → east wing belt

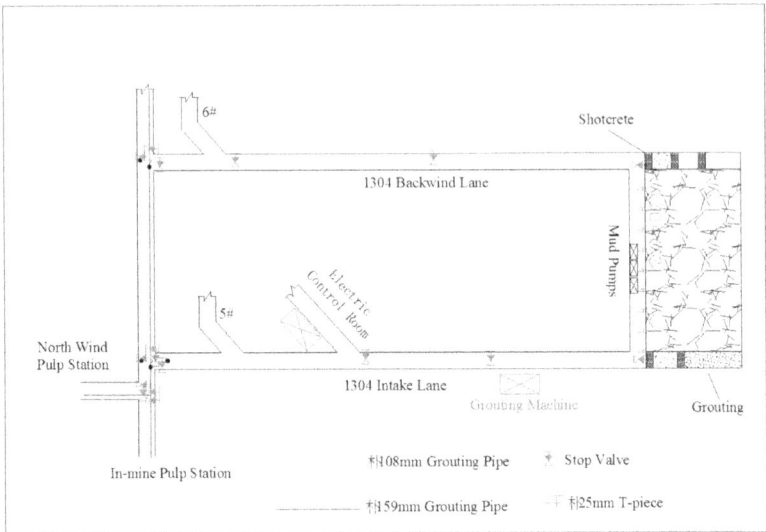

FIGURE 6.19   Diagram of long-distance delivery of coagulant with mud pump injection.

stone gate ($\Phi159\times6$) $\rightarrow$ first mining area return wind uphill ($\Phi159\times6$) $\rightarrow$ first mining area return wind uphill 6# link ($\Phi108\times5$) $\rightarrow$ 1304 backwind lane ($\Phi108\times5$). The system of long-distance delivery of coagulant with mud pump injection is shown in (Figure 6.19).

6.8.2   Uniform Pressure Fire Protection

Under the condition that the air volume of the working face is sufficient, the air volume should be reduced, and then the pressure difference between the working face and the lower end can be reduced, along with reductions in the width of the oxidation area and the air leakage from the mining area. The wind volume can be reduced in time once mining at the working face has ceased. The stopping line should be closed in time after the workface has stopped, followed by implementation of regional pressure equalization in the air mining area and reduction of the pressure difference between the two ends of the stopping line.

6.8.3   1304 Fire Prevention Technology During Track Following Hollow Excavation

During 1303 comprehensive workings back mining, slurry filling was carried out in the air mining area on the side of the transport, a broken coal bag wall was built every 50 m, and the air mining area between the two walls was filled with slurry to create a good fireproof condition for the 1304 intake lane along the air boreholes. The 1304 intake lane along the air boring period, the fireproof hole of the air mining area was constructed along the air tunnel, and the fireproof hole was used to fill the air mining area with pressure injection thickening colloid in order to reduce the air leakage in the mining area and prevent natural fire accidents. The1303 working face fire prevention measures during retrieval were shown below.

1. A wall was constructed at the end of the backwind lane to block wind leakage. Because the 1304 intake lane along 1303 backwind lane digging, so, in 1303 comprehensive work surface back mining period, in 1303 backwind lane end of the wall of broken coal bags to reduce the air leakage in the mining area, every 50 m to build.
2. The 1303 backwind lane buried pipe grouting. The 1303 working face back mining, in the backwind lane side pre-buried pipe for grouting, to fill the mining area to plug the leakage. A diagram of the 1304 intake lane fireproof drilling arrangement plan is shown in (Figure 6.20).
   1. Improvement of the grouting system:

      Improvement of the grouting pipe (Φ108 mm) from the return wind uphill of the first mining area of the east wing to the 1304 intake lane, during the digging period of the 1304 intake lane was performed. General excavation work was conducted to install the grouting pipeline. The installation of pipeline lagged behind the digging work face by 100 m. A Φ50 mm three-way valve was issued every 20 m, and the pipeline was hung along the belt to the gang; the hanging height was 0.3 m on the bottom plate.
   2. Construction of fireproof drilling
      a. 1303 open eyelet:

         In the 1304 intake lane corresponding to the 1303 open eyelet position to the east 10 m construction a set of drilling, perforation point are located in 1303 backwind lane south 10, 20, 30, 40, 50 m 3 coal roof plate under 1.0 m. The whole length of the borehole was sealed with Φ40 mm steel pipe, and the innermost end was sealed with a 1.0 m long flower pipe.

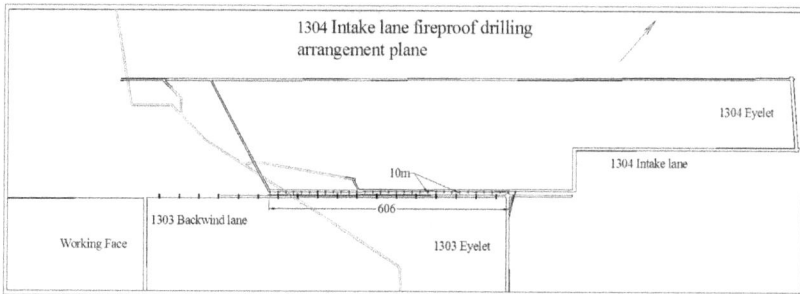

FIGURE 6.20   Diagram of the 1304 intake lane fireproof drilling arrangement plan.

b. 1303 backwind lane along the empty side:

From the intersection of track 1304 and the coal transport road, 600 m to the east is the roadway along the hollow side, and therefore, according to requirement for fire prevention work, fire prevention boreholes needed to be constructed. There were 60 holes in total, with one hole every 10 m, and each hole was 1.0 m above the top plate of the 1303 backwind lane, located in the middle of the top plate of the 1303 backwind lane. The drill hole was sealed with Φ40 mm steel pipe along the whole length, and the innermost end was sealed with a 1.0 m long flower pipe.

3. Gluing fireproof
    a. 1303 open eye injection rubber fire prevention:

    1304 track construction was used through 1303 open eyelet of 5 holes, to the open eyelet pressure injection gel for filling and plugging. Each hole injection volume was 50 m³, with a total pressure injection of 250 m³.
    b. 1303 backwind lane along the empty side of the injection of rubber fire prevention:

    Using the 60 holes constructed along the hollow side of the 1304 track, press injection of colloid into the hollow area of the 1303 backwind lane was performed, with filling and plugging, and the volume of each hole was 40 m³, providing a total press injection of 2400 m³. A large injection system for pressure injection of colloid was employed. A diagram of the 1304 intake lane through the 1303 backwind lane along the empty stay lane drilling arrangement is shown in (Figure 6.21).

4. Fire monitoring and detection
    a. After the construction of the borehole, the borehole was checked for temperature in time, and the temperature probe was installed in the borehole. The temperature was measured twice a week, and if the temperature exceeded 35°C, gas samples were taken for analysis and temperature curves were calculated.

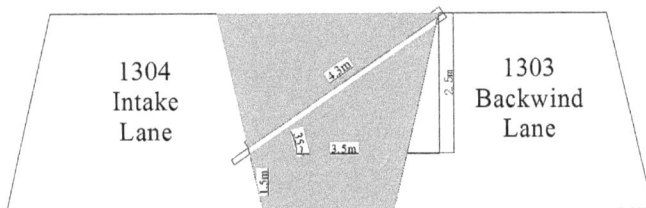

FIGURE 6.21 Diagram of the 1304 intake lane through the 1303 backwind lane along the empty stay lane drilling arrangement.

b. Monitoring beam pipes were installed in each drill hole in order to monitor the gas condition in the mining area. For drill holes with a temperature over 35°C, samples were taken and analyzed twice a week, and gas curves such as CO, CO2, C2H6, and C2H4 were established for analysis and assessment. Before the excavation construction, the bundle pipe was installed at the opening point, and then extended forward with the excavation.

c. During the excavation of the 1304 track, a CO sensor was installed at the headway and 10 m east of the intersection of 1304 track and coal transportation lane, so as to monitor the CO gas situation at the headway, return air flow, and provide timely data on abnormalities.

d. A full-time tile inspector was employed. An infrared thermometer was used to measure the temperature of the gang part of the roadway along the air side of the 1304 intake lane, twice in each shift. If the temperature of the gang part along the air side exceeds 30°C, it must be reported immediately, and the temperature of the drill hole near the abnormal point needs to be measured in order to analyze and assess the possible risk of spontaneous combustion.

6.8.3.1   Fireproof Treatment of the 1303 Mining Area

1. Fire prevention treatment of the 1303 old cuthole: before production at the working face, fire prevention drilling was constructed at the 1304 intake lane corresponding to the 1303 old cuthole, and the old cuthole was grouted and filled. The total grouting consisted of approximately 2000 m³.

2. The 1304 intake lane along the empty side of the fire prevention management: in order to effectively perform fire prevention in the 1303 backwind lane mining air area, the 1304 intake lane corresponded to the 1303 old cuthole west of every 10 m high and low layout of a drill hole, the full length of the hole sealing Φ 40 mm steel pipe, along the empty side of the grouting (glue) treatment.

3. Fire prevention drilling construction was completed, and the fire prevention drilling temperature was measured twice a week to check. When the temperature exceeds 30°C, a drilling temperature probe and beam tube should be installed and checked every two days. When the temperature is greater than 35°C, a timely injection of colloid is required.

6.8.3.2   Fire Monitoring

1. During drilling construction, monitoring of on-site construction should be conducted by personnel using an infrared thermometer to measure the temperature of the holes drilled.

2. It is important to focus on monitoring the 1304 track along the empty side, the end of the track, and the return corner of the working face, especially when the working face is pushed to the old lane position. In the process of pushing through the old lane, monitoring is vital. Changes in carbon monoxide or abnormalities must be reported immediately.

3. The use of 1304 intake lane along the empty side of the CO sensor with real-time monitoring of CO gas in the incoming air flow should be implemented. If the sensor detects the presence of carbon monoxide gas, it must be reported immediately and the emergency plan should be put into effect.

4. In the small cutting eye intake lane end at the pre-buried beam pipe and temperature sensor to monitor the gas and temperature situation in the mining area. A bundle pipe was placed at the return air flow on the low-pressure side of the west end of 1304 intake lane.

5. After construction of the drill hole, a good monitoring beam pipe was placed along with a temperature sensor, and the temperature inside the drill hole was tested once every 2 days. A gas analysis should be conducted for drill holes with a temperature exceeding 30°C, and change curves should be calculated.

6. In the borehole near the 1304 working face, a beam pipe and temperature probe were installed with a system 20–50 m ahead to monitor the gas and temperature changes in the old lane, with real-time monitoring.

6.8.4 FIRE PREVENTION TECHNOLOGY DURING THE WITHDRAWAL OF THE WORKING FACE

6.8.4.1 Fire Prevention Technical Program during the Mining Shutdown

1. Perfect pipeline system
   a. One shift before the workface is closed, establish an air supply and water supply system, set a six points three-way valve every 20 m, and connect it with the wind supply and water supply pipe of the track and transport, forming a double-way air supply and water supply system.
   b. During the closure period of the working face, a grouting system should be set up, with a Φ50 mm three-way valve placed every 20 m and connected with the grouting pipes of the track and transport, forming a double-way grouting system.

2. Construction drilling
   a. Bracket adjustment: One shift before cessation of mining on the working face, adjust the bracket spacing, with each bracket spaced no less than 100–150 mm apart, so as to construct inter-frame drilling.
   b. Drilling construction: Each stent was constructed with two injection holes, divided into high and low holes. One high hole was opened at the stent column, 5.7 m above the cut top line of stent 1304, with an opening angle of 71° and a hole depth of 6.0 m, with a full-length sealed 1 inch steel pipe and a 1.0 m long flower pipe at the

innermost end. Another hole was opened at the intersection of the top beam of the bracket and the tail beam, 4.0 m after the penetration 1304 frame and 7.4 m above the bottom plate in the broken coal body. The opening angle was 42°, the hole depth was 5.4 m. The full length was sealed with a 1 inch steel pipe, and the innermost end was a 1.0 m long flower pipe.

3. Glue injection fire prevention:

According to the working face of 1304 and the relationship of the surrounding roadway, the following fireproofing order was determined: fireproofing of the working face first, followed by fireproofing of the air side along the track, and finally fireproofing of the 1303 stopping line.

a. Working face fire prevention

Gluing volume: According to the volume of the broken coal body at the back of the bracket, the gluing volume of each bracket was 20 m^3, i.e., the gluing volume of each hole was 10 m3, and the total gluing volume of the working face was 2780 m^3.

Gluing process:

Due to the large amount of glue injection on the working surface, in order to facilitate rapid glue injection fire prevention, with universal injection of inter-frame drilling, the grouting machine was used to prepare the baking soda solution and increase the flow of the glue injection. That is, one grouting machine was stored at each end of intake lane and backwind lane to inject glue at the working face, and each grouting machine was responsible for 120 m of glue injection work. A diverter with one in (Φ50 mm) and five out (Φ25 mm) was connected at the Φ50 mm valve of the grouting pipe at the working face, and holes were drilled between the diverter and the stand. Ten holes were connected each time: 8 holes were injected, with 2 holes as spares.

After using the grouting machine to make a general injection of the holes between the racks, two sets of injection equipment were set aside at the working surface to make up the glue according to the drilling temperature, i.e., to create the glue for the holes whose temperature exceeded 35°C.

The order of gum injection: because there was less equipment on the side of the intake lane, and the intake lane was along the empty roadway, we first used the grouting machine at the end of the intake lane for gum injection treatment, and then used the grouting machine at the end of the backwind lane for gum injection treatment.

b. 1303 stopping line gluing fire prevention:

The slurry pipe was reserved for the 1303 stopping line and 3 high drilling holes outside the stopping line were used. Press injection of compound colloid to the 1303 stopping line was for fire prevention, press injecting 1000 m^3.

  c. Grouting and fireproofing of 1304 cut-off line:

After the closure of the 1304 stopping line, the slurry filling was done using the reserved slurry pipe to the stopping line, and 3000 m³ was injected.

6.8.4.2 Closed Fire Prevention Technology of the Stopping Line

1. Rail smooth side closure:

A 500 mm thick brick wall was constructed to close the 1304 stopping line at the end of the 1304 track, and a metal net was hung on the outside of the wall onto which a 0.5 m layer of thick cement mortar was sprayed. A slurry pipe on the wall to the 5# joint loading yard in the first mining area was reserved for filling the stopping line. A 500 mm thick brick wall 3 m inside the opening point was constructed, and a metal net was hung on the outside of the wall onto which a 0.5 m thick layer of cement mortar was sprayed, and a slurry pipe on the wall was reserved for filling the stopping line. The slurry pipe was reserved for filling the roadway between the two sealing walls.

2. Backwind lane side closure
  a. Construction of the cement barrel column was continued. In the 14 m between the existing barrel column and the coal wall of the stopping line, construction of two rows of barrel columns with a spacing of 500 mm (edge to edge) was continued.
  b. Each row of barrel pillars were hung with metal mesh on the inner side for slurry spraying. To ensure that the metal mesh was hung firmly, small wooden strips were installed on the barrel pillars to fix the metal mesh.
  c. The metal mesh of the outer (north) row of barrel pillars must be connected to the metal mesh of the coal wall at the end of the transportation, in order to ensure that the stopping line is completely blocked.
  d. A detection tube should be reserved. The monitoring tube was extended diagonally upward to 0.5 m below the top plate of the 1304 stopping line, and fix the outer end of the monitoring tube with iron wire. A monitoring bundle pipe measuring 10–15 m should be reserved in the stop line and integrated into the monitoring system.

3. Fire prevention plan
  1. Fire prevention drilling:

When the barrel pillar is drilled along the air side of 1304 transport line, a Φ40 mm × 2000 m steel pipe was inserted into the outer row of barrel pillars every 2 barrel pillars, and the end of the steel pipe was positioned at the top plate between the two rows of barrel pillars. A total of 5 fireproof holes were drilled.

  2. Perfecting the grouting pipe:

When the working face bracket was retracted to 130# frame before, two Φ40 mm steel pipes were left in advance: one of the steel pipe outlet was located 10 m south of backwind lane end, and the other was located 5 m south of the backwind lane end.

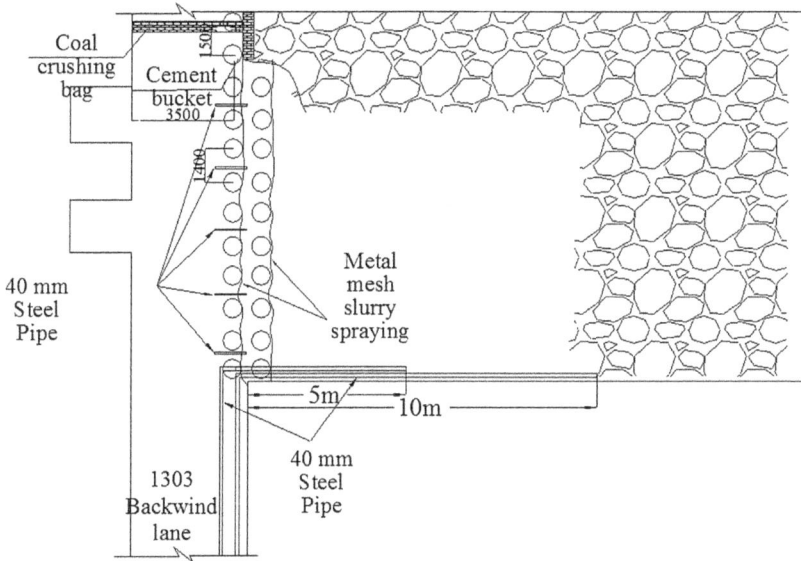

FIGURE 6.22   The 1304 backwind lane end injection holes and grouting pipe arrangement.

3. Grouting, glue injection, nitrogen injection fire prevention
   a. The Φ40 mm steel pipe was reserved in advance and placed between the cement barrel columns to fill the gap between the two rows of barrel columns with baking soda and water glass colloid. A total of 50 m^3 of colloid was injected.
   b. Two Φ40 mm steel pipes reserved in the stopping line were used to fill the gap in the mining area with 150 m^3.
   c. The grouting pipe reserved along the track was used to fill the void area with grouting 3000 m^3.
   d. Three nitrogen injection pipes reserved along the hollow stay lane in the backwind lane were used to inject 1800 m^3 of nitrogen into the closed area along the hollow stay lane. The 1304 backwind lane end injection holes and grouting pipe arrangement is shown in (Figure 6.22).

6.9   APPLICATION EFFECTS

The comprehensive fire prevention project of 1304 comprehensive discharge face was implemented in strict accordance with the formulated fire prevention technical measures, and the monitoring data showed that the adopted comprehensive fire prevention technical measures were suitable and effective.

An infrared thermal imaging camera was used to probe from the 1304 intake lane through the work surface to the backwind lane. For the intake lane, the 1303 backwind lane was probed 120 m along the empty stay lane section. The investigation results showed that in the 1304 intake lane, infrared thermal image detection showed that the temperature in the 1303 backwind lane 120 m along the empty stay lane section was normal, at about 27–28°C; the working surface and backwind lane section temperature was generally high, with a temperature range of 31–32°C.

The findings of our scientific analysis demonstrate that the comprehensive fire prevention technical measures developed for the 1304 comprehensive release face were effective. The successful fire extinguishing techniques presented herein allowed mining of the working face to resume safely.

REFERENCES

1. Guan HY, Ganderen JLV, Tan YJ, et al. (1998) Environment investigation and research on spontaneous combustion of coalfield in north China. Beijing: China Coal Industry Publishing House.
2. Wang ZH, Wang C, Yin JH (2015) Strategies for addressing climate change on the industrial level: affecting factors to CO_2 emissions of energy-intensive industries in China. Natural Hazards 75:S303–S317.
3. Song Y, Zhang M, Dai S (2015) Study on China's energy-related CO_2 emission at provincial level. Natural Hazards 77:89–100.
4. Adamus A (2001) Technical note: review of nitrogen as an inert gas in underground mines. Journal of the Mine Ventilation Society of South Africa 54 (3):60–61.
5. Cudmore JF (1988) Spontaneous combustion of coal and mine fire. International Journal of Coal Geology 9 (4):397–398.
6. Gao GW (1999) The present state and prospects for nitrogen fire prevention and fire elimination technology in coal mines of China. Journal of China Coal Society 24 (1):48–51.
7. Liu YX, Wu PJ (1997) Mechanism and application of thick loess fluid pouring for prevention of spontaneous combustion of residual coal in goaf. China Safety Science Journal 7 (1):36–39.
8. Singh RV, Tripathi DD (1996) Fire fighting expertise in Indian coal mines. Journal of Mines, Metals and Fuels 44 (6):210–212.
9. Wang XS, Zhang GS (1990) Prevention and control of mine fire. Xuzhou: Press of China University of Mining and Technology, pp. 14–43.
10. Wu C (2000) Test of chemical suppressants for fire prevention in mines with sulphide ores. Mineral Resources Engineering 9 (2):255–264.
11. Xu JC, Guo XM, Deng J (2000) Technology of mine fire direct extinguishing with temperature resistance high water gel for seam fire. Coal Science and Technology 28 (3):4–6.

7 Application of Safety Triad in Process Safety

Trent Parker, Ryan Shen, and Qingsheng Wang
Department of Chemical Engineering, Texas
A&M University, College Station, Texas

CONTENTS

7.1 INTRODUCTION: PROCESS INCIDENTS AND SAFETY TRIAD

A large number of incidents have occurred in process industries across the world. Many have resulted in injuries, deaths, economic losses, or environmental damage to the surrounding communities. Notable incidents include the Williams Olefins plant explosion in 2013 in Geismar, LA that resulted in 2 deaths and 114 injuries (U.S. Chemical Safety and Hazard Investigation Board, 2016) as well as the BP fire and explosion in 2005 in Texas City, TX that resulted in 5 deaths and 80 injuries (U.S. Chemical Safety and Hazard Investigation Board, 2007). In response to these incidents, the late Dr. M. Sam Mannan proposed the "safety triad," which

DOI: 10.1201/9781003140382-7

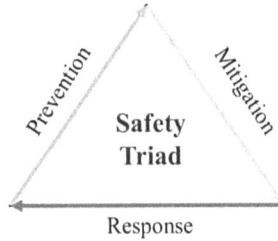

FIGURE 7.1   Safety triad. (O'Connor et al., 2019.)

presents the three core aspects of an effective safety system (O'Connor et al., 2019). Similar to the fire triangle, which depicts the three ingredients necessary for most fires, the safety triad depicts the three aspects of an effective safety system (Parker et al., 2019). These layers as shown in Figure 7.1, are prevention, mitigation, and response. In this safety triad, deficiencies in any of the factors can lead to a possible incident. Furthermore, the safety triad can be applied to incidents that have been caused by a variety of initiating events, as well as within onshore and offshore facilities.

7.2   PREVENTION

While Dr. Mannan strongly believed that all three parts must be given proper consideration for the success of the triad, he also believed that the prevention aspect should be implemented before the mitigation and response measures are considered. Examples of prevention measures are the development of accurate, well-written operating procedures and the implementation of safety management systems.

7.2.1   MITIGATION

For incidents that occur despite the implementation of prevention measures, mitigation measures are taken to lessen any adverse effects of an incident. These effects may include property damage or loss of life. Examples of mitigation measures are sensors and alert systems, backup power sources, and the separation of people from hazards.

7.2.2   RESPONSE

When incidents occur, appropriate response measures should be taken in order to minimize the adverse impacts of the incident on the environment, property, and human life and health. Examples of response measures include evacuation, shelter, and firefighting. Furthermore, following the prescribed response to an incident, there are later stages such as recovery, repair and cleanup, as well as investigation and reporting of the incident to reduce the likelihood of future recurrences.

7.3 INTEGRATION OF THE SAFETY TRIAD WITH OTHER SAFETY ANALYSIS APPROACHES

7.3.1 BOWTIE ANALYSIS

The bowtie analysis method is used to identify and analyze causal relationships in high-risk scenarios. As the name suggests, the term *bowtie* describes the shape of the diagram when the threats and responses are laid out. The goal of the bowtie method is two-fold; first, the diagram provides a visual representation of possible incidents that may occur for a given hazard. Second, it provides a representation of the control measures that can be instituted to minimize these incidents. A sample bowtie diagram is provided in Figure 7.2.

The bowtie format presents hazard, event, and response data as follows: the top-center node shows the *hazard*, which is the basis of the diagram and under this is the *top event*, which is the moment when control over the hazard is lost. At this point, no damage has occurred, but it is imminent. The threats are represented as nodes on the left side of the diagram and are possible causes of the top event. On the right side of the diagram are the *consequences*, which result from the top event occurring. The *barriers* shown on the left side of the diagram are preventive measures that limit or prevent the top event materializing. Barriers on the right side of the diagram are recovery barriers, which serve to prevent or mitigate the consequences of the top event. Additional factors referred to as *escalation factors*, can result in failure of the barriers. These are depicted in the diagram as lines connecting the escalation factors to their relevant barriers. Barriers that can prevent or mitigate escalation factors are also represented in the diagram in relation to their respective factors. The safety triad can be utilized in conjunction with the bowtie method to identify the necessary components to construct an effective safety system. As such, bowtie diagrams can be created for each of the prevention components to identify and address specific deficiencies. For hazards related to failures of the prevention components, bowtie diagrams can be created to identify practical mitigation components. Furthermore, hazards from possible failures of the mitigation components can be utilized in additional

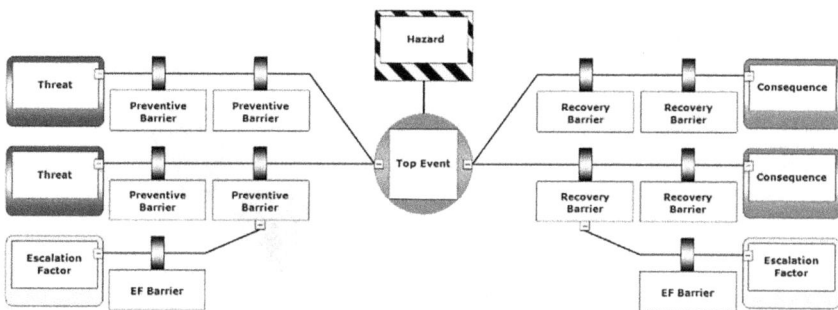

FIGURE 7.2 Sample bowtie analysis diagram. (CGE Risk, 2021.)

bowtie diagrams to identify appropriate response measures. Thus, the safety triad can be applied within the bowtie analysis method to establish a more effective safety system.

7.3.2 LAYERS OF PROTECTION ANALYSIS

An alternate risk analysis method is the layers of protection analysis (LOPA) method. Under this approach, a set of undesired incidents such as fires, explosions or chemical releases are outlined and the consequences of these events are estimated. Then, independent layers of protection that can be implemented are identified. Controls and safety systems are assigned protective credits based on their ability to independently control or mitigate the specified hazards. For a layer of protection to be considered an independent protection layer (IPL), its performance must not be affected by the failure of any other protection layer, or affected by the conditions that caused an alternate protection layer to fail. These IPLs include basic process control systems, pressure relief devices, and alarms with defined operator responses. LOPA analysis is utilized to analyze the probability of an incident occurring both with and without a given layer of protection. These IPLs are represented as progressively larger ovals and a LOPA sample is shown in Figure 7.3.

Similar to the bowtie method, the LOPA approach can be used in conjunction with the safety triad to identify necessary system components. For example, each of the prevention components of the safety triad can be incorporated into a LOPA diagram to determine their independence versus a process hazard. Then, each of the mitigation components can be incorporated into an additional LOPA

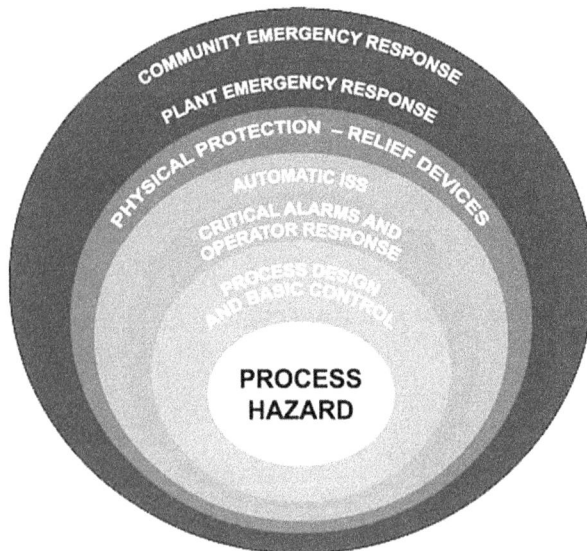

FIGURE 7.3 Sample LOPA diagram. (Willey, 2014.)

diagram to determine their independence. Lastly, based on the independence of the mitigation components, the response components can be incorporated into another LOPA. The results of the LOPA diagrams of the three components can be incorporated into one overall LOPA diagram to introduce the additional guidance that the safety triad provides.

7.3.3   HAZARD AND OPERABILITY ANALYSIS

Hazard and operability (HAZOP) is a systematic method of identifying possible hazards in a work process. Similar to the Bowtie and LOPA diagrams, the HAZOP consists of determining consequences associated with deviations from normal process parameters where it may create a hazard. To do this, a guide word is first identified for possible process deviations, such as no, higher, or more. Then, deviations that correspond to the guide word are listed. Possible causes of these deviations are then listed in the HAZOP, as well as their consequences. Lastly, actions to be taken to prevent or mitigate the consequences of these deviations are listed, as shown in Figure 7.4.

As with other process safety related diagrams, the HAZOP can be used in conjunction with the safety triad to comprehensively identify necessary safety system components. For this, guide words can be generated for each of the three safety triad components. Then, possible deviations associated with these guide words for each of the three components can be developed, as well as the causes, consequences, and necessary actions. This provides a robust method of evaluating the components of the safety system.

Preliminary HAZOP on Reactor - Example

Guide Word	Deviation	Causes	Consequences	Action
NO	No cooling		Temperature increase in reactor	
REVERSE	Reverse cooling flow	Failure of water source resulting in backward flow		
MORE	More cooling flow			Instruct operators on procedures
AS WELL AS	Reactor product in coils			Check maintenance procedures and schedules
OTHER THAN	Another material besides cooling water	Water source contaminated		

23

FIGURE 7.4   Sample HAZOP analysis diagram. (Hartwell, 2019.)

7.3.4 Safety Integrity Level Analysis

Safety integrity level (SIL) is used to determine the risk-reduction levels provided by various safety functions. For this method, the safety integrity level of each safety function is determined based on the probability of failure on demand (PFD) for that function. For example, for low demand operations, SIL 1 corresponds to a PFD of 0.1–0.01, while SIL 4 corresponds to a PFD of 0.0001–0.00001. Methods of assigning SIL values include risk matrices, as shown below in Figure 7.5. In this matrix, hazards with higher associated risks are assigned safety functions with higher SIL levels.

The safety triad can be used in conjunction with the SIL analysis method to identify necessary safety system components. To do so, safety system components for prevention, mitigation, and response can be organized based on their PFD and thus assigned SIL values. Then, these components can be depicted using a risk matrix to ensure that effective safety components are in place to address the hazards associated with the system.

7.3.5 Failure Modes and Effects Analysis

Failure modes and effects analysis (FMEA) is used to determine possible failures when designing a system. For this method, failure modes for a given system are identified and listed. Then, the severity of the hazards of the failure modes as well as their probabilities of occurrence are identified. Furthermore, the probability of detection for each of the failure modes is also determined, as shown in Figure 7.6. This allows appropriate safety measures to be implemented for each of the failure modes.

The FMEA analysis method can be used in combination with the safety triad to identify and implement the components of an effective safety system. Each failure mode identified in the FMEA can be categorized into one of the three safety triad components. From this, the severity and probability of occurrence

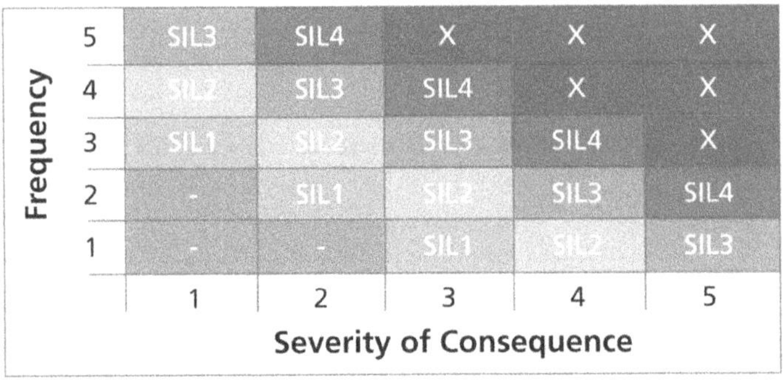

Frequency	1	2	3	4	5
5	SIL3	SIL4	X	X	X
4	SIL2	SIL3	SIL4	X	X
3	SIL1	SIL2	SIL3	SIL4	X
2	-	SIL1	SIL2	SIL3	SIL4
1	-	-	SIL1	SIL2	SIL3

Severity of Consequence

FIGURE 7.5　Sample SIL analysis with risk matrix. (Benmerrouche & Lee, 2015.)

FAILURE MODE & EFFECTS ANALYSIS (FMEA)				Date: 1/1/2018
Process Name: Left Front Seat Belt Install Process Number: SBT 445				Revision: 1.3

Failure Mode	A) Severity Rate 1-10 10=Most Severe	B) Probability of Occurence Rate 1-10 10=Highest Probability	C) Probability of Detection Rate 1-10 10=Lowest Probability	Risk Preference Number (RPN) AxBxC
1) Select Wrong Color Seat Belt	5	4	3	60
2) Seat Belt Bolt Not Fully Tightened	9	2	8	144
3) Trim Cover Clip Misaligned	2	3	4	24

FIGURE 7.6   Sample FMEA diagram. (Slack, 2015.)

and detection for each of these failure modes can be identified. This allows the robustness of the entire safety system to be assessed, with additional components added if necessary.

7.3.6   JOB SAFETY ANALYSIS

Job safety analysis (JSA) is used to identify hazards associated with each step in a task. For this analysis, each of the steps within a task are listed. Then, for each step, potential hazards are identified. For each of these hazards, recommended procedures are listed to either eliminate or reduce them, as shown in Figure 7.7.

The safety triad can serve as a useful tool to complement the JSA approach. For each step within the JSA, the safety triad can be utilized to identify whether there are robust prevention, mitigation, and response components in place to address

Basic Job Steps	Potential Hazards	Recommended Procedure
1 / Dig hole for tree.	Strike underground utilities	Make sure underground utilities have been located and marked.
2 / Move tree to hole location.	Lifting, back injuries	Always use a ball cart or tractor with trained operator to move tree to hole location.
3 / Place tree in hole.	Lifting, back injuries, damage to tree	Roll tree into hole or use two people to place tree into hole, depending on tree size.
4 / Remove wire cage.	Sharp edges of metal, cuts, lacerations	Wear gloves, use proper wire cutter
5 / Remove twine and burlap.	Cuts, lacerations with sharp cutting objects	Always cut away from body, wear gloves

FIGURE 7.7   Sample JSA diagram. (Kjellen & Albrechtsen, 2017.)

hazards. This can then guide the recommended procedures that are prepared for hazard elimination or reduction. Thus, the combination of the safety triad with JSA results in more comprehensive solution than JSA in isolation.

7.3.7   FAULT TREE ANALYSIS METHOD APPLICATIONS

Fault tree analysis (FTA) is a top-down approach to identifying component-level failures that result in a system-level failure occurring. For this method, the primary (component-level) failures are represented as circles in a diagram, which are connected to "and/or" gates. In the case of "or" gates, either outcome presents a primary failure in the output event which is represented by the rectangle above the gate. In the case of "and" gates, both primary failures must occur simultaneously to produce the output event. These output events are connected to additional "and/or" gates up to the location of the top output event, which is the system-level failure as shown in Figure 7.8. The FTA diagram illustrates the effects and interdependencies of each component-level failure on the system level failure, which leads to implementation of the systems that prevent occurrences of component-level failures. The safety triad can be used in combination with the FTA method to determine the robustness of each component of the triad. For this, the initiating events can be represented as component-level failures on the FTA diagram.

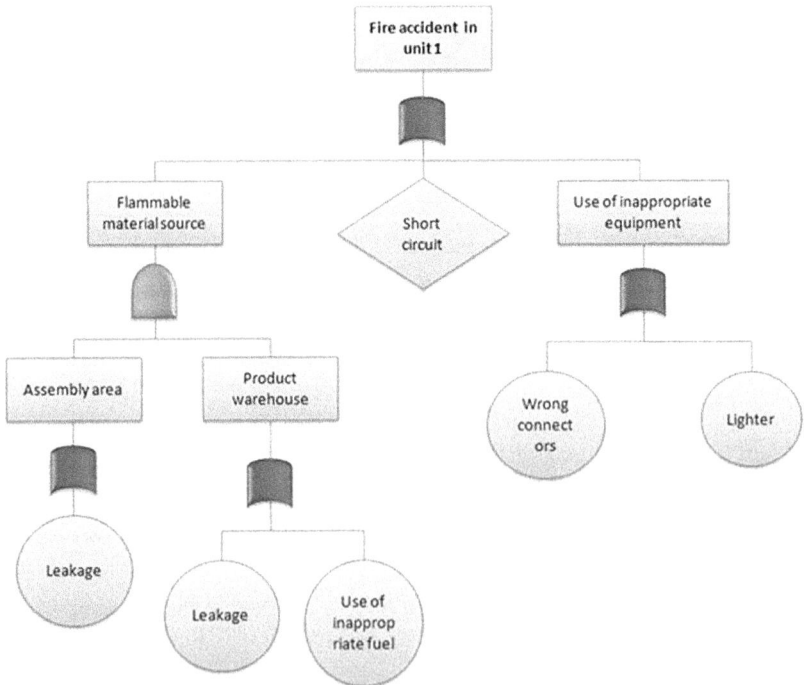

FIGURE 7.8   Sample FTA diagram. (Pedroni & Zio, 2012.)

The prevention, mitigation, and response components can then be identified to address the progressively higher-level output events within the diagram. Thus, using the safety triad in conjunction with the FTA ensures that each output event identified is properly addressed.

7.4 APPLICATION OF THE SAFETY TRIAD IN PROCESS SAFETY

7.4.1 Safety Triad and the Arkema Chemical Plant Fires

On August 25, 2017 Hurricane Harvey, a Category 4 storm, made landfall. Harvey resulted in a 500-year flood in the greater Houston area and 107 deaths, costing an estimated $125 billion in damage. One of the most significant industrial incidents stemming from Hurricane Harvey was the Arkema chemical plant fires pictured in Figure 7.9. These resulted from the combustion of organic peroxides that had been stored in refrigerated trailers. Nine trailers containing the peroxides were involved, with three combusting spontaneously due to decomposition of the peroxides. Another six were intentionally ignited with controlled burns. In total, more than 350,000 pounds of organic peroxides were combusted that resulted in 21 people requiring medical attention (U.S. Chemical Safety and Hazard Investigation Board, 2018).

The concept of the safety triad was used to identify the factors that contributed to the severity of this incident. Deficiencies in the safety triad primarily involved the prevention and response layers. The common failure point in the prevention layer was flooding. This failure mode was likely unexpected because flooding on the scale produced by Harvey was statistically very unlikely to occur during the expected life cycle of the plant. Response layer deficiencies included the decision to allow Highway 90 to remain open, despite being located within the evacuation zone.

FIGURE 7.9 Arkema chemical plant fire. (CBS News, 2017.)

This resulted in 21 people being exposed to decomposition products. This application gives an example of how the concept of safety triad can be applied to the investigation of a reactive chemical incident stemming from flooding caused by climate extremes. Because of the lack of awareness of potential flood risks in the process safety planning, flood insurance maps and related studies were not able to identify this issue, for this facility. None of Arkema's safeguards for electrical power failure met with company or industry standards in the event of flooding on this scale. Using the safety triad, the deficiencies in the three different layers can be systematically identified, even for an initiating event that will rarely occur, such as extreme weather. The results derived from these analyses can be utilized to enhance the process hazards analysis (PHA) and improve the resilience of the safety management system which comprehensively addresses all three layers of the safety triad (U.S. Chemical Safety and Hazard Investigation Board, 2018).

7.4.2 Safety Triad and the DuPont La Porte Facility Toxic Chemical Release

On November 14, 2014, four workers were killed by the release of approximately 24,000 pounds of highly toxic methyl mercaptan at the DuPont chemical manufacturing facility in La Porte, TX (U.S. Chemical Safety and Hazard Investigation Board, 2015; U.S. Chemical Safety and Hazard Investigation Board, 2019). The initiating event of this incident was determined to be cold weather; however, due to a series of safety system deficiencies involving the prevention, mitigation, and response layers, this cold weather resulted in a domino effect that caused a catastrophic incident. Deficiencies in the prevention layer in this incident mainly involved failure to identify the possibility of the formation of solid methyl mercaptan clathrate hydrate in a feed pipe during cold weather. This meant that there was no effective prevention measure to prevent this incident. For the 24 hours preceding the toxic release, the average temperature had been approximately 40°F and under 55°F for the previous three days in the Houston area. When temperatures fell below 52°F, water and methyl mercaptan formed the solid, ice-like methyl mercaptan clathrate hydrate. The reaction unit could not be restarted because the solid hydrate had formed in the methyl mercaptan feed pipe that led to the reaction section. However, there were no preventive measures to avoid the formation of this solid hydrate, such as insulation. Moreover, during the early troubleshooting stages, operators were unaware of the cause of the feed pipe blockage, and there were no standard operating procedures established to address this issue. Finally, because of the employees' lack of awareness of the possible hazards involved, methyl mercaptan escaped from the waste gas vent header into the atmosphere. This filled the manufacturing buildings with highly toxic vapors, killing the shift supervisor and three operators. Deficiencies in the mitigation layer of this incident primarily involved the inability of the facility to contain the toxic release or vent it to a less hazardous location. Specifically, for methyl mercaptan, no facility within the manufacturing building was equipped to contain the leak and direct it to an incinerator for destruction. Furthermore, although two ventilation fans were installed on the roof, they were inoperative during the release and lacked the capability to control the large quantity of gas required to maintain an acceptable human exposure level. Therefore, the undesirable impacts of methyl mercaptan exposure were not reduced effectively. Deficiencies in the response layer of this incident were mainly concerned with a lack of sensors that would have been able to detect the toxic release, triggering an effective emergency response plan. Alarms indicating high methyl mercaptan concentrations were only located in the control room, with none to alert fieldworkers. Furthermore, even operators who heard the alarms in the control room did not realize a release had occurred as they had previously detected methyl mercaptan gas release solely by odor. The emergency response team (ERT) was insufficiently prepared for this kind of incident. When the ERT was called upon to respond, they did not understand that a major toxic chemical release had occurred and thus, were unable to provide sufficient information to 9-1-1 emergency responders, even after they discovered the magnitude of the incident. Furthermore, in the emergency plan, there were insufficient secondary and tertiary control mechanisms for when a designated individual was unable to respond. In this incident, the loss of the shift supervisor meant that no one in the emergency plan could be assigned roles to disseminate the hazard information. Considering the deficiencies identified here in terms of the safety triad, the following safety measures could be proposed to address the issues (Parker et al., 2019). In the prevention layer, heat tracing or some other process should be implemented to prevent the formation of hydrates and detailed, well-written operating procedures should be developed regarding how to safely dissociate hydrates to prevent methyl mercaptan release. In the mitigation layer, a ventilation system with sufficient or excess capacity, or even an open building structure should be considered when processing toxic materials. In the response layer, a comprehensive emergency response plan should be developed, with all involved personnel trained on the plan. Moreover, efforts should be made to integrate personnel, facilities, equipment, and communication systems within the ERT to ensure that proper incident mitigation and responses occur in the future; one system, the Incident Command System (ICS) is recommended for use (Wang et al., 2012). This application gives an example of how the concept of the safety triad can be applied to track the initiation, development, and termination of a toxic gas release incident in the chemical process industry, and propose corresponding safety strategies that would minimize negative impacts on the emergency responders, operators, general public, facility, and environment. As described in this process, if any of the three layers of the safety triad are active and effective, the catastrophic consequences typically associated with major process incidents can be prevented. Furthermore, for prevention, mitigation, and response in the safety triad, the earlier the corresponding safety measures can be activated, the more efficient they are in reducing any knock-on effects. The lessons learnt from this incident can help promote the required improvements in process hazard analyses (PHAs) and prevent future similar occurrences.

7.4.3 Safety Triad and MXene Synthesis and Processing

MXenes are a recently discovered family of 2D transition metal carbides and carbonitrides. They consist of $M_{n+1}X_nT_x$ layers, where M is an early transition metal, X is carbon and/or nitrogen, and T is a combination of several terminal groups (i.e., −F, −OH, −O, −Cl, etc.) (Rout et al., 2019). They exhibit various unique properties, such as high conductivity, good mechanical properties, and hydrophilicity, so that they can be used for a wide range of applications including electronic devices, sensors, reinforcement for composites, and energy storage materials (Bondavalli, 2018). Currently, the production of MXenes is limited to laboratory scales, however there is a growing interest from industry in scaling up production. Scaling up production of MXene is not a simple process and may result in catastrophic incidents, loss of life and property. The correct process must be followed precisely and all potential hazards and risks must be taken into consideration. To reduce the number and severity of incidents associated with the synthesis and scale-up of MXene, the safety triad was applied to identify possible hazards with MXene production. The following is a discussion of the prevention, mitigation, and emergency response plans for the correct synthesis and postprocessing of MXene (Lakhe et al., 2019). Production of $Ti_3C_2T_x$ in a lab was selected for this study due to its well-known characteristics. This is a multi-step process shown in Figure 7.10, including MAX phase synthesis from raw materials, etching of the MAX phase to MXene clay, exfoliation to MXene nanosheets, and finally, post-processing of MXene.

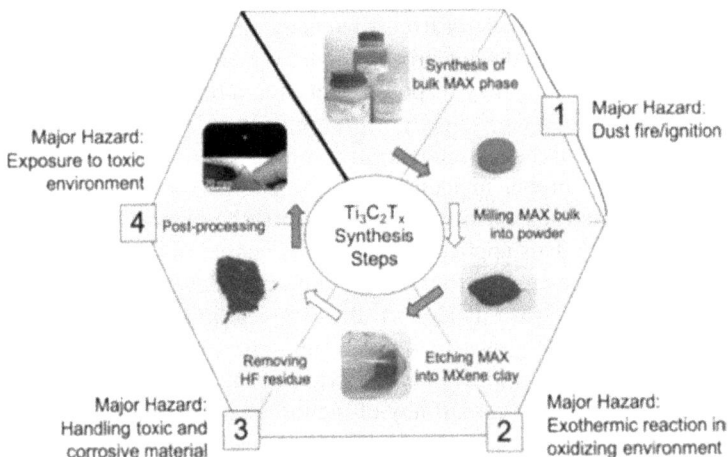

Step 1: Synthesis and milling phase of Ti_3AlC_2
Step 2: Etching Al from the Ti_3AlC_2 using a strong oxidizing agent such as HF
Step 3: Washing with the oxidizing agent (HF)
Step 4: Drying and post processing

FIGURE 7.10 $Ti_3C_2T_x$ synthesis from raw materials. (Lakhe et al., 2019.)

To systematically perform a hazard identification and risk mitigation analysis, different hazards associated with MXene production were evaluated by implementation of the safety triad. Major hazards were identified first, which were: spontaneous combustion, dust explosions, runaway reactions, and toxic chemical exposure. Thereafter, using bowtie analysis, the root causes of these hazards and possible major consequences were assessed. From these identified hazards, corrective/preventive measures for the laboratory scale production process were proposed. From this, inherently safer designs (ISDs), engineering controls, administrative controls, and personal protective equipment were found to be effective safety measures. However, as proposed by Lakhe et al. (2019), implementing ISD or engineering controls can be challenging in a laboratory setting, and administrative controls such as supervision, proper training, safe work practices, manuals, and procedures should be prioritized. Moreover, mitigation measures and emergency plans can be used to reduce the impact of undesirable events where there is a failure or lack of preventive measures. These risks can be situations such as incorrect placement of an extractor hood, testing sensors and alarm systems beforehand, and establishing evacuation plans and communicating this to staff. Communicating the expected hazards and risks to individuals and colleagues that work directly or indirectly with the process, or even just in the facility improves emergency responses and recovery plans. This application gives an example of how the concept of the safety triad can be applied to improve the safety processes in MXene production. This safety approach can scale-up, from large laboratory batches to commercial production. The same concept can be applied to any other novel laboratory process, by taking the means of loss prevention into consideration and implementing the required preventive measures. However, process safety needs to be considered in the development process, prior to actual implementation of a scale-up, as pre-planning will ultimately result in an inherently safer, simpler, and more economical production process.

7.4.4   Safety Triad and the Tianjin Incident

On August 12, 2015, a devastating warehouse fire and explosion occurred at the Ruihai International Logistics Co., Ltd. (hereafter, Ruihai Company) facility located at the Tianjin Port in China. This incident directly caused 165 deaths, of which 104 were firefighters, 8 missing persons, and 798 injuries. A total of 304 buildings, 12,428 commercial automobiles, and 7,533 containers were damaged, resulting in 6.866 billion RMB in direct economic losses. Moreover, a large number of hazardous chemicals at the scene produced poisonous and harmful substances that posed a dire threat to the surroundings and local residents (Fu et al., 2016; Zhou and Fan, 2017). The sequence of these events is shown in Figure 7.11.

A bowtie diagram was also prepared for this incident analysis and is presented in Figure 7.12. The top event is the spontaneous combustion of nitrocellulose, which ignited other nitrocellulose in the vicinity and released a large quantity of heat. Then the nitrocellulose flame spread extensively, reaching an area where ammonium nitrate was stored, which facilitated the final catastrophic explosion.

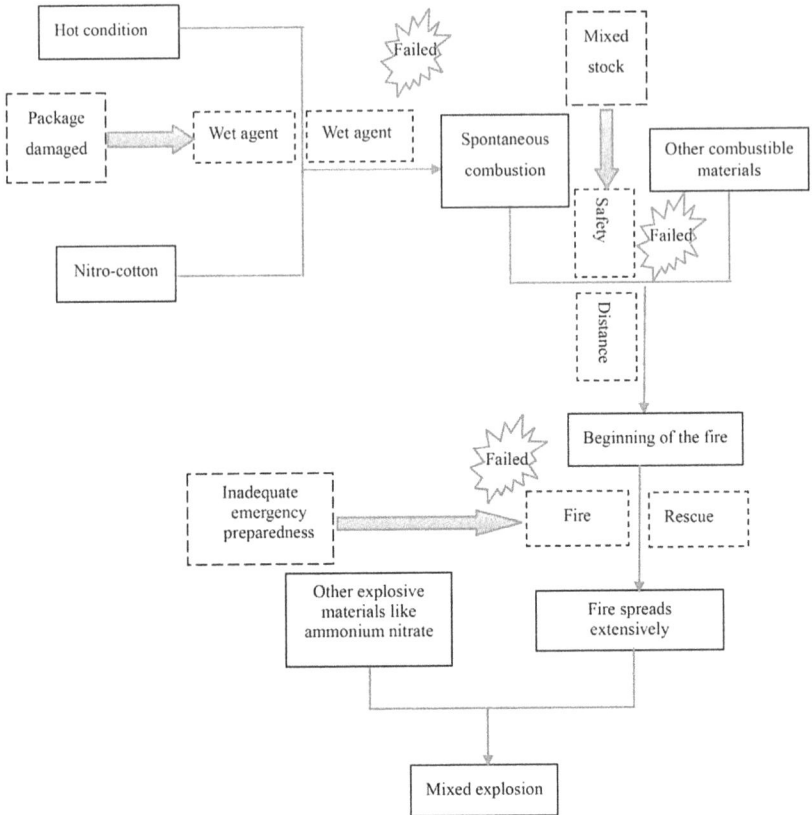

FIGURE 7.11 Event sequence diagram of the port fire and explosion in Tianjin, China. (Zhou and Fan, 2017.)

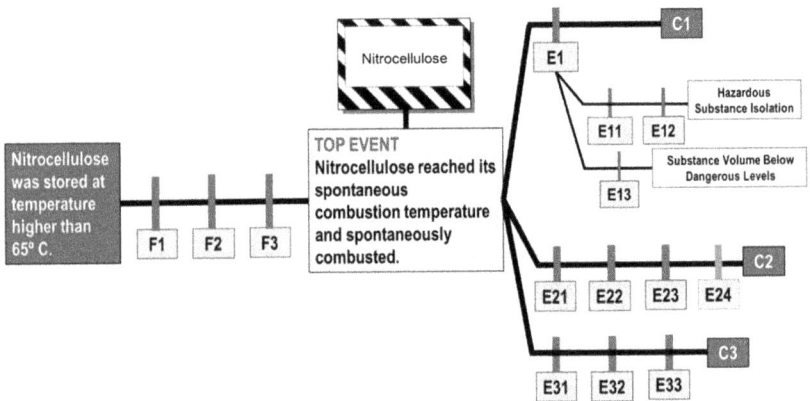

FIGURE 7.12 Bowtie analysis of the port fire and explosion in Tianjin, China. (Chen et al., 2019.)

The flame and shock waves resulting from the explosion detonated yet more ammonium nitrate stored further away, as well as other hazardous chemicals including potassium nitrate, calcium nitrate, magnesium, and sodium sulfide.

These were all being held in the Ruihai Company storage yard. In this event, the secondary explosion was much more powerful than the first. To prevent and mitigate the top event from escalating into further undesirable consequences, both active and passive barriers were installed, labeled E11–E13. However, all of the barriers failed as a result of shortcomings in individual and organizational behavior. For the response component of the safety triad, no training was provided to employees regarding handling of dangerous substances or in how to respond to potential emergencies involving such substances. Emergency responders also lacked training on how to handle nitrocellulose, which subsequently led to the explosion. Finally, the lack of an appropriate emergency plan or timely information regarding the state of the disaster led to a subsequent failure where the top event escalated into a disaster. Without effective mitigation measures onsite, the fire continued to burn, increasing the firefighters' exposure to very high temperatures with inadequate protection. The lack of compliance with emergency response requirements resulted in the deaths and injuries of many responders along with extensive damage to the surrounding community (Chen et al., 2019).

Notes
Red rectangles imply a barrier failed, green rectangles imply a barrier was successful.

F1: Ethanol or water should be used as a damping agent in nitrocellulose and should be sealed in thermoplastic.

F2: Hazardous chemicals should be unpacked, transported and unloaded according to regulations.

F3: Specialists should supervise on-site and check seals.

E1: Storage areas should be designed and managed to prevent secondary ignitions of ammonium nitrate (including E11–E13).

E11: Provide adequate separation distances among combustible materials and flammable substances.

E12: Hazardous chemicals should be stored in purpose built and properly equipped warehouses.

E13: The amount of hazardous chemicals stored should be restricted.

C1: Fire spreads to the ammonium nitrate containers. The first, and then second explosions occur. Subsequent fires cause extensive property damage.

E21: Firefighters need to be called to manage the escalating issue.

E22: Neighboring enterprises and communities need to be notified. Mutual protection safety agreements should be executed. Evacuation of on-site personnel should start in an organized way as soon as possible.

E23: There should be adequate safety zones between the warehouse and important public buildings and transportation facilities.

E24: Firefighters should organize evacuation of on-site personnel after a situation assessment.

C2: The accident caused mass casualties to workers.

E31: Access for firefighters and emergency responders must be maintained.

E32: Hazardous and highly hazardous materials must be officially registered.

E33: Firefighters must be adequately trained and equipped to handle hazardous chemical incidents.

C3: Firefighters suffer heavy casualties.

This case is another example of how the bowtie analysis system can illustrate a sequence of on-site events that lead to a severe incident. It identifies the direct failures in terms of the safety triad, with regard to the prevention, mitigation and response categories. This particular incident led to both a large number of deaths, and severe property damage. By determining the sequence of events using the bowtie analysis, and learning from failures in each of the three safety triad components that led to the incident, evidence-based guidance can be developed to prevent the future occurrences.

7.5   ADDITIONAL APPLICATIONS OF THE SAFETY TRIAD

The concept of the safety triad as a whole is to identify and address deficiencies in safety systems. This paper has investigated the triad as it relates to incidents in the chemical processing industry. It was shown that the triad can serve as a viable method to comprehensively identify deficiencies in safety systems, and thus reduce the likelihood of incidents. Applications of the safety triad to other types of incidents within more generic processing industries should be investigated in order to determine the extent of applicability of the triad to incident prevention and investigation.

REFERENCES

Benmerrouche, M., & Lee, R. (2015). Safety Integrity Level and Redundancy Requirements for the Top-Off Safety System. doi:10.2172/1493227

Bondavalli, P. (2018). *Graphene and Related Nanomaterials: Properties and Applications.* Elsevier. Amsterdam, The Netherlands.

CBS News. (2017, September 01). Arkema Chemical Plant Fire [Digital image]. Retrieved from https://www.cbsnews.com/news/flames-erupt-at-arkema-chemical-plant-flooded-by-harvey-in-crosby-texas/

CGE Risk. (2021, June 1). The Bowtie Method. Retrieved from https://www.cgerisk.com/knowledgebase/The_bowtie_method

Chen, Q., Wood, M., & Zhao, J. (2019). Case study of the Tianjin accident: Application of barrier and systems analysis to understand challenges to industry loss prevention in emerging economies. *Process Safety and Environmental Protection, 131,* 178–188. https://doi.org/10.1016/j.psep.2019.08.028

Fu, G., Wang, J., & Yan, M. (2016). Anatomy of Tianjin Port fire and explosion: Process and causes. *Process Safety Progress, 35*(3), 216–220. https://doi.org/10.1002/prs.11837

Hartwell, J. (2019, November 21). HAZOP - Hazard and Operability Study. Retrieved September 30, 2020, from https://www.iqasystem.com/news/hazop/

Kjellén, U., & Albrechtsen, E. (2017). In *Prevention of Accidents and Unwanted Occurrences: Theory, Methods, and Tools in Safety Management* (pp. 363–370). CRC Press, Taylor & Francis Group. Oxfordshire, England.

Lakhe, P., Prehn, E. M., Habib, T., Lutkenhaus, J. L., Radovic, M., Mannan, M. S., & Green, M. J. (2019). Process safety analysis for Ti3C2T x MXene synthesis and processing. *Industrial & Engineering Chemistry Research*, 58(4), 1570–1579. https://doi.org/10.1021/acs.iecr.8b05416

O'Connor, M., Pasman, H. J., & Rogers, W. J. (2019). Sam Mannan's safety triad, a framework for risk assessment. *Process Safety and Environmental Protection*, 129, 202–209. doi: 10.1016/j.psep.2019.07.004

Parker, T., Shen, R., O'Connor, M., & Wang, Q. (2019). Application of safety triad in preparation for climate extremes affecting the process industries. *Process Safety Progress*, 38(3), e12091. https://doi.org/10.1002/prs.12091

Pedroni, N., & Zio, E. (2012). Uncertainty analysis in fault tree models with dependent basic events. *Risk Analysis*, 33(6), 1146–1173. doi:10.1111/j.1539-6924.2012.01903.x

Rout, C. S., Late, D., & Morgan, H. (Eds.). (2019). *Fundamentals and Sensing Applications of 2D Materials*. Woodhead Publishing. Cambridge, England.

Slack, N. (2015). Failure Mode and Effect Analysis. *Wiley Encyclopedia of Management*, 1–2. doi:10.1002/9781118785317.weom100151. Wiley Publishing. Hoboken, New Jersey.

U.S. Chemical Safety and Hazard Investigation Board. (2007, March 20). BP America Refinery Explosion. Retrieved from https://www.csb.gov/bp-america-refinery-explosion/

U.S. Chemical Safety and Hazard Investigation Board. (2015, September 30). DuPont La Porte, Texas Chemical Facility Toxic Chemical Release: Interim Recommendations. Retrieved from https://www.csb.gov/assets/1/20/dupont_la_porte_interim_recommendations_2015-09-30_final1.pdf?15526

U.S. Chemical Safety and Hazard Investigation Board. (2016, October 19). Williams Olefins Plant Explosion and Fire. Retrieved from https://www.csb.gov/williams-olefins-plant-explosion-and-fire-/

U.S. Chemical Safety and Hazard Investigation Board. (2018). Extreme Weather, Extreme Consequences: CSB Investigation of the Arkema Crosby Facility and Hurricane Harvey. Retrieved from https://www.csb.gov/assets/1/20/csb_arkema_exec_summary_08.pdf?16265

U.S. Chemical Safety and Hazard Investigation Board. (2018, May 24). Arkema Inc. Chemical Plant Fire. Retrieved from https://www.csb.gov/arkema-inc-chemical-plant-fire-/

U.S. Chemical Safety and Hazard Investigation Board. (2019, June). Toxic Chemical Release at the DuPont La Porte Chemical Facility. Retrieved from https://www.csb.gov/dupont-la-porte-facility-toxic-chemical-release-/

Wang, Q., Ma, T., Hanson, J., & Larranaga, M. (2012). Application of incident command system in emergency response. *Process Safety Progress*, 31(4), 402–406. doi: 10.1002/prs.11538

Willey, R. J. (2014). Layer of protection analysis. *Procedia Engineering*, 84, 12–22. doi:10.1016/j.proeng.2014.10.405

Zhou, A., & Fan, L. (2017). A new insight into the accident investigation: A case study of Tianjin Port fire and explosion in China. *Process Safety Progress*, 36(4), 362–367. https://doi.org/10.1002/prs.11891

8 Forensic Analysis of Some Types of Industrial Explosions

Xiaoliang Zhang, Tingguang Ma,
Huiping Liu, and Zhangrui Liu
Department of Safety Engineering, Shanghai Institute
of Technology, Shanghai, People's Republic of China

CONTENTS

8.1 BACKGROUND OF INDUSTRIAL EXPLOSIONS

With the acceleration of industrialization in China, there has been a rapid increase in the use of conventional processing and manufacturing of chips and other high-end intelligent devices. The application of new materials has served to further enhance these manufacturing processes, which has inevitably led to the widespread use of flammable gases, explosive and flammable liquids, as well as industrial powder dielectric materials. The utilization of these materials has grown in proportion with the scale of the processes that are employed, which has also driven up the risk of

DOI: 10.1201/9781003140382-8

explosion. Some highly integrated processes may simultaneously make use of three types of material during manufacture. Occasionally, dust and gas may be stored to form a mixed or "hybrid" mixture, and when combined with a flammable liquid may also become cloud-forming, or may form an "aerosol" if powder or liquid particles are dispersed in the gas. As a consequence of these phenomena explosive hazards are more complex and more serious. Following the occurrence of an accident involving an explosion, the cause of the accident will be investigated, which will include the role of combustible industrial chemicals, damage assessment, laboratory identification, scene analysis, and accident prevention methods.

8.2 INDUSTRIAL MEDIUM CHARACTERISTICS

Combustible dusts, like flammable gas or combustible liquid vapor, not only have a lower explosive limit, a lower ignition energy, a low ignition temperature, and a higher explosive sensitivity but also have a large heat value on combustion, a higher pressure on explosion, a saturated vapor pressure due to the complexity of modern industrial production processes. However, combustible gas, combustible vapor, and combustible dust each have their own unique characteristics. In this chapter, three common and representative characteristic parameters of industrial chemicals are discussed. Combustible gas and vapor are classified into IIA, IIB, and IIC, and common characteristic parameters include the minimum ignition energy (MIE), the ignition temperature, and the heat of combustion, among others.

The common, typical combustible gas characteristic parameters are listed in Table 8.1. Other characteristic parameters of flammable vapor include flash point, ignition temperature, saturated vapor pressure, relative air vapor density, etc.

The common and representative characteristic parameters of combustible vapor are shown in Table 8.2. The characteristic parameters for explosive dust include median diameter, MIE, maximum explosion pressure, maximum explosion pressure rise rate, minimum ignition temperature (MIT) of dust cloud, MIT of dust layer, and minimum explosion limit, etc. Common and representative characteristic parameters of combustible vapor are listed in Table 8.3.

TABLE 8.1
Characteristic Parameters of Common Explosive Gases

Classification	Fuel Name	Minimum Ignition Energy (mJ)	Ignition Temperature (K)	Heat of Combustion (KJ/mol)
IIA	Methane	0.33	811	890.31
	Propane	0.31	723	2217.8
IIB	Ethylene	0.096	698	1411
IIC	Acetylene	0.02	578	1298.4
	Hydrogen	0.02	673	284

TABLE 8.2
Common Characteristic Parameters of Combustible Vapors

Classification (Flash Point)	Name	Ignition Temperature (K)	Saturated Vapor Pressure (Kpa)	Vapor Density (kg/m³)
<301 K	Ethanol	636	5.33(292 K)	1.59
	Benzene	771	13.33(299 K)	2.80
	Acetone	738	56.20(313 K)	2.00
301–318 K	Acetic acid	736	1.52(293 K)	2.07
	Turpentine oil	526	2.167(324.4 K)	<1
319–393 K	Phenol	988	38.722(293 K)	3.24
>665 K	Dioctyl adipate	508	0.32(473 K)	0.922

TABLE 8.3
Characteristic Parameters of Common Explosive Dust

Classification	The Name of the Dust	Median Diameter (μm)	Minimum Ignition Energy (mJ)	Maximum Explosion Pressure (MPa)	Maximum Explosion Pressure Rise Rate (Mpa*m/s)	Maximum Ignition Temperature of Dust Cloud (K)	Minimum Ignition Temperature of Dust Cloud (K)	Lower Explosive Limit (g/m³)
Inorganic particles	Aluminum powder	35	<1	0.73	128.6	923	543	40
	Magnesium powder	104	300	0.74	7.05	883	>723	25
Particulate organic matter	Analgin powder	/	1–10000	0.19–1.21	0.3–33.7	220–890	220–450	15–500
Particulate organic matter	Analgin powder	21	/	0.94	23	1033	>723	250
Mixed dust	Aluminum-ferroalloy powder	24	/	1	13.4	963	/	523
	Copper silicon alloy powder							

8.3 COMBUSTION AND EXPLOSION DYNAMICS OF INDUSTRIAL CHEMICAL MEDIUM

Gas explosions can be classified into two types. (1) Physical explosion formed due to the high pressure caused by the rapid expansion of gas under heat without enough time for ventilation. (2) Chemical explosion which occurs when a runaway reaction generates too much heat. For a generic fuel $C_aH_bO_cS_d$ burning in air, the stoichiometric reaction is stated as:

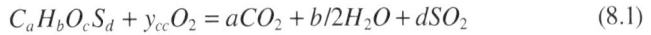

$$C_aH_bO_cS_d + y_{cc}O_2 = aCO_2 + b/2H_2O + dSO_2 \qquad (8.1)$$

$$y_{cc} = a + b/4 + d - c/2 \qquad (8.2)$$

$$y_{ch} = 1/(1 + 4.77y_{cc}) \qquad (8.3)$$

where y_{cc} is the stoichiometric air-to-fuel ratio and y_{ch} is the fuel concentration in air.

If the fuel fraction is above y_{ch} or the reaction is fuel-rich, incomplete combustion will occur. From the perspective of industrial explosion protection and explosion suppression, complete combustion gives off more heat than incomplete combustion, with a high combustion rate, a high explosion pressure, and a high pressure rising rate, so it is more dangerous. In the event of a disaster, there will be more serious effects. Therefore, in actual production, mixed gases should be avoided as far as possible so that the ranges of parameters that could lead to a combustion explosion are never reached. The most dangerous level of a flammable gas in industrial production is 1.1–1.3 times its stoichiometric concentration.

For vapor explosions, if a combustible liquid whose temperature is much lower than its flash point is pressurized and released into the atmosphere as a spray of small droplets in the air, the spray is ignited and spreads the flame like a premixed gas, forming a uniform explosive vapor cloud. The flash point of a flammable liquid refers to the lowest temperature at which the vapor released by the flammable liquid ignites when it comes in contact with air under atmospheric pressure, in the presence of an ignition source. When the fire source is sufficiently effective, the flash point is the lowest temperature at which the vapor and air mixture above the liquid surface can propagate the flame. If the combustion occurs in a constant volume, the pressure will rise. The constant volume adiabatic temperature rise can be expressed as:

$$T_2 - T_1 = \Delta E / C_V \qquad (8.4)$$

where ΔE is the free combustion heat per mole at constant volume; C_V is the average specific heat capacity of the gas system at constant volume; T_1 is the initial mixed gas temperature (before combustion); T_2 is the final temperature of the combustion product and nitrogen (from air) after complete combustion.

Assuming that the vapor is in an ideal state, the corresponding adiabatic pressure rise of constant volume can be estimated by using the equation of state.

$$\frac{P_2}{P_1} = \frac{n_2 T_2}{n_1 T_1} \qquad (8.5)$$

For dust explosions, as with gas explosions, there is a rapid chemical reaction between oxides (such as oxygen in the air) and combustibles. However, the detonation process of dust explosion is different from that of gas explosion, which is a molecular reaction, while dust explosion is a surface reaction, since dust particles are several orders of magnitude larger than molecules. In dust explosions, the following processes occur:

Heat conduction near the surface leads to a rise in the surface temperature;
The fuel near the surface of dust particles decompose under heat conduction and vapor is released;
The released vapor is mixed with air to form an explosive mixture of vapors, which will explode when they meet the fire source.

The heat generated by the burning flame promotes the decomposition of dust and continuously releases flammable gases so that the flame continue to spread. It can be seen that although dust explosion is the reaction between a solid phase and a gaseous phase, it is a gas-phase explosion in the final analysis and can be regarded as the storage of combustible gas in dust itself. The temperature rise of particle surface during the explosion is a condition, and heat transfer plays an important role in the explosion process. This is also the reason why a dust explosion requires greater ignition energy than a gas explosion. The motion trajectory in the DPM model can be obtained by integrating the motion equation of particles. Based on the inertia and force balance received by the dispersed particles, the motion equation of the dispersed phase particles can be obtained (taking the x direction in the rectangular coordinate system as an example):[1]

$$\frac{du_p}{dt} = f_D\left(u - u_p\right) + \frac{g_x\left(\rho_p - \rho\right)}{\rho_p} + f_x \tag{8.6}$$

where f_x is the additional acceleration term (force per particle mass); $\frac{g_x(\rho_p-\rho)}{\rho_p}$ is the resultant force of buoyancy for a single particle; $f_D\left(u - u_p\right)$ is the resistance of particle mass per unit; u is the velocity of continuous phase and particle phase; ρ_p, ρ is the density of the continuous phase and particle phase.

For the explosion of flammable gases, flammable vapors, and combustible dusts, the calculation of explosive equivalent is indispensable in the technical analysis of accident investigation. For the calculation of explosive equivalent, the following formula can be used:

$$\Delta P = 0.84\frac{\sqrt[3]{W}}{R} + 2.7\left(\frac{\sqrt[3]{W}}{R}\right)^2 + 7\left(\frac{\sqrt[3]{W}}{R}\right)^3 \tag{8.7}$$

where ΔP is the peak overpressure of shock wave (kgf), W is the explosive TNT equivalent (kg), and R is the distance from the detonation source (m).

8.4 GAS EXPLOSION

8.4.1 Gas Fuel Characteristics

The combustion heat of flammable gas is defined as the heat released when flammable gas and oxygen undergo a complete combusting reaction. If the amount of fuel at the explosion site can be estimated, the possible different fuel combinations can be compared by the intensity of combustion. For a gas-burning fire, the heat release rate is more important. The combustion heat explains why sometimes the damage to the compartment is unrelated to the heat generated by the fuel at that time[2].

The combustion releases heat, and the heat released raises the temperature of combustion products (flue gas) to a level called the flame temperature. The maximum flame temperature is the maximum temperature that the flue gas can reach. Different flammable gases have different maximum flame temperatures. For example, the maximum flame temperature of methane is above 2273 K, and the adiabatic oxyacetylene flame temperature is about 3273 K. The natural gas burning in oxygen can reach about 3673 K. The LPG-oxygen flame temperature is close to 2273 K.

Another important characteristic of a gas fuel is its color. The flammable gases are mainly inorganic combustible gases and hydrocarbons. There are some unique colors, such as fluorine gas (light yellow-green); chlorine gas (yellow-green); nitrogen dioxide (red-brown). Compared with dust, the colors of flammable gases cover a broader spectrum.

Ash and smoke also vary in color and possess different features. Flue gas covers all substances produced in combustion or pyrolysis, and are termed combustion products. Among them, any combustion product that can be seen in the air is called smoke (exhaust gas), which is a mixture of suspended soot, liquid particles, and gas produced by combustion. Common flame colors include hydrogen burning in the air, where the flame is light blue; hydrogen burning in chlorine gives off a pale-white flame; carbon monoxide and methane render a light blue flame.

8.4.2 Gas Ignition Characteristics

Once a gaseous fuel is mixed with air in a proper proportion and encounters a proper ignition source, the mixture will burn. At this time, the flammable range of fuel in air, vapor density, ignition energy, and ignition temperature is very important. In contrast, the flash point and the boiling point are not relevant[2].

MIE is lower than that of dust. This is mainly because solid fuels need to break their chemical bonds so that hydrocarbon molecules can freely react with oxygen. Therefore, a considerable amount of the ignition energy is used to decompose dust particles into gaseous fuels. This is why the energy required to ignite flammable gases and vapors is much lower than that of combustible dust.

MIT is only relevant when the ignition volume is considered. According to the Arrhenius theory, the temperature in the "ignition volume" cannot rise above the ambient temperature only through oxidation, which generates heat slowly. However, the critical temperature can be exceeded by pilot ignition. In addition, MIT caused by the ignition source will decrease as the "ignition volume" increases. In reality, the shape of the "ignition volume" will also play a part in ignition [3].

The lower explosion limit (LEL) determines the lowest concentration of flammable gas explosion. The mixture of flammable gas and air can only burn within a certain concentration range. If the gas concentration is lower than the LEL, the fuel concentration is considered too lean and will not ignite or explode in a flame; if the fuel fraction is higher than the upper explosion limit, the gas concentration is considered too high so it will not explode, but it can burn. If fresh air is added after burning for some time to reduce the concentration of flammable gas within the explosion limit range, it may explode. Most gases have a narrow explosive range. For gases with a wide range of explosion limits (such as hydrogen, carbon monoxide, carbon disulfide, etc.), when the gas concentration is within the explosion limit, the flame can ignite and explode. In a closed system, the explosion range and flammable range are basically the same. The combustion limit is related to the temperature, pressure, oxygen content, and the diameter of the container. As the temperature, pressure, and oxygen content rise, the lower explosive limit will decrease. For common flammable gases, if the oxygen concentration in the mixture is reduced to 6%–14%, combustion or explosion will be avoided; in addition, the smaller the diameter of the container, the narrower the explosive range and the lower the risk. There are other factors at work in the non-closed system, which are more complicated.

Flammable gas explosion in a closed system is one of the main forms of industrial hazard. Many observations and studies have been carried out to investigate the propagation of the flame. The flame divides the spherical container into three zones: burned zone, burning zone, and unburned zone; the flame in a cylindrical container is more complicated. At the beginning, the flame spreads outward from the center. As the distance of propagation increases, the flame is deformed due to cooling and friction near the wall, and it still spreads outward from the wall. When there is an obstacle in the container, the flame propagation changes greatly. As the flame approaches the obstacle, the gas flow at the lower part of the flame front is blocked, making the flame lag behind, while the upper part of the flame will accelerate the gas flow due to the smaller flow area. As the flame front crosses the obstacle, the flame front is deformed. Once it has passed the obstacle, it deforms sharply, and the burning rate is significantly increased[1].

8.4.3 TECHNICAL ASSESSMENT

The incomplete combustion products produced by the combustion of materials are combustible. Therefore, when gases, vapors, dusts, and other incomplete combustion products mix with air and encounter an ignition source, they may cause an explosion. In the accident investigation analysis, this type of accident can be analyzed based on the different combustion products. Some burning gases will show a certain color. For example, flames of hydrogen, carbon monoxide, and methane are light blue when burning in air; if the hydrogen is burning in chlorine, the flame is pale. In the case of on-site video recording, the color of the burn/explosion can be used to identify the potential fuel of the accident. Furthermore, because the MIE of the gas is lower than that of the dust, the temperature range of the on-site process can be compared with the MIE of the gas and the MIE of the dust to predict the type of accident. Accordingly, the above-mentioned MIT and LEL values can be compared with the temperature and concentration values on site. If the temperature and concentration values are within the explosive range, the accident type is determined to be a gas explosion.

Under normal circumstances, it is necessary to carry out on-site sampling and testing to identify elements contained in the combustion products, and infer the possible ignition source based on the evidence from the accident site. Generally, Gas Chromatography and Mass Spectroscopy (GCMS) is relied on to detect the species for forensic analysis. Gas chromatography produces different reactions with the solid phase in the chromatographic columns according to the different characteristics of the gas. The mobile phase is taken out of the chromatographic column at different rates, and then captured by the detector. The final output is an electrical signal. The peak value of the chromatogram can be used to determine which substance the detected gas is. At the same time, gas chromatography can determine the content of macromolecular gas products in explosives. In order to establish whether the cause of the explosion is the burning of grease or organic matter, infrared spectroscopy can effectively detect whether the product after the explosion contains organic functional groups, and then the specific type of the substance can be determined by laboratory analysis.

Secondly, by investigating the accident site, checking the damage of the explosion (such as building walls, glass, and other major markers, as well as the distance between these markers and the accident center), the damage degree and damage level can be confirmed, which can then be used to estimate the accident explosion equivalent TNT level. The samples retrieved from the site are analyzed in laboratories, including the raw materials and intermediates used during the production process, and products found after the explosion. GCMS is then applied for qualitative analysis to determine the elements or functional groups present in the products through the obtained spectrum. The type of fuel involved in the ignition can be inferred from the results.

In order to avoid the same accident, it is necessary to simulate different accident scenarios, carry out a reconstruction of the accident, and put forward opinions and suggestions on how to avoid the same type of incident.

Case Study: An oxygen cylinder explosion accident:

An oxygen cylinder exploded when an operator was cutting steel plates in a mold company, and he died on the spot. According to witness records on the scene, the worker was changing the gas cylinder at the time of the incident, and a large amount of oily and sandy mixed substances were found at the bottom of the cylinder after the explosion, while the use of the oxygen cylinder is not allowed to appear in the cylinder.

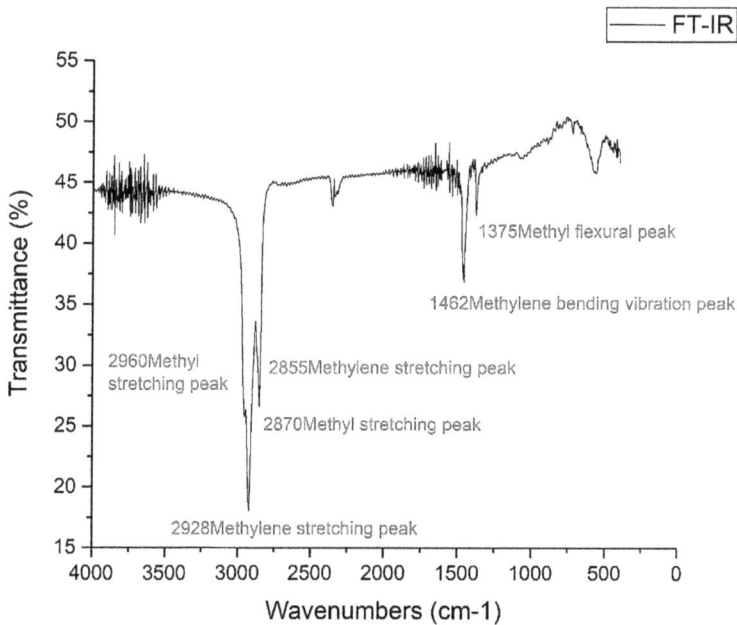

FIGURE 8.1   Infrared spectrum of steel residue samples.

It was established that the explosion occurred in the process of opening the cylinder, so the possibility of combustible gas flashback was ruled out. By sampling and analyzing using laboratory infrared spectroscopy, a spectrum was obtained, as shown in Figure 8.1 It was determined through functional group comparison that the medium was a type of grease, and it was therefore preliminarily concluded that the primary ignition fuel was likely grease. The on-site investigation found that the fixing measures of the liquefied gas cylinders were insufficient and they were thrown 10 meters away. The valve of the liquefied gas cylinder was damaged, while the overall cylindrical structure was complete, and fragments were scattered on the road 50 meters away.

An analysis of the scattered fragments (Figure 8.2) revealed that the cylinder wall thickness was not significantly thinner and no significant local plastic deformation (bulge phenomenon) was observed. Thus, the explosion fragments were characterized as plastic explosive, which is typical for a chemical explosion. The above laboratory analysis is summarized in the steps outlined below. The collected samples were first analyzed by infrared spectroscopy. Then the sample taken at the bottom of the oxygen cylinder contained only CH bonds, which may have been the product of the explosion of esters. Thus, the ignition medium was confirmed to be grease. Turning on the oxygen valve sharply creates frictional heat, which could ignite any oily substance in the cylinder, causing a chemical explosion. A BOW-TIE model for the entire event was established through scenario analysis. In contrast to traditional risk identification methods, this model is

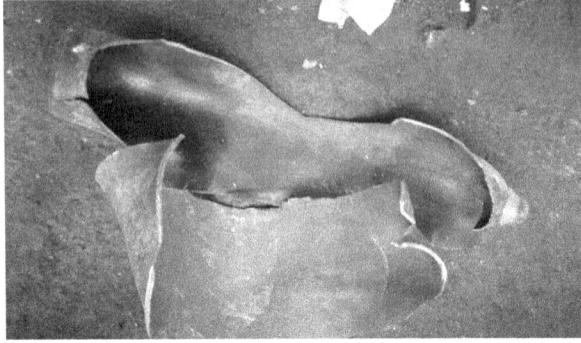

FIGURE 8.2   Photo of oxygen cylinder fragments after explosion.

established for the entire event through scenario analysis. For an oxygen cylinder explosion, it is vitally important to eliminate any possibility of explosion caused by the burning of grease and the possibility of backfire.

As far as this incident is concerned, the grease-burning accident caused by frictional heat was the result of the simultaneous presence of grease, an ignition medium, and a high-heat ignition source. The safety of the gas cylinder was not checked regularly, the gas cylinder and the grease-containing substance were stored together, the accessories of the gas cylinder were contaminated by grease, there was grease present on the personnel in contact with the gas cylinder, the gas cylinder and the grease substance were transported at the same time. Any of these five situations alone, or several of them, may have resulted in grease being present in the cylinder. At the same time, when the oxygen content of the gas cylinder is small, it will cause a small-scale fire, but when the oxygen content of the gas cylinder is large, an explosion may occur. The specific BOW-TIE model diagram is shown in Figure 8.3.

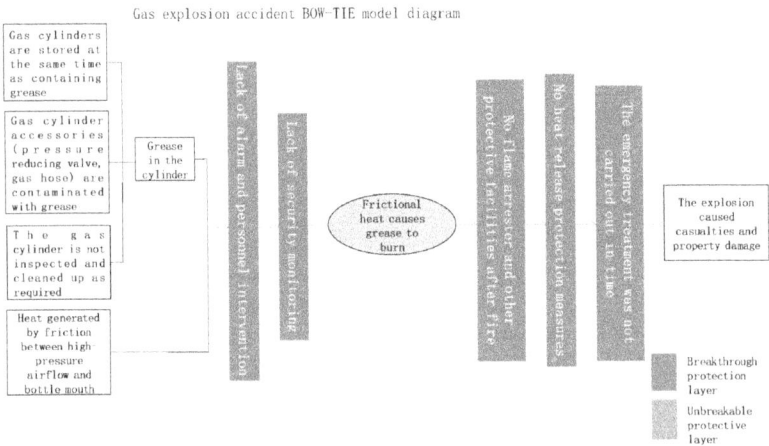

FIGURE 8.3   BOW-TIE model diagram for the oxygen cylinder explosion accident.

8.5 LIQUID VAPOR EXPLOSION

8.5.1 FUEL VAPOR CHARACTERISTICS

The heat of combustion can be used to calculate the theoretical adiabatic flame temperature of a vapor explosion, and to estimate the explosive equivalent if the mass of the fuel involved in the explosion can be roughly estimated. The burning speed of flammable liquid can be used to indicate the speed of fuel burning, and its speed is related to the fuel type, concentration, pressure, temperature, and other conditions. In general, the more carbon, hydrogen, sulfur, phosphorus, and other combustible elements in a fuel, the faster it burns. In addition, the starting temperature of the fuel, the heat of combustion, the concentration of inert gas, and so on, also affect the combustion rate of the fuel.

Vapor pressure is the characteristic that differentiates vapor fuels from other fuels. Vapor pressure is the pressure in space created by the evaporation of molecules from the surface of a liquid into the air, if evaporation takes place in a closed system and at a certain temperature, when enough steam is vaporized into the air, it will reach an equilibrium with the liquid. At this point, the pressure of the gas in space generated by the vapor is termed the saturated vapor pressure. The higher the saturated vapor pressure of the flammable liquid, the more readily the substance will evaporate and thus reach the explosive limit, possibly leading to an explosion. For example, at 308 K, the vapor pressure of acetone is 30.67 kPa, while that of benzene is 12.69 kPa, and that of ethanol is 7.51 kPa. Therefore, acetone has the highest risk, followed by benzene, and ethanol has the lowest risk.

The closer the vapor density is to the air density, the more significant the mixing effect is. Vapor with a high vapor density will sink downward when it leaks, and will spread horizontally like a liquid when it meets the ground, while vapor with a low vapor density will rise upward when it leaks, and will accumulate in the high plane like smoke when it meets the roof. Therefore, vapors and air mixtures have various concentration layers, some of which may be within the explosive (combustion) limit and are easily ignited[2]. For example, the vapor density of methane is 0.55, the vapor density of methanol is 1.11, and the vapor density of pentane is 2.48[4]. Therefore, methanol leakage will mix with the air more evenly, while methane and pentane leakage will form aggregation on the roof and the ground, thus producing a stratification phenomenon (if there is no external disturbance).

8.5.2 VAPOR IGNITION PROPAGATION CHARACTERISTICS

Flash point can be used to evaluate the fire risk of combustible liquids. The flash point of a flammable liquid is the lowest temperature at which the liquid produces flammable vapor. This does not mean that the vapor can spontaneously ignite at the flash point temperature, but rather that it can be ignited by a flame, a small arc, or some other local heat source.

The ignition point is the lowest temperature at which a continuous flame can be produced continuously on the surface of a liquid, as opposed to the point at which instantaneous flashover can be produced. Usually, it is several degrees above the rated flash point. Ignition points may be more useful than flash points in assessing the impact of combustible liquids on the fire environment.

Ignition temperature is very important when considering combustible liquid fire. It refers to the temperature at which a flammable liquid is ignited without the influence of any external ignition source. In fact, all fires start locally, where the temperature is high and the steam/air ratio is appropriate. This area may be very small, but importantly, the temperature at this very small point in space exceeds the ignition temperature, and it is not difficult to achieve such high temperatures locally, such as with friction, impact, etc., which produce small but very hot sparks capable of triggering an explosion.

Each flammable vapor has MIE. It is the energy that must be transferred to the fuel to initiate its initial REDOX reaction. MIE is very low for most flammable vapors, such as hydrocarbon vapors with MIE of 0.25 mJ. The ignition energy depends on the concentration of the vapor, and is lowest when the vapor concentration is stoichiometric or ideal.

The ignition temperature and explosion limit of flammable vapor can be measured using a flammable vapor explosion limit tester. An explosion limit tester can only be used to measure and evaluate the reaction characteristics of substances during heating and combustion under laboratory conditions. The test results cannot be directly used to describe fire risk or the fire risk of substances under real ignition conditions, but it can be used as an important reference factor for fire risk assessment. Under different environmental conditions, the combustible liquid vapor explosion limit value also varies.

For example, Shanghai Institute of Technology's Institute of Industrial Safety and Explosion Protection has studied the explosion characteristics of methanol/air mixtures with different oxygen content at an initial pressure of 0.3 MPa–0.75 MPa and an initial temperature of 423 K. The upper explosion limit increases as the initial pressure rises (as shown in Figure 8.4), and according to the model

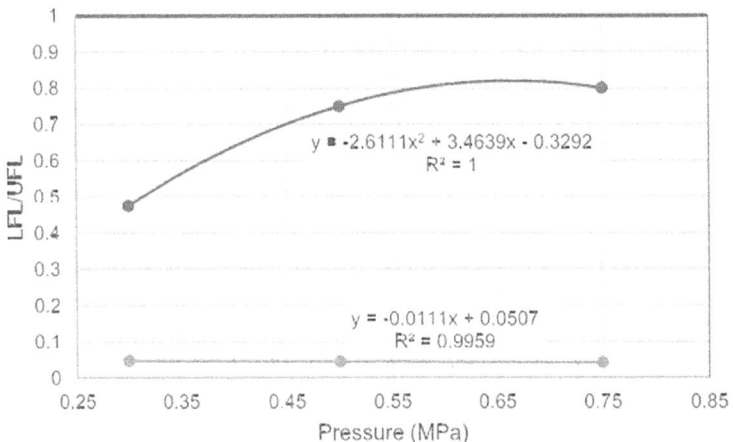

FIGURE 8.4   Variation of explosion limit with pressure.

prediction and experimental research on the explosion limit under different oxygen contents,[5] a relationship between the initial pressure of dust and the explosion limit was found. For different types of dust, the explosion limit changes with the initial pressure, which plays a key role after the dust explosion, based on the results of the technical analysis.

8.5.2.1   Accident Technical Appraisal

In the accident technical appraisal of a vapor gas explosion, the explosive chemical medium must be determined first. Gas chromatography is usually used to analyze the residual vapor in the explosion. After the vapor explosion, there may be unignited fuel and incomplete combustion products. Analyses of these residual substances, combined with the on-site data, can essentially determine the explosive medium. Another aspect of the analysis that provides valuable information is flame color. Some materials have a special flame color. Ethanol and methane, for instance, produce a pale blue flame when burned in air, sulfur burned in an oxygen flame is blue-purple, and hydrogen in a chlorine flame is a pale color. A recording of the scene of an explosion by an eyewitness or live video can definitively exclude certain substances and can sometimes be used as an auxiliary means to determine the explosive medium. After determining the explosive medium, it is necessary to find the ignition source. Common ignition sources of steam explosion include open fire, static electricity, metal impact spark, and electrical spark, among other causes. By obtaining a detailed understanding of the process and site situation, and applying the elimination method, it is possible to determine the ignition source.

When describing a fire scene, the damage can be determined indirectly by observing the damage at the explosion site, such as the distance of fragments from the explosion, the breaking and melting of glass, the propagation of shock waves, and injuries sustained by personnel. If it is necessary to know the reaction rate and power, the heat release rate, heat release, and explosive equivalent of the fuel should be determined. In the laboratory, a cone calorimeter can be used to determine the heat release rate and total heat release of combustible liquid fuels. Using the measurements of the heat of combustion, ignition time, and so on, the explosion equivalent can be calculated with Equation (8.8):

$$W_{TNT} = \alpha \cdot W \cdot Q_V / Q_{TNT} \qquad (8.8)$$

where α is the steam cloud explosion efficiency factor, generally 3% or 4%; W is the material mass; Q_V is the heat of combustion; and Q_{TNT} is the explosion heat of *TNT*, generally $4.52 \times 10^6 J/kg$.

Due to incomplete combustion of the explosive medium, the laboratory identification and analysis of the residues at the explosion site may further facilitate identification of the explosive medium. For example, if it is initially determined that the explosion medium is methane, and then excess CO and CO_2 are identified in the explosion tail gas, it can be further determined that the explosion medium is alkane vapor. The large amount of CO in the explosion tail gas is one of the characteristic products of alkane vapor explosion.

After determining the explosive medium and ignition source, it is necessary to determine the cause of the accident and establish how the explosive medium appeared (leakage and evaporation). Usually the leakage of flammable liquid is caused by metallographic corrosion of safety accessories, containers, etc. The ignition source needs to be elucidated. Next, the accident scene should be reconstructed, followed by a systematic analysis. Finally, targeted preventive measures based on the scene restoration of the combustible liquid fire accident should be proposed.

8.5.3 CASE ANALYSIS

A company's EPS foam production workshop had three fire incidents in the preexploitation process (two of the three fire points were located in the conveying pipeline and one was located at the silo inlet), resulting in varying degrees of property damage (Fig 8.5). The company's foaming process uses intermittent prefoaming methods, which can be divided into several processes such as feeding, foaming, and maturation. The foaming process requires steam at about 150°C to heat the EPS particles. After heating, the EPS particles are softened, and the volatile agent (pentane) contained in it escapes and expands. The foamed EPS particles then pass through a pre-expander and a fluidized bed. The fan blows particles along the conveying pipeline to the silo, where they are allowed to stand for 2 hours for maturation treatment. The matured particles enter the weighing silo and enter the next process. After verification and communication on site, two raw materials are used for the intermediate pre-foaming process, the basic composition of which is polystyrene, though other components are also present. Black EPS particles (hereinafter referred to as black particles) can be divided into polystyrene (90%–95%), pentane (4%–7%), hexabromocyclododecane (<1%), and carbide

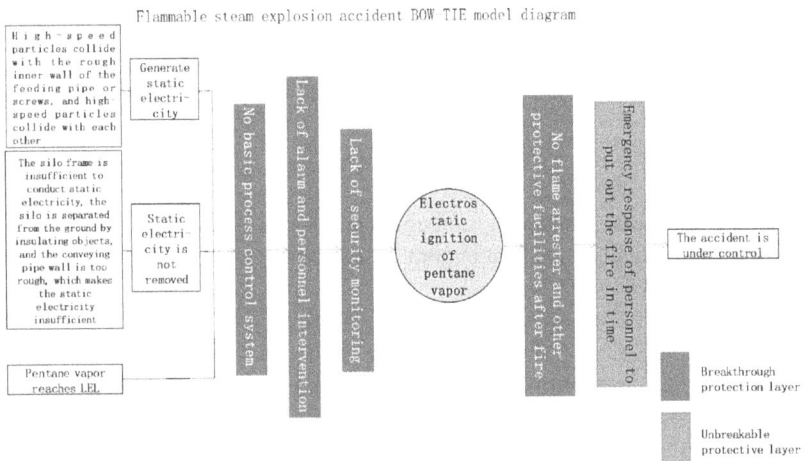

FIGURE 8.5   BOW-TIE model of pentane vapor explosion accident.

(0.01%–5%). White EPS particles (hereinafter referred to as white particles) can be divided into polystyrene (93%–97%) and pentane (5%–7%).

In this case, the media that may cause ignition include pipes, silo mesh, polystyrene foam, and pentane gas. After the field investigation and testing, the pipe and the silo mesh could be eliminated as the initial ignition medium (the pipe is made of metal, which can exclude the initial ignition). According to the specific location of the three fires and the description of the three accidents by employees who were present, the silo mesh was the first potential medium to be excluded. Regarding the possible causes of the fire, the three accidents were not considered to be caused by initial ignition of other media or fuel bunker net cloth, so the possible initial ignition medium was determined to be polystyrene particles or pentane gas. In order to confirm the ignition medium, two dust layer ignition temperature tests were carried out for EPS particles in the laboratory, and the test temperatures were 423 K and 473 K, respectively. The test results were as follows: the EPS particles were heated to 423 K, resulting in foaming accompanied by pungent gas (pentane). No combustion was observed. After cooling, the particles agglomerated and hardened. The EPS particles were heated to 473 K, and molten liquid appeared, accompanied by pungent gas (pentane). No combustion was observed, and the particles agglomerated and hardened after cooling. Polystyrene particles could thus be excluded as the ignition medium. As mentioned above, EPS particles contain pentane, a foaming agent. During the process of particle foaming, pentane is vaporized from the foam hole, forming a surface pentane vapor. During the ignition temperature test of the EPS particle dust layer, the concentration of volatile gas on the particle surface was also tested. During the test, a box cover was installed on the outside of the test device, and the test point was located at the four corners of the box body. The test results in the first 5 minutes of the experiment and the first 10 minutes of the experiment are shown in Tables 8.4 and 8.5.

After the equipment was in a stable state for a period of time, the concentration of pentane gas was measured at the exhaust hood of the fluidized bed and the feed

TABLE 8.4
Test Results of the First 5 Minutes of Heating

Test point	1	2	3	4
Peak value (ppm)	2230	2690	1500	1880

TABLE 8.5
Test Results of the First 10 Minutes of Heating

Test point	1	2	3	4
Peak value (ppm)	2390	2910	3180	2980

inlet at the top of the silo. The field test results showed that the short-term pentane gas concentration at the exhaust hood of the fluidized bed and the feed inlet of the silo could reach more than 5000 ppm. Pentane is a class A flammable liquid with a boiling point of 309.1 K, a flash point of 225 K, and an explosion limit of 1.4%–7.8%. MIE is 0.49 mJ when the concentration of pentane reaches 2.55%. The lower limit of explosion is about 14000 ppm, and the lower limit of combustion is slightly lower than the lower limit of explosion. Therefore, the concentration of pentane gas in a confined space can reach the lower limit of combustion and explosion in an instant. In addition, according to the accident record, the flame was observed to be light blue at the time of the fire. A light blue flame is typically produced in the combustion of alkane substances. Therefore, it can be inferred that the ignition medium was the pentane gas volatized from EPS particles.

The company's three fire incidents were analyzed by on-site detection and investigation in order to eliminate the possible causes of accidents, such as hot surface, flame and hot gas, sparks generated by machinery, electrical equipment, stray current, lightning, electromagnetic wave, ionizing radiation, ultrasonic wave, and shock wave. The possible ignition sources were chemical reaction and static electricity. As mentioned above, EPS particles turn into white foam pellets after pre-foaming, and their properties are stable. Secondly, the pentane gas in the air delivery pipeline with a single medium is relatively stable and is not prone to decomposition. Therefore, it can be concluded that the ignition source was a non-chemical exothermic reaction. Static electricity was the most likely ignition source for three reasons. First, EPS particles that have been pre-bubbled are blown under the positive pressure of the fan. Collisions and friction occur between the foaming balls as well as between the foaming balls and the conveying pipeline, which tend to generate static electricity. This accords with the condition of electrostatic generation of powder; second powder electrostatic electrification and its resistivity (rho) are as follows: 1) the rho < 108 $\Omega \cdot m$, don't consider the electricity; 2) 108 < rho < 1012 $\Omega \cdot m$, the possibility of causing current is small, requires grounding; 3) the rho > 1012 $\Omega \cdot m$, electrification, requires special protective measures. The resistivity of polystyrene powder for 1017–1019 $\Omega \cdot m$, for high insulating material, readily produces static electricity, though it is not readily released. In addition, the three fire points had a large resistance, which intensified collisions. Static electricity can be easily produced by a metal nail in the pipe (which can produce tip discharge), a fast flow rate of feed inlet, increased collision frequency, and other problems. Therefore, the ignition source was determined to be static electricity.

As pentane gas volatiles in the transport pipeline were ejected out of the feed port after electrostatic ignition and deflagration, two of the three accidents burned out the silo, causing certain economic losses. However, in one of the accidents, due to the small amount of pentane vapor, the accident damage was relatively light.

The laboratory's reconstruction of the accident process revealed that the cause of the fire was the concentration of pentane gas reaching the lower explosive limit within a short time, and the pentane gas was ignited by static electricity, which in turn caused other substances to ignite, leading to intensification of the flame combustion.

The high-speed particles collided with the inner wall of the feeding pipe and the screws in the pipe to generate static electricity, and the static electricity of the transportation pipeline and the silo was not discharged in time, which resulted in accumulation of static electricity. Also, pentane vapor was generated during the EPS particle foaming process. When the lower explosive limit was reached, the combination of the two caused static electricity to ignite the pentane vapor. If the operator discovers and extinguishes the fire in time, only the materials in a small area will catch fire. If the fire is not detected and extinguished in time, it will cause a fire in the workshop or even cause the pentane vapor cloud to explode.

8.6 DUST EXPLOSION

8.6.1 Characteristics of Dust Fuel

The intensity of a dust explosion and its consequences largely depend on the fuel characteristics of the dust involved in the explosion reaction. The higher the combustion heat of dust, the lower the lower limit of its explosive concentration. Once an explosion occurs, it will be high temperature and high pressure, and the explosive power will be great, such as occurs in explosions caused by coal powder, carbon powder, sulfur powder, and so on.

The particle size of the dust is closely related to the specific surface area. The finer the particle size, the greater the specific surface area of the dust, the greater the degree of dispersion in the air, the longer the suspension time, the lower the explosion limit, and the greater the explosion risk. However, when the particle size is lower than a certain value, as the particle size decreases, the LEL is basically unchanged, and sometimes increases instead, such as occurs with magnesium powder. Generally, when the particle size of combustible dust is larger than 400 μm, an explosion is less likely to occur[1].

Dust moisture content is the percentage of moisture that can be dried and removed on the surface of the dust, in the pores and capillaries. The lower the moisture content of the dust, the more likely it is to explode. When the dust humidity exceeds 30%, it will not readily detonate. Dust explosion is also related to the volatile matter contained in it.

8.6.2 Ignition Propagation Characteristics of Dust

Dust explosion is a multiphase deflagration process, which is much more complex than gas explosion. Many factors affect the possibility of dust explosion and the severity of consequences after ignition. Smoldering is an ignition propagation characteristic of dust that is different from combustible gas and combustible liquid. Generally, smoldering may occur on the surface of the dust layer with no air circulation or exposure to external heating flow. The key features of the smoldering process of dust combustion are the porous characteristics of dust particles, such as particle size, morphology, particle-to-particle porosity, particle humidity, air humidity, particle group, and particle layer.

The limit oxygen concentration of a dust cloud refers to the minimum oxygen content required when a dust cloud explodes[6], and its impact on the dust explosion limit can be ascertained by filling in different volume ratios of nitrogen to adjust the oxygen. Generally speaking, under normal temperature and pressure, when the oxygen volume fraction is lower than 8%, the organic dust/air mixture will not explode[7].

MIE refers to the minimum spark energy that can cause the dust cloud to burn or explode. Compared with gas explosion and combustible vapor explosion, MIE required for dust explosion is higher, usually above tens of millijoules, and ignition of hot surfaces occurs less readily[6]. MIT of dust comprises two parts: the minimum ignition temperature of the dust layer (MITL) and the minimum ignition temperature of the dust cloud (MITC). The larger the dust particle size and the higher the moisture content, the higher the MIT of the dust, and the lower the risk of fire and explosion of this kind of dust[8]. Generally, the MITL of white powder on wheat is 553 K and the MITL of brown coal powder is 503 K. If the measured MITL of dust is lower than this temperature, the lowest ignition temperature of the dust therefore belongs at the lower end of the scale[9]. The LLEL of a dust cloud is the lowest concentration of a dust cloud in which the deflagration flame can continue to spread in the dust cloud under the action of a given energy ignition source. The lower the LEL of dust cloud, the greater the risk of explosion[10]. The LEL of general industrial dust is between 15 and 60 g/m^3, and the upper explosion limit is 100 times greater than its lower limit. The extremely wide range of dust explosion is also one of the important differences between gas explosion and combustible vapor[11].

The maximum explosion pressure is the maximum value of the explosion pressure of a certain dust determined by a series of tests under various dust concentrations. The maximum pressure rise rate is the maximum slope of the explosion pressure curve with time. In order to obtain the maximum explosion pressure and the maximum explosion pressure rise rate of a certain kind of dust, the concentration of the test dust needs to be changed, generally between 300 and 2500 g/m^3. In the range of lower dust concentration, the maximum explosion pressure and maximum pressure rise rate of dust rises rapidly with the increase of dust concentration. After exceeding a certain critical concentration, the maximum explosion pressure and maximum pressure rise rate are maintained at a higher value in a wider concentration range[11]. For dust, the explosion pressure is about 0.8–1 MPa.

8.6.3 TECHNICAL APPRAISAL

In the technical appraisal of combustion and explosion, first of all, the type of industrial media should be determined. Depending on the production process of the factory involved in the accident and the remaining substances on site, combined with the on-site traces formed after the explosion, investigators should determine whether the explosive substance is a combustible dust. Generally, there is no obvious explosion center (point) at the explosion fire site caused by combustible dust explosion. There are fewer ejections, large chunks, unevenness, and a short ejection distance. This is similar to the scenario that is generated after the explosion of combustible mixed gas, and therefore it is important to make a careful differentiation between the two scenarios [12]. The focus should be on the flame color (for example, when sulfur powder burns blue) and the number of explosions (dust explosion may cause a second explosion). The type and source of the combustible dust should be established. Secondly, the extent of site damage needs to be assessed, which includes collection of data related to the location, area, range of the explosion, the longest distance of explosive projectile, and collapse of the building. Then, laboratory identification and analysis on the residual dust on site should be performed, as well as analysis of the moisture content, limiting oxygen concentration, and explosive characteristics of the combustible dust involved in the accident (MIE, MIT, LEL, maximum explosion pressure, and rise rate), to provide evidence of the cause of the explosion. Having established the cause of the accident and obtained the results of the accident site survey, based on laboratory thermal safety equipment, combined with the dynamic behavior of dust particles, dust particles clear indicators of the characteristics and impact objectives. Based on the theory of the "particle element method", professional modeling tools such as FLACS, a computational fluid dynamics (CFD) tool, can be used to simulate the influence of various factors, such as moisture content, particle size, ambient temperature, humidity, and porosity on the characteristics of dust cake accumulation. Such tools can be used to form rules, "reverse-engineer" the accident process, reconstruct the real fire/explosion scene, and systematically analyze the disaster mechanism of the accident. Finally, preventive measures and opinions on safety improvement should be put forward to mitigate future risk of dust explosion accidents.

8.6.4 Accident Case

A dust explosion accident occurred in a pharmaceutical factory during the powder mixing operation, resulting in two deaths and one serious injury. The equipment was also seriously damaged. The main raw materials of the plant are potassium clavulanate and microcrystalline cellulose. Thus, flammable and explosive substances may be mixed during the powder mixing operation of the end process. Potassium clavulanate is also termed clavulanic acid. Its molecular formula is $C_8H_8KNO_5$ and it has a molecular weight of 237.25. It is a β-lactamase inhibitor. In addition, it is a white or slightly yellow crystalline powder, slightly odorous, and readily attracts moisture. Potassium clavulanate is easily soluble in methanol, slightly soluble in ethanol, and insoluble in ether. Microcrystalline cellulose is a free-flowing crystalline powder (non-fibrous particles), insoluble in water, dilute acids and most organic solvents, and slightly soluble in 20% alkali solutions. In the production process, potassium clavulanate is obtained through biological fermentation and then sieved through a 40-mesh filter screen. After sieving, it is mixed with microcrystalline cellulose in a rotary double-cone mixer at a speed of 25 r/min. The powder charge is 200–500 kg. According to the analysis of the main process conditions and the analysis of residual substances at the accident site, the accident was identified as a dust explosion caused by mixed dust

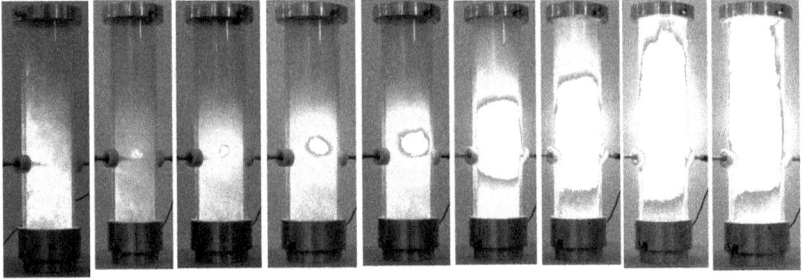

FIGURE 8.6 Comparison of high-speed photography images of flame propagation of florfenicol in drug dust.

composed of potassium clavulanate and microcrystalline cellulose (hereinafter referred to as mixed dust)[9,13,14]. Based on preliminary research and laboratory analysis, the flame propagation characteristics and thermal safety parameters of different drug dusts vary greatly, as shown in Figures 8.6 and 8.7[15]. Therefore, for different types of drug dust, the complete thermal safety parameters must be obtained to gain a comprehensive understanding of the accident catastrophe mechanism. For potassium clavulanate dust, an electric drying oven and an analytical balance were used to determine that the moisture content of the mixed dust was 1.6%. Experiments with different volume ratios of nitrogen to adjust the oxygen revealed that the maximum allowable oxygen content required to prevent the mixed dust from igniting was 14.6%; the lowest ignition temperature of the mixed dust layer measured by the MITL tester was 473 K; the lowest ignition temperature of the mixed dust cloud measured by the Godbert-Greenwald furnace was 643 K; the 20 L spherical explosion test device determined that the LEL of the mixed dust was 55 g/m^3, the maximum explosion pressure was 0.22 MPa, the maximum pressure explosion rise rate was 7.64 MPa/s, the explosion index was 2.07, and the dust explosion intensity level was St 1; the maximum explosion pressure varied with dust concentration, as shown in Figure 8.8.

Using a differential thermal scanning thermal analyzer, the fusion endothermic peak of the mixed dust was about 462 K and the heat release was 633 J/g,

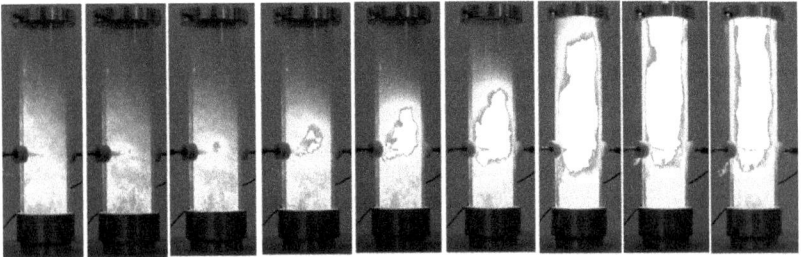

FIGURE 8.7 Comparison of high-speed photographic images of flame propagation of drug dust tilmicosin.

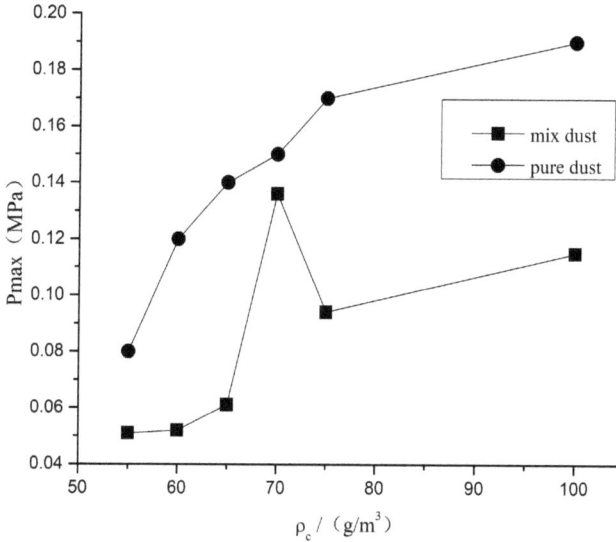

FIGURE 8.8   Trend chart of maximum explosion pressure with dust concentration.

indicating a high level of explosion risk. According to laboratory research, the thermal dynamic behavior of different drug dusts during the heating process varies considerably. For example, the differential thermal scanning analysis of florfenicol and tilmicosin is displayed in Figure 8.9[16].

The mixed dust was analyzed using capillary gas chromatography and it was found that the mixture contained 93 ppm acetone gas. In addition, based on the on-site analysis of the ignition source, static electricity was identified as the primary ignition source for fire and explosion in the process of mixing potassium clavulanate dust. The BOW-TIE model was used to analyze the accident. Static electricity was determined to be the main cause of the entire accident. The manufacturing equipment involved in the accident did not employ monitoring measures

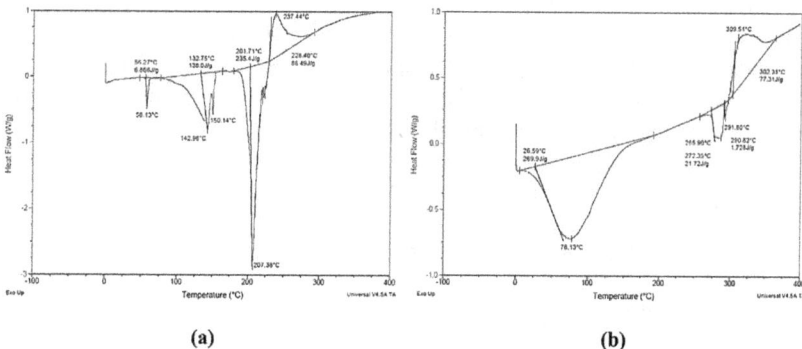

(a)   **(b)**

FIGURE 8.9   DSC thermograms of florfenicol (a) and tilmicosin (b) dust.

Dust explosion accident BOW-
TIE model diagram

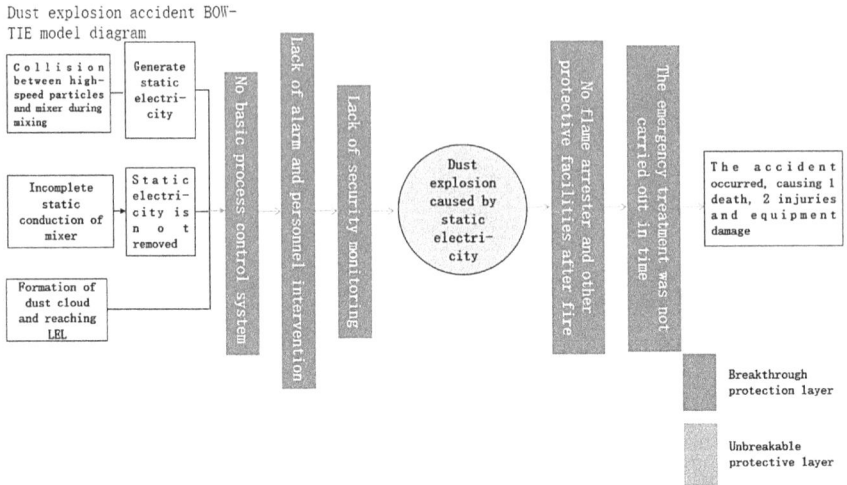

FIGURE 8.10   BOW-TIE model of dust explosion accident.

such as static electricity removal and process control, which indirectly led to further explosion accidents. After the accident, the correct emergency measures were not taken, which led to a large accident loss. Detailed results are shown in Figure 8.10.

For the prevention of dust explosion accidents at industrial sites the following prevention and control measures should be implemented. First, dust humidity should be increased as dust clumps do not disperse readily and the water vapor can absorb heat to prevent the dust from catching fire. At the same time, it is important to pay attention to the product quality. Second, inert gas should be applied to reduce the oxygen content of the dust, so that the organic dust/air mixture will not explode. Third, the heating surfaces, bearings, and rotating parts of the workshop machines need to be regularly checked to prevent the temperature from reaching the lowest ignition temperature of the dust layer. Fourth, to avoid dust being raised during the process of mixing powder, grinding, and packaging, particular attention should be paid to the greater impact of the release pressure of the packaging on the lowest ignition temperature of the dust cloud. Fifth, mixing of combustible gas or vapor in the dust/air mixture from the source of the process should be avoided to reduce the possibility of lower explosion limits.

REFERENCES

1. Mingshu Bi, Gang Li, Xianfeng Chen, Guogang Yang. Gas and dust explosion prevention engineering. Chemical Industry Press, China, 2017.
2. John D. Dehaan, Aiping Chen, Xiaonan Xu. Kirk fire investigation. Chemical Industry Press, China, 2006.
3. Rolf K E. Explosion hazards in the process industries. University of Bergen, Norway, 2016.

4. Weifan Zhang. Safety manual of common dangerous chemicals. Chemical Industry Press, China, 1994.
5. Zhang Xiaoliang, Li Hao, Yang Luying, Ding Dongmei, Ma Tingguang. The explosion parameters of methanol under variable pressures and 423. Journal of Loss Prevention in the Process Industries, 2020, 64(02):104079.
6. Weiguo Cao, Junjie Zheng, Yuhuai Peng, Feng Pan. Experimental study on explosion characteristics and flame propagation process of corn starch dust. Explosive Materials, 2016, 45(01):1–6.
7. Zhaolin Gu. Wind blowing dust: near-surface turbulence and gas-solid two-phase flow. Science Press, China, 2010.
8. Gang Li, Xiaoyan Liu, Shengjun Zhong, Junxiang Dang. Experimental study on the lowest ignition temperature of grain-associated dust. Journal of Northeastern University, 2005(02):145–147.
9. Xiaoliang Zhang, Henggen Shen, Peihui Zhao. Determination of explosion parameters of potassium clavulanate and microcrystalline cellulose mixed powder. Fire Science and Technology, 2009, 28(11):800–802.
10. Congzhang Zhou, Ruiping Zhang, Yongfang Yu. Discussion on the lower limit concentration of dust cloud explosion. China Safety Science Journal, 1995(03):39–42.
11. Yige Wang. Research on Maximum Explosion Pressure and Maximum Pressure Rising Rate in Dust Explosion. Fire Science and Technology, 2011, 30(03):193–195.
12. Jincheng Yan. On-site investigation and cause identification of combustible dust explosion fire–Also on the art of investigation and inquiry. Fire Science and Technology, 1995(03):40–42.
13. Xiaoliang Zhang, Henggen Shen, Peihui Zhao, Shufen Chen. Study on the explosion accident of drug mixing powder and preventive measures. China Safety Science Journal, 2010, 20(01):67–71+181.
14. Xiaoliang Zhang. Research on the explosive characteristics of $C_8H_8KNO_5$ (potassium clavulanate) mixed dust in the transportation pipeline in the pharmaceutical industry. Donghua University, China, 2012.
15. Qian Shen. Study on the explosive characteristics of florfenicol and tilmicosin drug dust. Shanghai Institute of Technology, China, 2016.
16. Xiaoliang Zhang, Qian Shen, Xiaobo Shen, Zhikai Zhang, Sunjie Xu, Shengjun Ye. Minimum ignition energy of medicinal powder – Florfenicol and Tilmicosin. Journal of Loss Prevention in the Process Industries, 2016, 39; 30–38.

9 Risk Assessment of Hydrogen Fueling Stations

Jo Nakayama and Atsumi Miyake
Institute of Advanced Sciences/Center for
Creation of Symbiosis Society with Risk, Yokohama
National University, Yokohama, Japan

CONTENTS

9.1 INTRODUCTION

In the quest for realizing sustainable and energy-efficient societies, hydrogen is a promising zero CO_2 emission fuel for a variety of vehicles such as cars, buses, trucks, motorbikes, and even ships. This environmentally-friendly characteristic of hydrogen can also significantly contribute to reductions in the emission of harmful fossil fuel-based pollutants such as hydrocarbons, NOX, and SOX. Importantly, hydrogen also has tremendous value as a renewable energy source. Hydrogen can be produced via electrolysis through energy generated from wind, solar power, or hydropower sources. In this regard, the cycle between hydrogen production and its use is ideal for sustainability. However, the practical use of hydrogen is impeded by the issues of regulation, cost, and technology. Consequently, developed countries, particularly Germany, the USA, and Japan, have attempted to solve the various problems underlying the establishment of hydrogen infrastructure. In this context, the Japanese government has invested significant efforts in the building of hydrogen infrastructure such as hydrogen fueling stations. In March 2019, the Japanese government outlined a strategic road map for the implementation of hydrogen and fuel cells toward realizing a "hydrogen society" (METI, 2020). The road map lays out guidelines to reduce the costs of fuel cells, hydrogen fueling stations, and hydrogen-fueled buses by 2025. In particular, as essential infrastructure for the commercialization of fuel cell vehicles (FCVs) and buses, the guidelines recommend that the government should increase the number of hydrogen fueling stations to 320 by 2025 and to 900 by 2030.

Currently, the storage, transportation, and supply of hydrogen are the biggest challenges toward realizing a hydrogen economy; it is not easy to safely and effectively treat large amounts of gaseous/liquified hydrogen owing to the various hazards associated with the gas. This has led to extensive research on hydrogen hazards such as flammability and metal-embrittlement characteristics in order to develop safer and low-cost hydrogen infrastructure. Against this backdrop, in this chapter we provide an overview of the risk assessment of hydrogen fueling stations. Firstly, we discuss the various types of hydrogen fueling stations. Secondly, we consider certain related information databases to focus on the lessons learned from incidents and accidents related to hydrogen fueling stations in Japan and the USA. Thirdly, we consider the qualitative risk assessment of hydrogen fueling stations. Finally, we present the results of a simulation-based safety investigation of hydrogen fueling stations.

9.2 HYDROGEN FUELING STATIONS

There are currently several designs of commercialized hydrogen fueling stations which are broadly categorized as stationary and mobile stations. Both types of stations have the following common equipment: a hydrogen compressor, 82 MPa-pressurized hydrogen-storage tanks, and dispensers for the delivery of hydrogen to FCVs. Regarding the differences between the two types of stations, mobile stations usually consist of trucks loaded with the aforementioned common equipment. These mobile stations are advantageous in the sense that FCVs can be supplied with hydrogen at temporary locations; however, the amount of hydrogen storage in a truck is limited.

Figure 9.1 and Table 9.1 present information on the various types of stationary stations. A typical station is classified as an off-site gaseous- or liquified-hydrogen-type

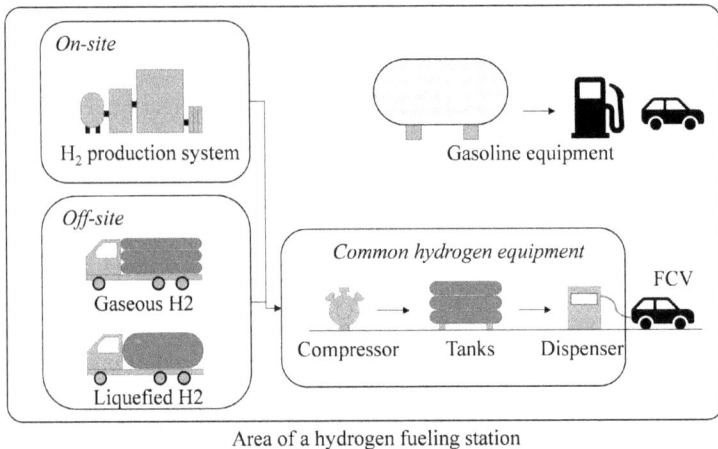

Area of a hydrogen fueling station

FIGURE 9.1   A simplified diagram of stationary hydrogen fueling stations.

TABLE 9.1
Types of Stationary Hydrogen Fueling Stations

	Off-site		On-site			
	Gaseous Hydrogen	Liquified Hydrogen	Hydrocarbon	Methylcyclohexane	Ammonia	Hydrogen Absorbing Alloy
Single	A	B	C	D	E	F
Hybrid	G	H	I	J	K	L

station (A or B). In the case of an off-site station, hydrogen is produced at other (nonlocal) hydrogen production sites and then transported to the station by trucks, whereas an on-site station contains both the common equipment and the hydrogen production system at the station itself. The presence or absence of a hydrogen production system is the only difference between on-site and off-site hydrogen fueling stations. It should be noted that off-site stations are the most popular type of station globally.

As regards station classification based on fuel production (relevant only to on-site stationary stations), hydrogen can be generated by the reaction of hydrocarbons (methane, propane, etc.), methylcyclohexane (MCH), ammonia, or hydrogen-absorbing alloys at the stations (C–F). Additionally, it is worth noting that some stations offer both hydrogen and gasoline systems at the same location and are referred to as hybrid stations (G–L). Hybrid off-site and on-site stations offer a critical advantage in terms of efficiency in supplying fuel to hydrogen and gasoline vehicles.

This basic understanding of the type of hydrogen fueling stations indicates the need to address the following safety issues: (1) process hazards corresponding to the storage of extremely high-pressure hydrogen (82 MPa), (2) hydrogen hazards such as flammability and embrittlement, and (3) other material hazards such as the flammability of propane and MCH. Therefore, it is necessary to perform risk assessment to identify risks and propose effective safety measures for the construction of hydrogen fueling stations. This move can also aid in the commercialization of FCVs.

9.3 INCIDENT/ACCIDENT ANALYSIS OF HYDROGEN FUELING STATIONS

Globally, the number of hydrogen fueling stations is significantly lower than that of gasoline stations. Therefore, very few incidents and accidents related to hydrogen fueling stations have been reported thus far. However, it is important to determine the common causes of accidents in order to implement safety measures. In this regard, Sakamoto et al. (2016b) analyzed the accidents and safety-related incidents occurring at on-site and off-site hydrogen fueling stations in Japan and the USA. Their analysis mainly used data from two databases: (i) the High Pressure Gas Safety Act Database created by the High Pressure Gas Safety Institute of Japan, and (ii) the Hydrogen Incident Reporting Database (HIRD) created by the Pacific Northwest National Laboratory, USA. Subsequently, they applied a classification method to categorize the reported incidents and accidents to determine common failures. Their method classified incidents and accidents into six categories: (1) Leakage I: leakage due to the damage and fracture of the main bodies of apparatuses and pipes (including welded parts), (2) Leakage II: leakage from flanges, valves, and seals (including degraded nonmetallic seals), (3) Leakage III: leakage due to other factors, e.g., human error and external impact, (4) Explosion and fire, (5) Burst and fracture, and (6) Others.

According to the classification shown in Figure 9.2, the 21 reported incidents and accidents in Japan from 2005 to 2014 can be categorized into 3 Leakage I incidents, 14 Leakage II incidents, 2 Leakage III incidents, 1 explosion incident, and 1 "bursting" incident. Moreover, the 22 incidents and accidents in the USA from 2004 to 2012 can be categorized into 4 Leakage I incidents, 6 Leakage II incidents, 3 Leakage III incidents, 5 "bursting" incidents, and 4 incidents attributed to other causes. Furthermore, two fire accidents triggered by Leakage I also occurred. Here, we note that the investigation report on one of these two fires was published by Harris and Marchi (2021). Analysis of the above data shows that the underlying cause of Leakage I incidents in Japan and the USA was mainly design error, that is, poor planning that did not consider material fatigue. Leakage II incidents were a frequent occurrence, caused by inadequate torque and inadequate sealing. Furthermore, human error was one of the main causes of Leakage III. Meanwhile, in the USA, one of the main causes of Leakage III was found to be

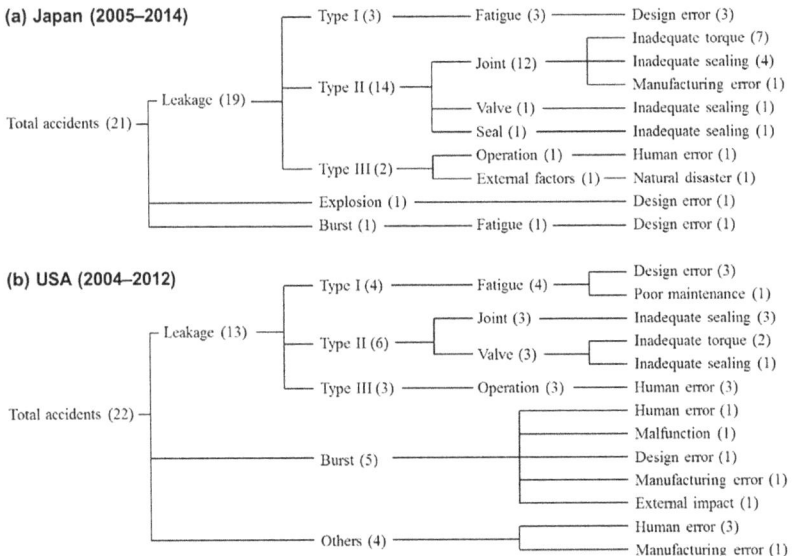

FIGURE 9.2 Tree diagram of incidents and accidents involving hydrogen fueling stations in (a) Japan and (b) the USA. (Sakamoto et al., 2016b.)

hydrogen leakage from a filling hose damaged by erroneously starting an FCV. It should be noted that these analyses can contribute to estimating safety issues and implementing corresponding safety measures. In particular, as regards the use of materials, hydrogen-leakage analysis can aid in the development of new design methods and materials to withstand high-pressure hydrogen environments. In addition, the information related to accidents associated with FCV users can be used to establish safety regulations related to self-service hydrogen fueling stations and safer site use planning.

Figure 9.3 illustrates the vulnerable equipment/components and locations generally associated with the incidents and accidents reported at hydrogen fueling stations in Japan and the USA. Most of these vulnerabilities were associated with the common equipment used in all types of hydrogen fueling stations, e.g., joints, compressors, and dispensers. This type of analysis of vulnerabilities aids in the identification of leakage risks and prioritizes equipment maintenance. The findings of Sakamoto et al. (2016b) show that most accidents and incidents involve small hydrogen leaks; however, some accidents can lead to major consequences such as jet fires. Furthermore, leakages mainly occur at the joints owing to inadequate torque and sealing. Other causes include design errors of the main components of the apparatus and human error. Therefore, it is very important to consider the usage environment in the design stage itself to adequately install and maintain the hydrogen fueling equipment.

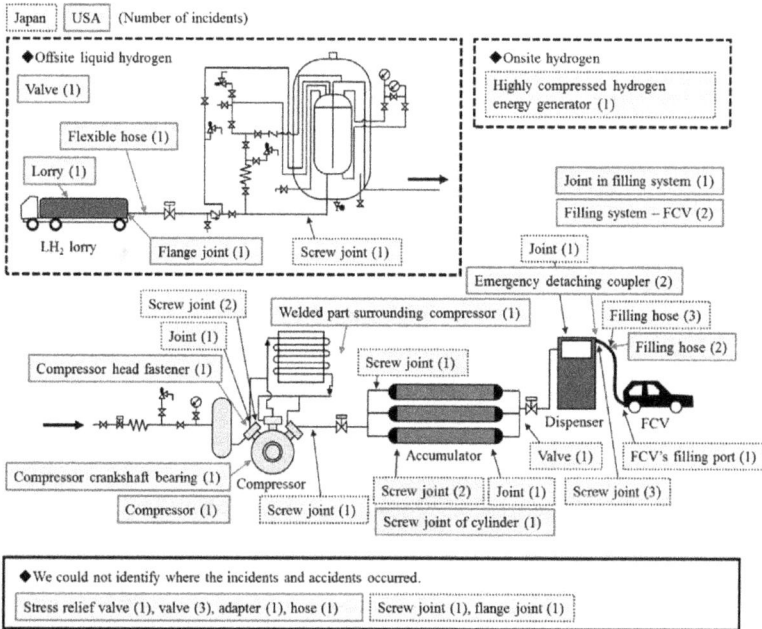

FIGURE 9.3 Details of positions and parts involved in incidents and accidents at hydrogen fueling stations in Japan and the USA. (Sakamoto et al., 2016b.)

9.4 CONCEPT OF RISK ASSESSMENT OF HYDROGEN FUELING STATIONS

We begin this section by noting that several methods of risk analysis and assessment have been developed by professionals in the field of process safety. Here, it is appropriate to note that the petroleum and petrochemical industries have grappled with safety issues for decades considering that they store, produce, and transport a large amount of dangerous substances such as explosives and acutely toxic and flammable substances.

Based on the lessons learned from fatal accidents, experience, and research, a general risk assessment procedure adapted to hydrogen fueling stations has been established over the years, as shown in Figure 9.4. The risk assessment procedure consists of hazard identification (HAZID), consequence analysis, frequency analysis, risk evaluation, and the recommendation of risk-reduction measures. HAZID is the important first step of risk assessment: if the hazards associated with chemicals and processes are not identified, accident scenarios and risks cannot be fully analyzed. Consequence analysis is the critical second step. Here, HAZID and consequence analysis are generally conducted by means of experiments and simulations. Next, the frequency of various scenarios is qualitatively or quantitatively analyzed using approaches such as brainstorming, event tree analysis, and fault tree analysis. Risk is qualitatively or quantitatively evaluated by using the results of the consequence and frequency analyses. In particular, qualitative risk assessment, which is a well-known brainstorming approach, uses a matrix to rank consequences and frequencies. On the other hand, quantitative risk assessment (QRA) generally uses two concepts: individual risk and societal risk (Center for Chemical Process Safety [CCPS], 2000).

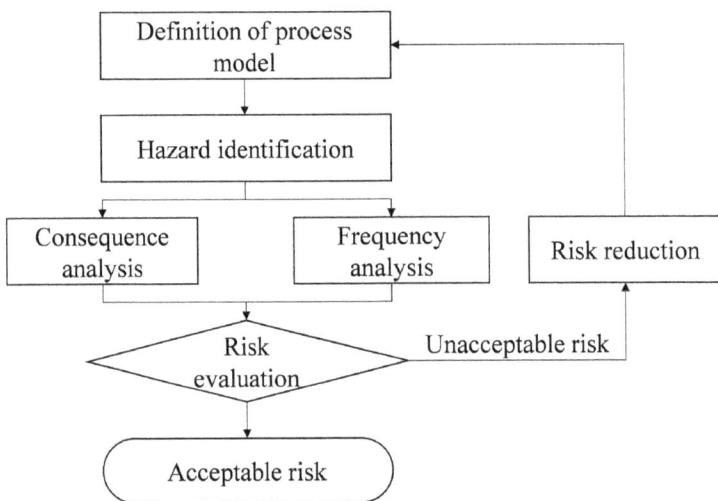

FIGURE 9.4   A general procedure of risk assessment.

The concrete methods of consequence analysis, frequency analysis, and risk evaluation depend on the analytical purposes and process stages. It is of fundamental importance to analyze risks from the earlier stages of the process life cycle (Srinivasan and Natarajan, 2012). Figure 9.5 shows the effectiveness of various risk reduction strategies across the process life cycle. It is clear that risk-reduction strategies generally consist of inherent, passive, active, and procedural measures. At the research stage, it is easy to eliminate and reduce hazardous chemicals and processes. The most important principle is inherent safety, which is a strategy underlined by four basic concepts: minimization, substitution, moderation, and simplification (Hendershot, 1997). At the later stages of process development, because the relevant chemicals and processes are already defined, it is easier to focus on design details and operations; and approaches concerning passive, active, and procedural measures become effective. For instance, hazard and operability studies (HAZOP) and failure modes and effects analysis (FMEA) are effective at the detailed design stage because process design conditions such as the temperature, pressure, and flow rate are clearly defined. This implies that risk assessment methods must be selected according to both the purpose and the process stage.

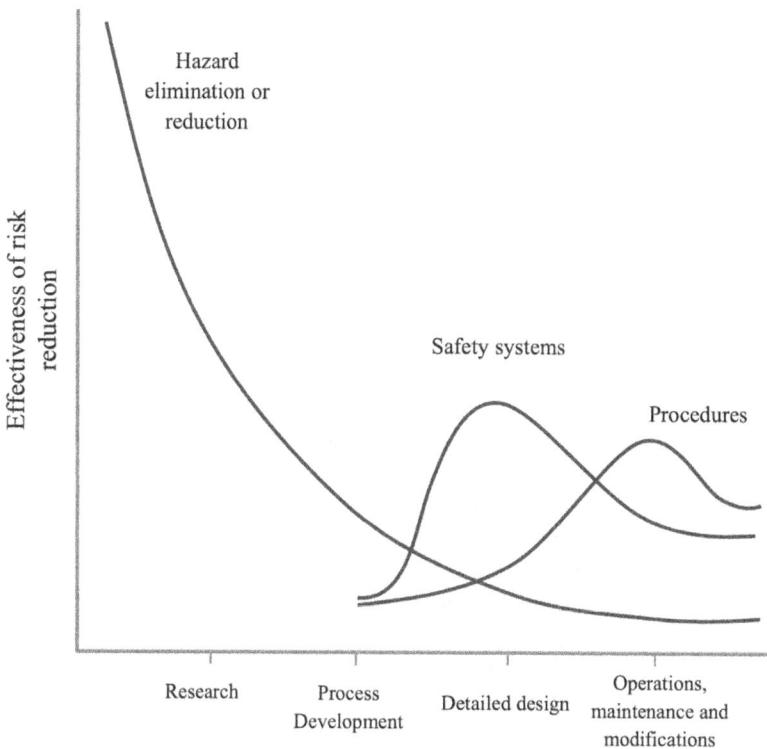

FIGURE 9.5 Effectiveness of various risk reduction strategies across process lifecycle. (Srinivasan and Natarajan, 2012.)

Many studies have focused on hydrogen hazards, risk analysis, and risk assessment of hydrogen fueling stations. Crowl and Jo (2007) conducted a preliminary assessment of the hazards and risks of hydrogen compared to methane and gasoline as traditional fuels. Hydrogen dispersion and explosion phenomena of high-pressure and liquified hydrogen were analyzed by computational fluid dynamics (CFD) simulation (Kikukawa, 2008a; Middha and Hansen, 2009a and Middha et al., 2009b; Baraldi et al., 2009; Kim et al., 2011; Ichard et al., 2012; Hall et al., 2014). Based on the relevant investigations, effective and practicable 2D and 3D simulation software such as FLACS (Gexcon, 2020) and HyRAM (Sandia National Laboratories, 2020) were developed. On the other hand, the uncertainty of accidents and the frequencies of hydrogen leaks were investigated using a conditionally autoregressive model (Kodoth et al., 2018) and a time-based Bayesian estimate method (Kodoth et al., 2020), respectively. For the construction of stand-alone compressed and liquified hydrogen fueling stations, qualitative and quantitative risk assessments have already been conducted (Kikukawa et al., 2008b and 2009; Kim et al., 2013; Casamirra et al., 2009; Lowesmith et al., 2014; Tsunemi et al., 2019). In addition, safety distance, which is an inherently safer measure, was analyzed using the quantitative risk analysis method (Matthijsen and Kooi, 2006).

9.5 HAZARD IDENTIFICATION STUDY

HAZID studies are one of the qualitative methods used for comprehensively identifying hazards and undesirable accident scenarios (Nakayama et al., 2015). It is important not to overlook accident scenarios because effective safety measures can be built into the relevant equipment as the various scenarios are identified. In this regard, HAZID studies specialize in detecting hidden scenarios that may cause catastrophic accidents, which are named worst-case scenarios. After qualitative risk analysis, a quantitative analysis is needed for the detailed determination of the consequences and likelihoods of the risk. However, not all risks can be investigated via detailed analyses because of prohibitive costs and the difficulties in acquiring long-term technical knowledge on gas/chemical dispersion, fire, and explosions via simulations. Therefore, a HAZID study is first conducted at the process development stage to roughly identify accident scenarios and rank risks for quantitative risk analysis. In general, an expert team performs a HAZID study to imagine and present every possible hazard and undesirable scenario during brainstorming sessions. For greater objectivity and logicality, certain guide words are chosen before conducting the HAZID study. These guide words depend on the location, site, and process information related to natural hazards, external event hazards, chemical hazards, and process hazards. With regard to hydrogen fueling stations, it is particularly important to refer to incident and accident databases regarding processes involving hydrogen and other dangerous substances.

In the above context, we refer to the HAZID study carried out by Nakayama et al. (2016) to identify the worst-case scenarios of a hybrid gasoline-hydrogen fueling station with an on-site hydrogen production system using MCH. Such a station has potential hazards related to gasoline, hydrogen, and the hydrogen

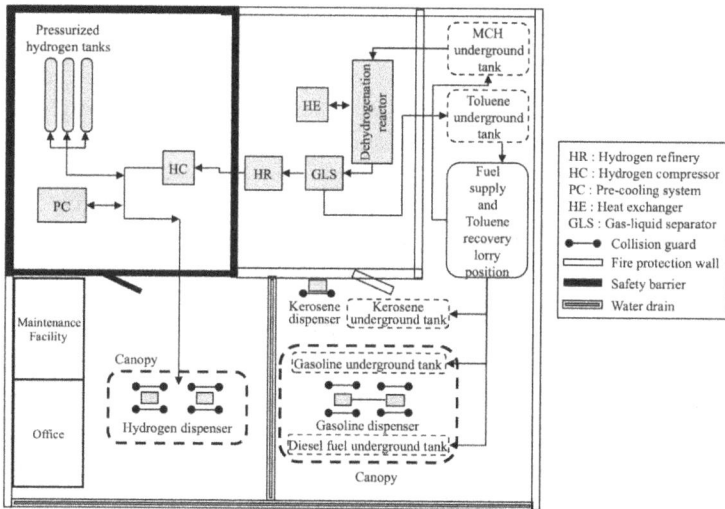

FIGURE 9.6   Hybrid gasoline-hydrogen fueling station model. (Nakayama et al., 2016.)

production system. In this regard, previous studies have determined accident scenarios involving compressed hydrogen and gasoline systems (Nakayama et al., 2015; Sakamoto et al., 2016ba). Consequently, HAZID studies have been conducted to consider the safety issues related to the coexistence of gasoline and hydrogen production systems.

The model of such a hybrid station is shown in Figure 9.6, wherein the gasoline and kerosene supply systems consist of underground tanks and dispensers. The hydrogen supply system consists mainly of a hydrogen production system, a hydrogen compressor, pressurized hydrogen tanks, a pre-cooling system, and dispensers. The hydrogen production system is further composed of a dehydrogenation reactor, a heat exchanger, a gas-liquid separator, and a hydrogen refinery. Here, we note that the pressure of the system is <1 MPa. In the dehydrogenation reactor, MCH decomposes into toluene and hydrogen in the presence of a catalyst at 300–400°C (Figure 9.7), and subsequently, toluene and hydrogen are separated in the gas-liquid separator. Toluene is then transported to an underground tank for recovery and resupply at an MCH hydrogenation plant. Meanwhile, hydrogen is refined to eliminate impurities at the hydrogen refinery and transported to the hydrogen compressor.

Considering the above context, researchers performed a HAZID study using guide words, risk criteria, and a HAZID study sheet (Nakayama et al., 2016). The

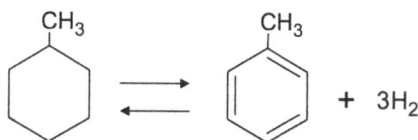

FIGURE 9.7   Methylcyclohexane dehydrogenation reaction. (Nakayama et al., 2018.)

HAZID study was also conducted from two perspectives based on the concept of the bow-tie diagram. The one perspective is that prevention, control, and mitigation measures are sufficient to address the problem at hand. This perspective simulates realistic circumstances and identifies safety measures installed in the station to prevent loss of containment and mitigate the consequences. The other perspective, however, is that only mitigation measures are effective, and thus, prevention and control measures are not considered. This perspective imagines incident scenarios after loss of containment, and it discusses whether mitigation measures are sufficient. These two perspectives can aid in determining the safety measures implemented in the station for risk reduction. One example scenario is described as follows: (1) "Combustible material" is selected as a guide word. (2) At a gas-liquid separator, toluene is observed to leak from a pipe failure caused by corrosion or fatigue. (3) The toluene pool gradually spreads, and toluene gas is dispersed in the atmosphere owing to vaporization. A vapor cloud explosion of toluene can occur if it is ignited, and the explosion can affect people and equipment outside the station. (4) If there are no safety measures in the system, the risk is high because the consequence of causing damage outside the station can be catastrophic, although the frequency of this occurrence in the lifetime of a hydrogen fueling station is occasional. (5) Various safety measures are already implemented in the system. (6) The levels of the explosion consequence and frequency are respectively reduced to major damage and "improbable" because of suitable material selection and the use of isolation valves, flame detectors, fire protection walls, and safety barriers. Therefore, the risk level is reduced to medium. In the HAZID study, 314 risks were identified based on scenarios with and without safety measures considering the coexistence of gasoline and hydrogen production systems. For example, a characteristic scenario is that a massive amount of gasoline leaks from an operating gasoline lorry, the resulting gasoline pool fire affects the hydrogen production system, and MCH, toluene, and/or hydrogen eventually leak out due to the damage caused by the thermal radiation of the pool fire.

Tables 9.2 and 9.3 list the risk matrices, which indicate the risk distributions to clearly visualize the risks. Almost all of the risks can be reduced to medium or

TABLE 9.2

Risk Matrix without Safety Measures (Nakayama et al., 2016)

		Frequency			
		1	2	3	4
Consequence	5	0	71	98	0
	4	0	63	82	0
	3	0	0	0	0
	2	0	0	0	0
	1	0	0	0	0

TABLE 9.3
Risk Matrix with Safety Measures (Nakayama et al., 2016)

		Frequency			
		1	2	3	4
Consequence	5	19	0	0	0
	4	95	27	0	0
	3	126	19	0	0
	2	26	2	0	0
	1	0	0	0	0

low risks by applying various prevention, control, and mitigation measures, such as the use of shutdown valves based on seismometers, seismic design, hydrogen detectors, and fire protection walls. The HAZID study also identified 46 high-risk scenarios. It should be noted that detailed investigations are needed to propose safety measures for risk mitigation and to accept risks that are practical from the viewpoint of the "as low as reasonably practicable" (ALARP) concept.

HAZID studies are an effective method to qualitatively analyze risks during preliminary risk analysis. Therefore, after HAZID studies, more detailed risk analyses such as CFD simulations, HAZOP, FMEA, and emergency response planning are needed at the detailed design stage for the construction of hydrogen fueling stations. If hydrogen fueling stations are expected to face thermal hazards such as thermal explosions and runaway reactions, it is also important to conduct thermal hazard analyses (Nakayama et al., 2018). In other words, for inherently safer hydrogen fueling stations, it is critical to rigorously analyze risks with qualitative and quantitative methods throughout the process lifecycle.

9.6 SIMULATION-BASED INVESTIGATION

From the results of previous HAZID studies of hydrogen fueling stations (Nakayama et al., 2015 and 2016), two critical scenarios can be identified to prevent the domino effect at hydrogen fueling stations. In this context, it is noteworthy that while several studies have focused on the risks of hydrogen fueling stations, very few have conducted investigations on the risks related to the domino effect. The domino effect (Reniers, 2010), which generally refers to a sequence of events that spatially and temporally escalate and propagate, can prove fatal to people, buildings, and the environment within and around the site in question. Therefore, studies on domino-effect risk are crucial to ensure the safety of hydrogen fueling stations. The first critical scenario was determined as a gasoline pool fire affecting a liquified-hydrogen storage tank at a hybrid hydrogen-gasoline fueling station (Nakayama et al., 2015). As regards this situation, Sakamoto et al. (2016a) analyzed the temperature and stress of the tank due to the temperature distribution resulting from the gasoline pool fire. Moreover, additional safety measures were suggested to reduce the safety distance.

The second critical scenario and related simulation-based safety investigation (Nakayama et al., 2017) involved a hydrogen fueling station with MCH-based hydrogen production, and the domino-effect scenario can be described as follows: (1) Toluene or MCH continually leaks from a pipe in the hydrogen production system. (2) A pool fire of toluene or MCH occurs in the system. (3) Thermal radiation from the pool fire affects the pressurized hydrogen tanks which usually store a total of 2000 L of hydrogen at 82 MPa. (4) A large amount of hydrogen in the tanks is rapidly released though a safety vent during the emergency. Here, we note that the resulting increase in temperature can structurally weaken the pressurized hydrogen-storage tanks; hence, if the safety-vent system is not adequately designed or operated, the tanks may rupture, causing a catastrophic accident. To analyze the domino-effect scenario, Nakayama et al. (2017) investigated the thermal radiation from the pool fires, the effects of thermal radiation on the tanks, and the variation in the hydrogen flow rate released through the safety vent by using PHAST 6.54 (DNV, 2021), ANSYS 16.0 (ANSYS, 2020), and SimulationX 3.6 (ESI Group, 2021), respectively. The simulation conditions are described in the literature (Nakayama et al., 2017).

As per the results of the pool fire analysis, the physical properties of toluene and MCH pool fires are listed in Table 9.4. In addition, based on the relationship between the representative thermal radiation of MCH and the downwind distance (1.0 m/s) at the height of 0 m, the MCH thermal radiation was determined as 4 kW/m^2 at 10.5 m, 9.5 kW/m^2 at 7.1 m, 25 kW/m^2 at 3.8 m, and 37.5 kW/m^2 at 2.7 m. Such radiation levels from pool fires can severely affect the tanks and pipes that store pressurized hydrogen. This also means that thermal radiation can damage the tanks as they are composed of carbon fiber reinforced plastics (CFRPs) which decomposes relatively easily with increasing temperature. Secondly, the thermal characteristics of the tank were analyzed. Figure 9.8 illustrates temperature distribution at the surface of a pressurized hydrogen tank after 30 min of exposure to a pool fire (a) without and (b) with protection from a steel enclosure. As shown, the maximum surface temperature of the tank nearest the pool fire in case (a) is 428°C, whereas that in case (b) is 224°C. It should be noted that the threshold

TABLE 9.4

Physical Properties of Pool Fires of Toluene and Methyl Cyclohexane

Material	Wind Velocity (m/s)	Pool Diameter (m)	Flame Height
Toluene	1.0	1.83	5.58
	3.0	1.82	5.56
Methylcyclohexane	1.0	1.59	5.53
	3.0	1.57	5.49

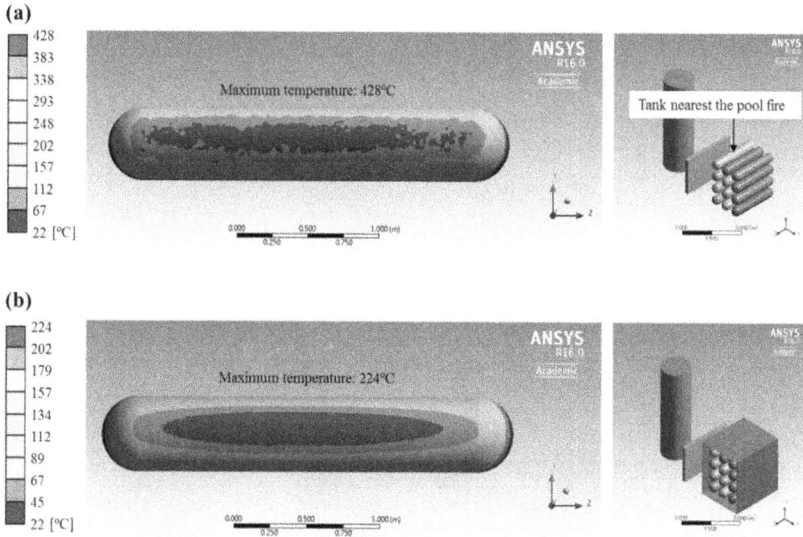

FIGURE 9.8 Temperature distribution on the surface of a pressurized hydrogen tank after a 30 min exposure to a pool fire (a) without and (b) with protection from a steel building. (Nakayama et al., 2017.)

surface temperature of a type-III CFRP tank is approximately 205°C. Thus, the tanks can rupture owing to thermal radiation, but a steel enclosure can effectively protect the tanks against thermal radiation to prevent and delay tank rupture. Furthermore, for greater safety during emergencies, it is important to evaluate and implement a safety valve system.

We should also consider the analysis of the hydrogen flow rate from the vent outlet and the pressure in the tanks during emergency hydrogen release. Figure 9.9 illustrates the characteristics of mass flow through vents of various sizes. It can be observed that the pressure and mass flow are dependent on the diameter of the safety valve: a smaller diameter corresponds to a slower pressure reduction rate and mass flow, which means that the risk of the physical explosion of the tanks damaged by the fires increases with time. Conversely, a larger valve diameter corresponds to a more rapid pressure reduction rate and a higher mass flow from the vent outlet. From a hydrogen hazard perspective, a larger valve diameter increases the risk of both hydrogen jet fires at the vent outlet and of hydrogen explosions in the atmosphere, although the risk of tank rupture is reduced. These results signify the importance of optimal vent design. Finally, we need to consider the results of simulation-based safety investigations in the domino-effect scenario as this aids in identifying relevant safety measures. Firstly, to prevent and mitigate MCH and toluene leakage, the pipe material should be suitably selected, and leak detectors and shutdown valves must be installed in the hydrogen production system. Secondly, in the case of a pool fire, the station layout should be designed considering not only economically efficient land use

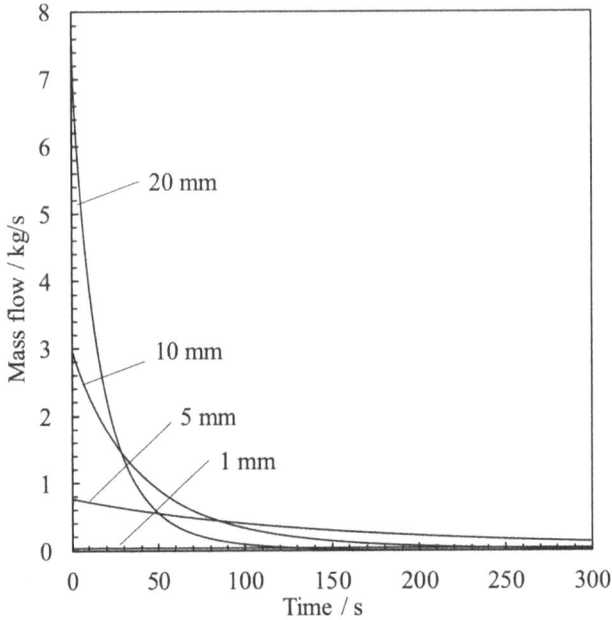

FIGURE 9.9   Mass flow from the vent outlet. (Nakayama et al., 2017.)

but also safety. The safety distance between equipment must be conservatively estimated, and the provision of fire extinguishing equipment can effectively prevent a cascading event. Finally, the material and thickness of the tanks as well as the housing building must be considered to reduce the effects of thermal radiation. Meanwhile, we note that the use of a safety vent system is a reasonable and practical measure to reduce the probability of tank rupture. Therefore, using CFD simulations, the design of the vent system should be optimized considering the station layout and hydrogen dispersion behavior.

9.7   SUMMARY

In this chapter, we presented an overview of the risk assessment of hydrogen fueling stations. In particular, we considered the HAZID study, which is a qualitative risk assessment method, and we subsequently examined a worst-case scenario of a typical hybrid hydrogen-gasoline fueling station using simulation tools. These methods focus on risks at the preliminary process design stage, which means that QRAs are subsequently needed to construct the stations and evaluate the effectiveness of the safety measures. Although we only considered the study by Nakayama et al. (2019) in investigating the safety issues underlying a typical hydrogen fueling station in this chapter, it is important to investigate security issues related to hydrogen fueling stations, which can be estimated based on the experiences encountered in the petroleum and petrochemical industries.

In conclusion, we note that in the quest for establishing a hydrogen society, it is necessary to conduct risk and safety assessments based not only on engineering but also on the social sciences; fostering the public acceptance of hydrogen-related technologies is crucial because the related technologies are new. Although hydrogen is a well-known chemical substance, new technologies using hydrogen may pose unfamiliar risks. Engineering risk assessment can reveal the risks, and the results can be applied to inculcate public acceptance. For example, Hienuki et al. (2019) suggested that exposure to hydrogen technology leads to trust in the technology, eventually leading to its acceptance. Along these lines, Ono et al. (2019) analyzed whether information on the risk or safety measures would change the public acceptance of hydrogen fueling stations in Japan. These results can also be applied for the safer design and operation of hydrogen fueling stations. Thus, the use of methods based on engineering and the social sciences is important to create an advanced and sustainable society.

ACKNOWLEDGMENT

Certain sections of this chapter were supported by the Council for Science, Technology and Innovation: Cross-ministerial Strategic Innovation Promotion Program; and the "Energy Carrier" program (Funding agency: JST) as well as the "Promotion Program for Scientific Fire and Disaster Prevention Technologies" by the Fire and Disaster Management Agency of the Ministry of Internal Affairs and Communication in Japan.

REFERENCES

ANSYS, https://www.ansys.com/ [Accessed 27 September 2020]

Baraldi D., Venetsanos A.G., Papanikolaou E., Heitsch M., and Dallas V. (2009). *Journal of Loss Prevention in the Process Industries* 22, 303–315.

Casamirra M., Castiglia F., Giardina M., and Lombardo C. (2009). Safety studies of a hydrogen refuelling station: Determination of the occurrence frequency of the accidental scenarios. *International Journal of Hydrogen Energy* 34, 5846–5854.

Center for Chemical Process Safety (CCPS). (2000). Guidelines for chemical process quantitative risk analysis. Wiley-AIChE, Hoboken, New Jersey.

Crowl D.A., and Jo Y.D. (2007). The hazards and risks of hydrogen. *Journal of Loss Prevention in the Process Industries* 20, 158–164.

DNV, https://www.dnv.com/ [Accessed 27 September 2021]

ESI Group, https://www.esi-group.com/products/system-simulation [Accessed 27 September 2021]

Gexcon, https://www.gexcon.com/ [Accessed 27 September 2020]

Hall J.E., Hooker P., and Willoughby D. (2014). Ignited released of liquid hydrogen: Safety considerations of thermal and overpressure effects. *International Journal of Hydrogen Energy* 39, 20547–20553.

Harris A.P., and Marchi C.W.S. (2012). Investigation of the hydrogen release incident at the AC transit Emeryville facility. Sandia Report SAND2012-8642, Sandia National Laboratories.

Hendershot D.C. (1997). Inherently safer chemical process design. *Journal of Loss Prevention in the Process Industries* 10(3), 151–157.

Hienuki S., Hirayama Y., Shibutani T., Sakamoto J., Nakayama J., and Miyake A. (2019). How knowledge about or experience with hydrogen fueling stations improves their public acceptance. *Sustainability* 11(22), 6339.

Ichard M., Hansen O.R., Middha P., and Willoughby D. (2012). CFD computations of liquid hydrogen releases. *International Journal of Hydrogen Energy* 37, 17380–17389.

Kikukawa S. (2008a). Consequence analysis and safety verification of hydrogen fueling stations using CDF simulation. *International Journal of Hydrogen Energy* 33, 1425–1434.

Kikukawa S., Yamaga F., and Mitsuhashi H. (2008b). Risk assessment of hydrogen fueling stations for 70MPa FCVs. *International Journal of Hydrogen Energy* 33, 7129–7139.

Kikukawa S., Mitsuhashi H., and Miyake A. (2009). Risk assessment for liquid hydrogen fueling stations. *International Journal of Hydrogen Energy* 34, 1135–1141.

Kim E., Lee K., Kim J., Lee Y., Park J., and Moon I. (2011). Development of Korean hydrogen fueling station codes through risk analysis. *International Journal of Hydrogen Energy* 36, 13122–13131.

Kim E., Park J., Cho J.H., and Moon I. (2013). Simulation of hydrogen leak and explosion for the safety design of hydrogen fueling station in Korea. *International Journal of Hydrogen Energy* 38, 1737–1743.

Kodoth M., Aoyama S., Sakamoto J., Kasai N., Khalil Y., Shibutani T., and Miyake A. (2018). Evaluating uncertainty in accident rate estimation at hydrogen refueling station using time correlation model. *International Journal of Hydrogen Energy* 43, 23409–23417.

Kodoth M., Aoyama S., Sakamoto J., Kasai N., Khalil Y., Shibutani T., and Miyake A. (2020). Leak frequency analysis for hydrogen-based technology using Bayesian and frequentist methods. *Process Safety and Environmental Protection* 136, 148–156.

Lowesmith B.J., Hankinson G., and Chynoweth S. (2014). Safety issues of the liquefaction, storage and transportation of liquid hydrogen: An analysis of incidents and HAZIDS. *International Journal of Hydrogen Energy* 39, 20516–20521.

Matthijsen A.J.C.M., and Kooi E.S. (2006). Safety distances for hydrogen filling stations. *Journal of Loss Prevention in the Process Industries* 19, 719–723.

Middha P., and Hansen O.R. (2009a). Using computational fluid dynamics as a tool for hydrogen safety studies. *Journal of Loss Prevention in the Process Industries* 22, 295–302.

Middha P., Hansen O.R., and Storvik I.E. (2009b). Validation of CFD-model for hydrogen dispersion. *Journal of Loss Prevention in the Process Industries* 22, 1034–1038.

Ministry of Economy, Trade and Industry (METI), Formulation of a New Strategic Roadmap for Hydrogen and Fuel, https://www.meti.go.jp/english/press/2019/0312_002.html [Accessed: 22 September 2020]

Nakayama J., Sakamoto J., Kasai N., Shibutani T., and Miyake A. (2015). Risk assessment for a gas and liquid hydrogen fueling station. *49th Annual Loss Prevention Symposium 2015, LPS 2015 - Topical Conference at the 2015 AIChE Spring Meeting and 11th Global Congress on Process Safety*, 138–150.

Nakayama J., Sakamoto J., Kasai N., Shibutani T., and Miyake A. (2016). Preliminary hazard identification for qualitative risk assessment on a hybrid gasoline-hydrogen fueling station with an on-site hydrogen production system using organic chemical hydride. *International Journal of Hydrogen Energy* 41, 7518–7525

Nakayama J., Misono H., Sakamoto J., Kasai N., Shibutani T., and Miyake A. (2017). Simulation-based safety investigation of a hydrogen fueling station with an on-site hydrogen production system involving methylcyclohexane. *International Journal of Hydrogen Energy* 42, 10636–10664.

Nakayama J., Aoki H., Homma T., Yamaki N., and Miyake A. (2018). Thermal hazard analysis of a dehydrogenation system involving methylcyclohexane and toluene. *Journal of Thermal Analysis and Calorimetry* 133, 805–812.

Nakayama J., Kasai N., Shibutani T., and Miyake A. (2019). Security risk analysis of a hydrogen fueling station with an on-site hydrogen production system involving methylcyclohexane. *International Journal of Hydrogen Energy* 44, 9110–9119.

Ono K., Kato E., and Tsunemi K. (2019). Does risk information change the acceptance of hydrogen refueling stations in the general Japanese population? *International Journal of Hydrogen Energy* 44, 16038–16047.

Reniers G. (2010). An external domino effects investment approach to improve cross-plant safety within chemical clusters. *Journal of Hazardous Materials* 177, 167–174.

Sakamoto J., Nakayama J., Nakarai T., Kasai N., Shibutani T., and Miyake A. (2016a). Effect of gasoline pool fire on liquid hydrogen storage tank in hybrid hydrogen-gasoline fueling station. *International Journal of Hydrogen Energy* 41, 2096–2104.

Sakamoto J., Sato R., Nakayama J., Kasai N., Shibutani T., and Miyake A. (2016b). Leakage-type-based analysis of accidents involving hydrogen fueling stations in Japan and USA. *International Journal of Hydrogen Energy* 41, 21564–21579.

Sakamoto J., Misono H., Nakayama J., Kasai N., Shibutani T., and Miyake A. (2018). Evaluation of safety measures of a hydrogen fueling station using physical modeling. *Sustainability* 10(11), 3846.

Sandia National Laboratories, Hydrogen Risk Assessment Models (HyRAM), https://energy.sandia.gov/programs/sustainable-transportation/hydrogen/quantitative-risk-assessment/hydrogen-risk-assessment-model-hyram/ [Accessed 27 September 2020]

Srinivasan R., and Natarajan S. (2012). Development in inherent safety: A review of the progress during 2001–2011 and opportunities ahead. *Process Safety and Environment Protection* 90, 389–403.

Tsunemi K., Kihara T., Kato E., Kawamoto A., and Saburi T. (2019). Quantitative risk assessment of the interior of a hydrogen refueling station considering safety barrier systems. *International Journal of Hydrogen Energy* 44, 23522–23531.

10 Comparative Assessment and Validation of USEPA Air Quality Modelling Techniques Using Pre-Post Point-Source Observations Near a Coal-Based Thermal Power Plant

Rahil Changotra
School of Energy and Environment, Thapar Institute of Engineering and Technology, Patiala, India

Amarpreet Singh Arora
School of Chemical Engineering, Yeungnam University, Gyeongsan, South Korea

Himadri Rajput, and Amit Dhir
School of Energy and Environment, Thapar Institute of Engineering and Technology, Patiala, India

CONTENTS

DOI: 10.1201/9781003140382-10

10.1 INTRODUCTION

Owing to advancements in technology, agriculture and industry, in combination with a rapidly growing population, air quality is declining worldwide. Increased traffic flows and emerging cities all contribute to increased waste, and where there are inadequate environmental norms or regulations in place, there are a wider range of pollutants being emitted. The public are completely reliant on a continuous supply of energy to perform their daily activities and work. Thus, to meet this rising energy demand, both private and government sectors have made huge investments in the field of energy generation, and primarily electricity production. At present, almost 60% of the global electricity production is produced using fossil fuels. In India, coal is the sole fossil fuel that exists in abundant quantities and is regarded as the primary fuel for electricity generation in thermal power plants (Mishra, 2004). In this regard, the total coal based energy contribution in India is 1.94 Lac MW (54.6%) of the 3.56 Lac MW total installed electricity production capacity (CEA, 2017). In thermal power plants, coal undergoes combustion and is converted into heat energy, regardless of the numerous health and environmental hazards. The coal combustion process in power plants leads to the discharge of several pollutants, viz. sulfur dioxide (SO$_2$), nitrogen oxides (NOx), carbon dioxide (CO$_2$), chlorofluorocarbons (CFCs), and a number of other trace gases, along with airborne inorganic particles in the form of suspended particulate matter (SPM) and fly ash. Indian coal has a high ash content, thus the poor quality combustion techniques contribute significantly to India's overall gaseous, pollutant and particulate matter emission problems. The chemical composition of the air is being altered due to the addition of all these emitted gases, particulates, and volatile substances, which are toxic to the environment and living beings. Table 10.1 shows the types of discharged air pollutants and their various effects on the environment. The prolonged and continuous emission of these primary pollutants from coal-based plants, i.e. NO$_x$ and SO$_x$, also have an adverse effect on the surrounding buildings, architectural structures, and historical monuments. Even metallic structures experience corrosive effects in the form of acid rain. It is also worth noting that the huge volume of CO$_2$ emissions (approximately 0.91–0.96 kg/kWh) from the thermal power plants also contribute toward the problems of global warming and climate change (Pokale, 2012). The World Health Organization (WHO) estimates that approximately 8 Lac deaths and 4.7 million people will experience serious health issues annually due to air pollution (WHO, 2002). The areas most affected by air pollution are the developing nations in Asia, as they report two-thirds of global casualties due to air

TABLE 10.1

Types and Effects of Air Pollutants Discharged into the Atmosphere

Major Air Pollutant	Associated Health Issues	Source(s) of Pollutant
Suspended particulate matter (SPM), Particulate matter ($PM_{2.5}$ and PM_{210})	Disturbs exchange of gases in the lungs causing respiratory illness	Combination of liquid and solid inorganic and organic materials containing nitrates, sulfates, ammonia, carbon, mineral dust, sodium chloride, and water
Ozone (O_3)	Induces infections in respiratory systems (pneumonia, colds), asthma; difficulty in breathing	Part of a photochemical smog generated by the interaction of air pollutants and sunlight
Nitrogen dioxide (NO_2)	Enduring intake is toxic; reduces lung function and causes bronchitis in asthmatic children	Part of $PM_{2.5}$ and O_3, found in nitrate aerosols, produced by burning fuels, vehicles engines and during electricity generation
Sulphur dioxide (SO_2)	Decreased pulmonary function within 10 minutes of exposure; causes respiratory inflammation including mucus secretion, coughing, bronchitis and asthma attacks; and eye irritation	Industrial activities and burning of fossil fuels
Carbon dioxide (CO_2)	Reduces brain and respiratory functions; reduces oxygen levels; causes vision defects	Burning oil, coal, and natural gases
Carbon monoxide (CO)	Reduces blood oxygen levels, slows reflexes, increases sleepiness and create confusion	Burning diesel, petrol, and wood; and smoking
Lead (Pb)	Damages nervous system in children	Petrol, diesel, lead batteries, paints, and coloring agents

Source: WHO.

pollution (Cohen et al., 2004). Therefore, the quantitative and qualitative evaluation of pollutants discharged from thermal power plants is of crucial importance to develop stringent measures and regulations due to their harmful impact on humans and the environment.

10.2   AIR QUALITY MODELING

Rapid industrialization and urbanization have added ample amounts of pollutants to the air we breathe and consequently deteriorated the air quality in developing countries, like India. To control, prevent, and abate air pollution in the country, the Government of India passed the Air (Prevention and Control of Pollution) Act in 1981. According to Section 2[a] of this act, 'air pollutants' are defined as 'any solid, liquid or gaseous substance as well as noise existent in the atmosphere with concentration that may tend to be harmful to human beings or plants or environment or property or any other living creatures' and Section 2[b] further defines 'air pollution' as 'the presence of any air pollutant in the atmosphere.' Under the provisions set out in the Air (Prevention and Control of Pollution) Act of 1981, the 1982 National Ambient Air Quality Standards (NAAQS) were proposed by the Central Pollution Control Board (CPCB) to protect ambient air quality, these standards were revised by the CPCB in November 2009. The NAAQS standard was developed in line with policy guidelines to regulate emissions of pollutants into the atmosphere and limit their impact on the environment through industrial and human activity.

The air quality of any location that deteriorates due to air pollutants must be determined by monitoring or by modeling to determine its concentrations and characteristics relative to ambient air. Monitoring data only presents the extant condition, and this is not always economical or practical as it does not facilitate accurate predictions regarding air quality. Alternatively, the dispersion model for air quality predicts the transport and fate of emitted pollutants in the air from any particular source. After emission of primary pollutants from the source, these pollutants are mixed and dispersed in the ambient air via physical and chemical processes such as mechanical/thermal turbulence and chemical reactions, leading to decreased atmospheric concentrations. Ground-level concentrations (GLCs) of any particular pollutant at a specific spatial location (i.e. receptors) can be predicted by air dispersion models that use a mathematical equation that incorporates numerous complex meteorological processes for the dispersion of pollutants. To predict the GLCs of any pollutant, air dispersion models include input parameters such as source inventory, meteorological data, and to some extent, information on the dimensions of structures of buildings and details of local terrain within the study area. A generalized overview of how the input information is utilized in computer-based air pollution models is shown in Figure 10.1. The two most widely used short-range (up to 50 km) air dispersion models intended to support the United States-Environmental Protection Agency's regulatory options are the Industrial Source Complex Short Term Model 3 (ISCST3) and the American Meteorological Society Environmental Protection Agency Regulatory Model (AERMOD). Both these steady-state models take into account the Gaussian plume equation (Equation 10.1) that relies on the assumption that pollutants disperse in the atmosphere according to a normal statistical distribution (Cooper and Alley, 2002; Holmes and Morawska, 2006). Accordingly, the concentration of pollutants is at a maximum at the release point i.e. stack, and decreases in both vertical and lateral directions according to the normal distribution, as shown in Figure 10.2.

$$C(x,y,z) = \frac{Q}{2\pi u \sigma_y \sigma_z} \times \left[\exp - \left(\frac{y^2}{2\sigma_y^2} \right) \right] \left\{ \exp \left(\frac{-(z-H)^2}{2\sigma_z^2} \right) + \exp \left(\frac{-(z+H)^2}{2\sigma_z^2} \right) \right\}$$

$$(10.1)$$

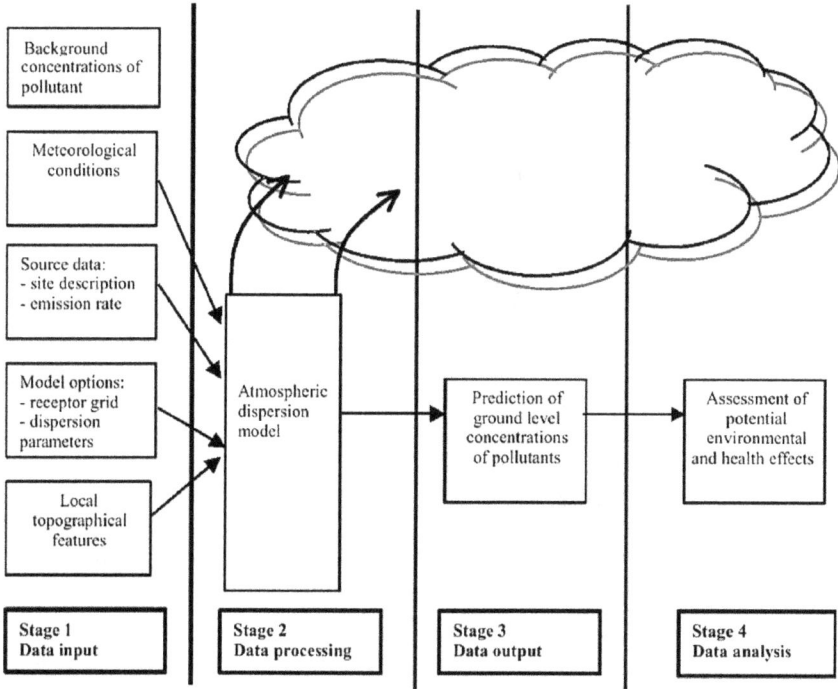

FIGURE 10.1   Overview of a generic air pollution modeling procedure. (MENZ, 2004.)

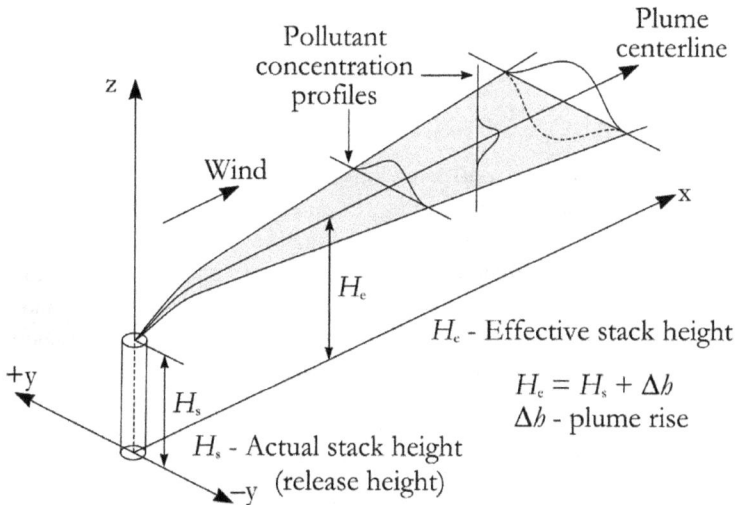

FIGURE 10.2   Representation of plume distribution according to the Gaussian equation.

where

    C is the steady-state concentration of a pollutant at a point (x, y, z), $\mu g/m^3$

    Q is the emission rate of a pollutant, $\mu g/s$

    σ_y and σ_z are the standard deviations of the lateral and vertical spread parameters, respectively

    U is the mean wind speed at the release height

    y is the horizontal distance from the plume centerline, m

    z is the vertical distance from the ground level, m

    H is the effective height of the stack ($H = h + \Delta h$), where h is the physical height of stack and Δh is the plume rise.

Thus, air dispersion models use numerical tools and mathematical equations to simulate the physicochemical processes affecting pollutants, followed by modifications allowing for their reaction and dispersion in the air. Both models are regulatory, and it is not possible to make changes in their algorithms, therefore these models could generate different results in the same circumstances. Since different models are likely to generate different results under different conditions, it would be interesting to evaluate and validate the magnitude and nature of these differences. The objective is to estimate the output from both models in the form of pollutant concentration at a particular receptor point by considering the point sources of the pollutant and the meteorological parameters in the study zone. Further, no detailed study of the prediction of ground-level concentrations at sensitive receptors using dispersion modeling has been published related to Thermal Power Plants in India. This chapter runs an inventory of high priority pollutants such as sulfur dioxide (SO_2), nitrogen dioxide (NO_2), and SPM as emitted from thermal power plant stacks and compares the outcomes of the alternate computer models, AERMOD, and ISCST3. The observed GLCs at nearby sites were recorded using sensitive receptors and the predicted concentrations from the models were compared with the observation data and validated for the study area.

10.3 DESCRIPTION OF STUDY AREA AND INPUT PARAMETERS

The study zone comprises a 30 × 30 km area in the Punjab, India including a thermal power plant as a primary source of emissions in the Rajpura province of Punjab. Eight receptor sites lie within the study area (Figure 10.3). Figure 10.3 depicts the study zone with a power plant (labeled *T*) and the receptor locations (labeled *A–H*). The major source of emission was the 1400 MW coal and supercritical technology-based thermal power plant established by the Punjab State Electricity Board, Government of Punjab, and commissioned by Nabha Power Limited on 8 December 2013.

    The general elevation and geographical positions of the power plant site are 271 m and 30°32′35″ to 30° 33′ 52″ N and 76° 33′ 41″ to 76° 35′ 04″ E, respectively. To study the influence of emissions from the thermal power plant on ambient air quality, eight sensitive receptors were selected on the basis of population density, prevailing wind parameters, topography of the area, and other local parameters

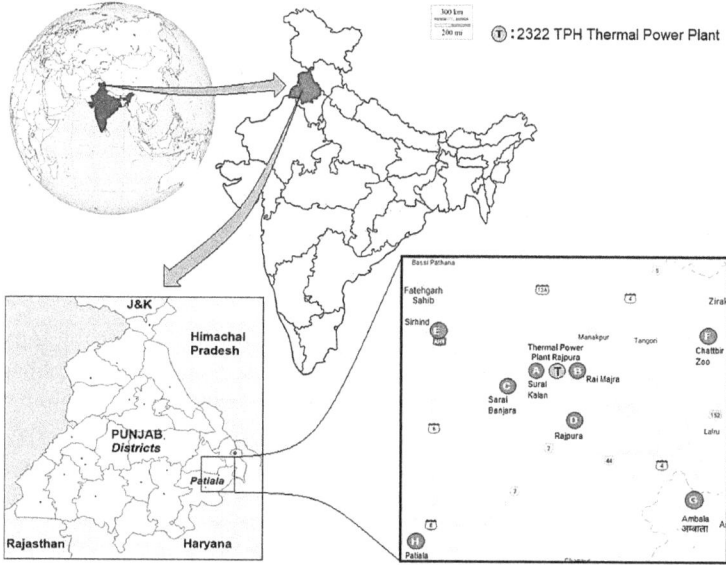

FIGURE 10.3   Study zone of 30 × 30 km area with the locations of thermal power plant and eight receptors.

within the study zone. The topographical area within the study zone was found to be flat with little-to-no variance in elevation. Table 10.2 displays the details of the eight receptor locations including direction and distance from the thermal power plant stack. The source inventory from the point sources (i.e. stacks) of the thermal power plant was prepared and compiled for the SO_2, NO_2, and SPM emissions. SO_2, NO_2, and SPM were taken as indicator pollutants for the source

TABLE 10.2
Direction and Distance of Sensitive Receptors from the Thermal Power Plant Stack

Receptor	Name	Direction and Distance from the Power Plant	Longitude	Latitude
1	Sural Kalan	W, 2 km	76°33′24.24″E	30°33′25.94″N
2	Rai Majra	E, 1.6 km	76°35′33.97″E	30°33′19.02″N
3	Sarai Banjara	SW, 5.7 km	76°31′09.96″E	30°32′32.75″N
4	Rajpura	S, 7.7 km	76°35′32.39″E	30°28′58.56″N
5	Sirhind	NW, 19.6 km	76°22′57.19″E	30°37′07.77″N
6	Chattbir Zoo	E, 22 km	76°47′19.24″E	30°36′02.94″N
7	Ambala	SE, 25.6 km	76°44′57.56″E	30°23′28.39″N
8	Patiala	SE, 29.4 km	76°22′14.39″E	30°21′13.13″N

TABLE 10.3
Emissions Data from the Stacks of the Power Plant in the Study Zone

S. No	Stack	Height of Stack (m)	Flue Gas Temperature (k)	Diameter of the Stack (m)	Emission Velocity (m/s)	Emission Rate (g/s)		
						SPM	SO_2	NO_2
1.	2322TPH	275	410	7.5	25	79.45	1151	1035
2.	2322TPH	275	421	7.5	24.7	79.34	1143	1027

inventory, due to their significant discharge rate during coal-based power production, and because the World Bank and Indian Standard emission guidelines have identified them as hazardous. Moreover, these pollutants are regularly monitored by several authorized bodies and their concentrations can be easily predicted by the given dispersion models. The emission parameters for the source inventory were calculated using a stack monitoring assessment performed during the peak working hours of the thermal power plant, with the help of a stack monitoring kit that complies with CPCB guidelines. The average emission characteristics of all point sources are shown in Table 10.3.

The meteorological parameters are important variables for any dispersion model as they play a crucial part in governing the dispersion of air pollutants in the air. Important meteorological conditions that control the diffusion and transport of pollutants are convective transport in horizontal direction (wind direction and speed), convective transport in vertical direction (mixing height and atmospheric stability), and topographical conditions of the study zone. Thus, to ensure accurate prediction of GLCs for the selected pollutants within the study zone, the mean hourly-average meteorological data for a duration of one year was retrieved from the installed Weather Monitoring Station (WatchDog weather station, Model 2900ET, Spectrum Technologies, USA).

Based on the source inventory and meteorological data, the dispersion analysis was performed using the ISCST3 and AERMOD models (Version 8.9.0) and was used to predict the daily average (24-h) GLCs of SO_2, NO_2, and SPM for the selected receptors within the study zone. Validation of models was achieved by comparing the predicted GLCs of pollutants from models at all the receptors with the actual observed concentrations. Since the operation of the power plant commenced in December 2013, the 24-h average background concentrations of all selected pollutants was monitored prior to and after the commencement of the operational activities in the power plant. Thus, it was possible to superimpose the daily average background concentrations of pollutants before the inauguration of the power plant over the predicted GLCs, and furthermore, validate this using the daily average observed concentrations. The results of both models were compared with each other, and then with the standard limits prescribed by the CPCB in NAAQS (CPCB, 2009).

The mean hourly-based meteorological data for a one year period was processed in the meteorological preprocessors of the AERMOD and ISCST3 systems, i.e. AERMET and PC-RAMMET, respectively. These meteorological preprocessors,

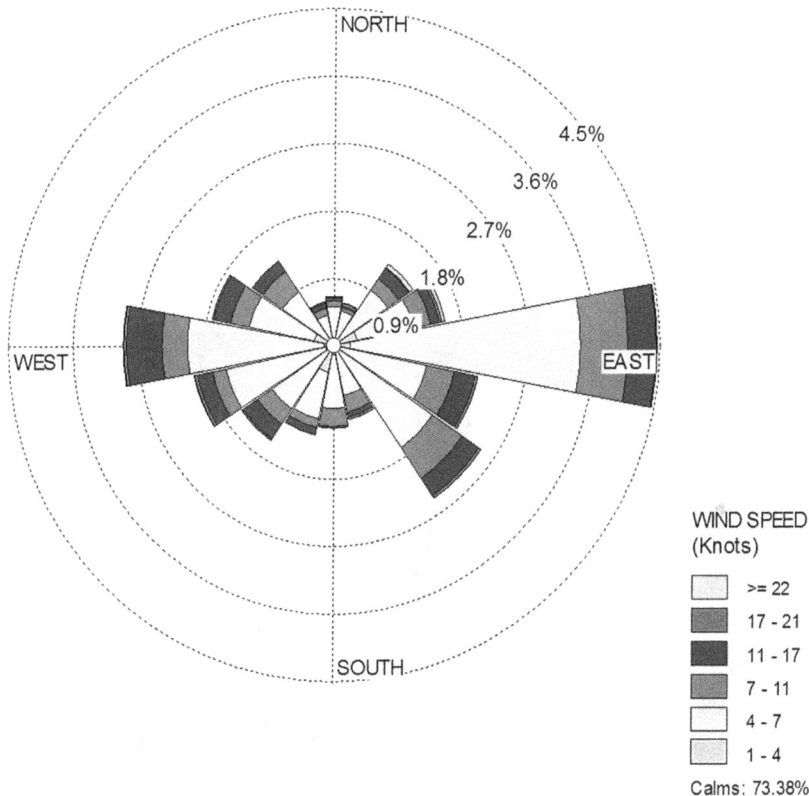

FIGURE 10.4   Wind rose diagram for the study period.

AERMET and PC-RAMMET, were used to create an output file that is used as a source file in AERMOD and ISCST3, respectively (US EPA, 1999; 2012). Standard surface parameter values were used viz. Bowen ratio, surface roughness, and albedo, based on the land use data for the relevant land types, in this case cultivable land 70%, and urban land 30%. Based on the meteorological data, a wind rose diagram was plotted using the meteorological pre-processors to determine the predominant wind directions and speeds within the study zone, for the study period. Figure 10.4 shows that the average wind speed was 1.52 m/s (1 Knot = 0.514 m/s) with a 73.3% frequency of calm winds, dominant wind directions were West, East, and South-east.

10.4 MODEL PREDICTIONS

With the help of Google Earth, the study zone (30 × 30 km), thermal power plant, and all receptors were identified on maps as shown in Table 10.2 and Figure 10.4. Due to the topographical situation of the study zone the flat terrain option was adopted, the building downwash option was not used during the model run.

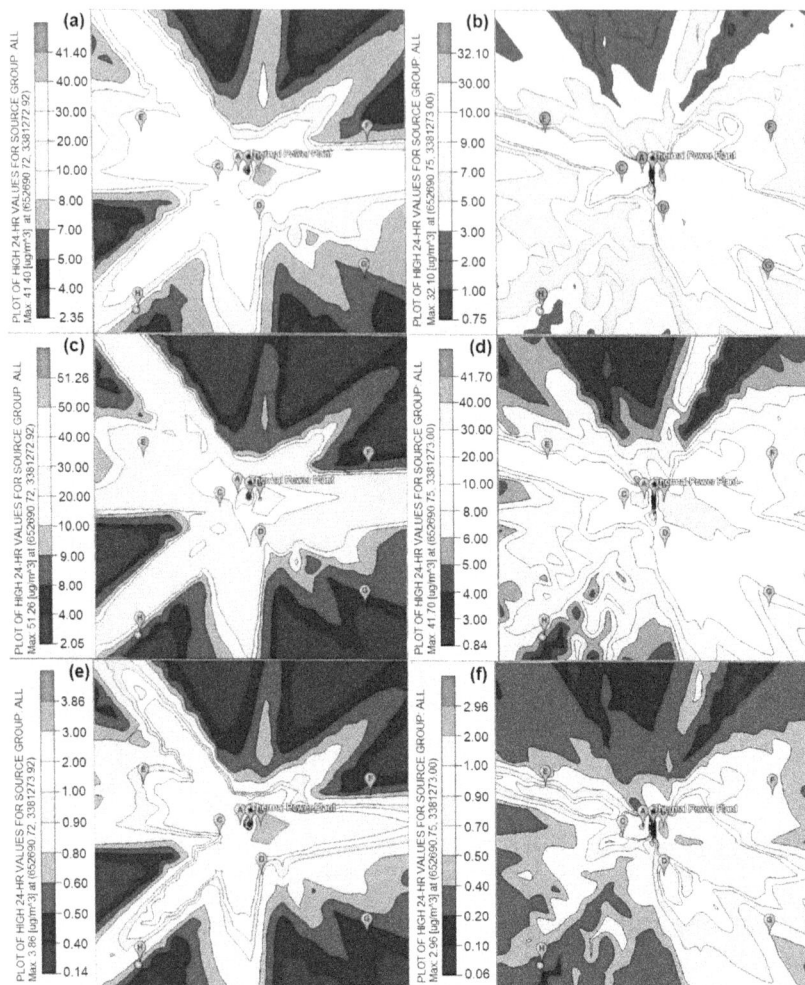

FIGURE 10.5 Contour maps of the 24-h (daily) average GLCs (in µg/m³) of (a), (b) NO₂; (c), (d) SO₂; and (e), (f) SPM; generated by the AERMOD and ISCST3 models, respectively.

After collecting and compiling the necessary source inventory and meteorological data required to simulate the dispersion of pollutants, the AERMOD and ISCST3 models were run successfully and the outcomes of the daily average SO_2, NO_2, and SPM GLCs were predicted with their corresponding contour maps.

Figure 10.5 (a) and (b) depicts the predicted daily average NO_2 GLC levels under AERMOD and ISCST3, respectively. AERMOD and ISCST3 predicted the highest NO_2 levels of 41.40 µg/m³ and 32.10 µg/m³, respectively, due to the presence of the thermal power plant. Consistent with the modeling results, the highest NO_2 concentrations were observed by both models at a same receptor i.e. Village Rai Majra (Receptor 2) adjacent to the plant site. Since the lifespan of

NO_x in air is 1–2 days due to wet and dry deposition effects (Baumbach, 1996), this assumption was disregarded during the models run. Moreover, the NO_x emissions due to the power plant were considered quite low due to the super-critical nature of the thermal power plant having a specialized low NO_x system. Also, it can be observed from Figure 10.5 (a) and (b) that the contour map generated by AERMOD for NO_2 emissions had more characteristics when compared to the ISCST3 contour map. The overall results of the maximum daily average NO_2 GLC from the power plant were below the daily standard concentration value (80 $\mu g/m^3$) prescribed by the CPCB/NAAQS/2009.

Figure 10.5 (c) and (d) depicts the daily average SO_2 GLCs predicted by both models due to the thermal power plant. AERMOD and ISCST3 predicted the maximum daily average SO_2 GLCs as 51.26 $\mu g/m^3$ and 41.70 $\mu g/m^3$, respectively. Similarly to the NO_2 predictions, the receptor most affected by SO_2 emissions was Receptor 2, and this was confirmed by both models. Though, the maximum daily average SO_2 concentrations from the thermal power plant were below the prescribed standard concentration value (80 $\mu g/m^3$) prescribed by the CPCB/NAAQS/2009, it should be noted that the power plant successfully retains a large amount of the generated SO_2 due to the high efficiency (~95%) of the flue gas desulphurization systems.

Figure 10.5 (e) and (f) depicts the daily (24-h) average SPM GLCs as predicted by the AERMOD and ISCST3 models. The highest daily SPM levels from the power plant were predicted to be 3.86 $\mu g/m^3$ and 2.96 $\mu g/m^3$ by AERMOD and ISCST3, respectively. These observations showed that SPM emissions from the thermal power plant did not exhibit any substantial effect on the receptors. One of the probable reasons for the low prediction levels under both models could be the exclusion of the dry or wet deposition aspect of the models. Furthermore, the high efficiency (99.9%) of electrostatic precipitator leads to reduced SPM emissions, such that regardless of coal utilization with a 34 wt. % ash content, SPM emissions are reduced. Moreover, no substantial influence of particulate matter on air quality is anticipated due to the prior installation of dust suppression systems at the coal stockyard and at all coal transfer and handling points. The predicted highest daily average SPM concentration from the power plant was very much lower than the daily prescribed limit (100 $\mu g/m^3$) prescribed by the CPCB/NAAQS/2009.

It can be noticed from the contour maps that AERMOD more accurately describes the distribution and characteristics of pollutants than ISCST3. The contour maps generated by AERMOD show that the impact of pollutants discharged from thermal power stacks can be observed at further distances than expected, whereas the impact of pollutants are mostly confined to a limited area, according to the contour maps generated by the ISCST3 models. These findings also suggested that emissions from the power plant were most obvious at the same receptor sites, as predicted by the models. This can be attributed to the fact that the AERMOD model uses Briggs equations under convective conditions, with wind and temperature gradients at the top of the stack and half-way to the rise of the plume. Conversely, the ISCST3 model considers Briggs equations with vertical wind speed and temperature gradient at the stack-top, under all stability conditions (US EPA, 2004).

According to the contour maps of both models, the greater dispersion of pol-lutants from the power plant stacks were confined to South-east and Easterly directions of the study zone which were also the predominant wind directions observed in the wind rose diagram. Even the un-polluted areas saw some effects from the pollutants due to long-range transport mechanisms. Hence, the wind parameters, primarily wind speed and direction could have played an important role in the displacement and dispersion of air pollutants and gases within the study zone.

10.5 COMPARISON OF AERMOD AND ISCST3

Although both models generated the similar contour map patterns, the predicted concentrations of selected pollutants differed at all receptors. Table 10.4 compiles the predicted SO_2, NO_2, and SPM concentrations under both models. According to Table 10.4, the predicted GLCs SO_2, NO_2, and SPM under AERMOD criteria were higher at all receptors, compared to the projections under ISCST3; this leads to overall higher expected pollutant levels from the thermal power plant under the AERMOD criteria.

ISCST3 projections tended to predict lower GLCs around the point source, when compared to AERMOD. As distance from the stack increased, the ISCST3 model predicted higher GLCs at only Receptor 7. These outcomes can be ascribed to the fact that both the models use different base algorithms in their respec-tive meteorological processors to generate the meteorological data. According to the model predictions, Village Rai Majra (Receptor 1) and Village Sural Kalan (Receptor 2) were highly affected regions, owing to their proximity to the power plant stack.

As discussed above, even though both models generated the same contour map patterns, the overall predictions of GLC differed quantitatively at all receptors.

TABLE 10.4
Maximum Daily Average GLCs of SO_2, NO_2, and SPM (in µg/m³) as Predicted by ISCST3 and AERMOD, due to the Thermal Power pPant

Receptor	SO_2		NO_2		SPM	
	AERMOD	ISCST3	AERMOD	ISCST3	AERMOD	ISCST3
1	38.71	33.41	28.17	23.43	2.69	2.02
2	51.26	41.7	41.4	32.1	3.86	2.96
3	28.14	22.14	27.37	23.18	2.41	1.97
4	29.36	24.74	26.43	24.76	1.91	1.57
5	26.64	20.17	25.78	22.59	1.52	0.92
6	8.97	8.29	9.25	8.51	0.77	0.62
7	7.97	8.15	6.9	6.93	0.48	0.61
8	8.16	5.19	5.02	4.77	0.67	0.39

The similarity in the contour maps could be due to the assumptions during the models as there are some common algorithms used by both. However, the actual difference in the results can best be observed during a detailed analysis of the GLCs. In general, the GLCs of the selected pollutants as predicted by AERMOD were higher in comparison to the ISCST3 model. Various factors are responsible for the difference in the predicted results of both models. For example, when the mixing lid was lower than the plume centerline under convective conditions, the ISCST3 model assumed there were zero GLCs during the model run. On the other hand, AERMOD reflects three mechanisms of plume behavior in the atmosphere; (a) a plume that enters the mixing lid and gently scatters in the stable loft layer; (b) an 'indirect plume' that is brought directly to the ground when it is trapped in an updraft on reaching the mixing lid and; (c) a 'direct plume' that is diverted back to ground in a downdraft (US EPA, 2004). The ISCST3 model faces possible under-predictions in its results due to its 'all or nothing' treatment of plumes, an issue that is avoided in the AERMOD model. AERMOD's approach to 'convective downdrafts and updraft behavior of plumes in a probability density function (PDF)' is one advanced feature AERMOD has over ISCST3, which establishes the higher prediction validity of the model (US EPA, 1995; 2004). Another reason for the difference in the results could be the involvement of surface parameters such as surface roughness, Bowen ratio, and albedo parameters in AERMOD, which are not considered by ISCST3 algorithms (Faulkner et al., 2012). While AERMOD computes the surface parameters in its meteorological pre-processors by considering the seasonal variations in the data, the ISCST3 meteorological pre-processor does not compute these and consider the surface features at the receptor sites (Silverman et al., 2012). Besides this, the AERMOD model uses Gaussian treatment in a vertical and horizontal direction for plume dispersion under stable conditions and a non-Gaussian PDF for plume dispersion in vertical direction, whereas, ISCST3 considers only Gaussian treatment for dispersion of plume in vertical and horizontal directions. As a result of using a non-Gaussian PDF for plume dispersion in unstable conditions, the AERMOD approach offers improvements over ISCTS3, and is closer to a representation of observed conditions (US EPA, 2003). Furthermore, AERMOD considers convective downdrafts and updrafts under unstable meteorological conditions (for example, greater turbulence) and under stable meteorological conditions (for example, low turbulence due to convention and buoyancy) it considers wind speed and temperature variations on the top of the stack. On the contrary, the involvement of convective turbulence is not considered in the ISCST3 model algorithms (Silverman et al., 2012). Hence, the updated algorithms in the AERMOD model lead to a more realistic and precise projection of conditions.

10.6   VALIDATION OF MODELING RESULTS

Validation of model performance is an important step in evaluating the predictive worth of a model. The best way to assess model performance is to compare the predicted values with observed values for a given receptor location, and then present the comparison results in a series of correlation coefficient graphs. The background concentrations of pollutants were monitored before the thermal power plant operations and the monitored values were superimposed with the predicted values from the models. Cumulative values were compared with the observed values of the pollutants after the commencement of the power plant. Thus, validation of model performance was performed for all selected pollutants, SO_2, NO_2, and SPM, at each receptor location separately.

10.6.1 Validation of NO_2 Model Performance

Table 10.5 shows the predicted and observed NO_2 concentrations at each receptor. According to the results, both models under-predicted NO_2 concentrations relative to the observed data at Receptors 1, 2, 3, 4, and 7. This may be a product of the model's inability to account for long-range NO_2 transports.

Additionally, these outcomes are somewhat expected due to the lack of consideration of building downwash effects in the models which would report higher concentrations of pollutants at nearby residential areas due to the strike of the plume on local buildings. During the modeling process, the effect of building-downwash was not taken into account due to the large study zone, and consequently, this might have led to the under-predicted NO_2 values in both models. Table 10.5 shows that both models over predicted NO_2 concentrations at Receptors 5, 6, and 8. The reason for the difference in NO_2 concentrations in the models, and at all receptors, could be the variance in background NO_2 concentrations prior to and after the commencement of the thermal power plant. In Figure 10.6(a), the AREMOD correlation coefficient is 0.852, higher than that of 0.797 for the ISCST3 model, shown in Figure 10.6(b). This indicates it is a better predictor of NO_2 concentrations for these receptors.

TABLE 10.5
Predicted and Observed Values of NO_2

Receptor	Background Concentration Before Power Plant Operation ($\mu g/m^3$)	Predicted GLCs of NO_2 ($\mu g/m^3$) AERMOD	ISCST3	Observed GLC After the Power Plant Operations ($\mu g/m^3$)
A	16.85	44.97	40.23	53.57
B	13.23	54.64	45.3	57.81
C	19.96	47.27	43.08	48.43
D	31.72	58.13	56.46	59.67
E	24.28	50.07	46.88	37.12
F	17.44	26.7	25.96	23.13
G	21.25	28.16	28.19	32.10
H	9.73	14.79	14.54	11.61

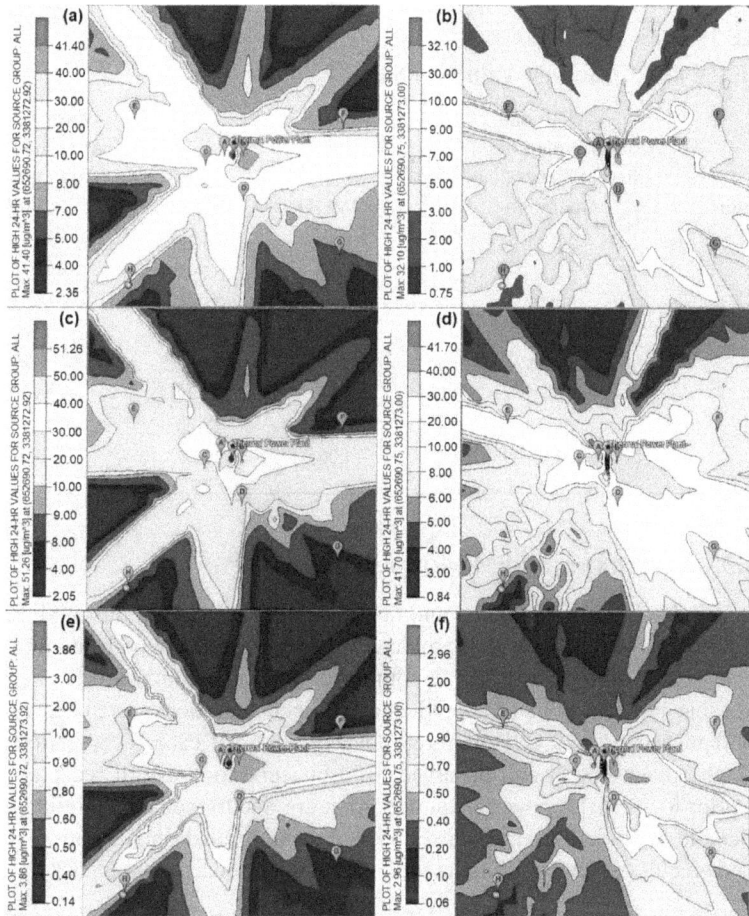

FIGURE 10.6 Comparison of observed GLCs and predicted GLCs of (a) NO_2 by ISCST3 and (b) NO_2 by AERMOD; (c) SO_2 by ISCST3 and (d) SO_2 by AERMOD; and (e) SPM by ISCST3 and (f) SPM by AERMOD.

10.6.2 VALIDATION OF SO_2 MODEL PERFORMANCE

Table 10.6 displays the predicted and observed concentrations of SO_2 at all receptors. The results generated by both models were close at most of the receptors after they were combined with background SO_2 concentrations.

However, both models over predicted the observed SO_2 concentrations at all receptor locations, which could be due to the lower background SO_2 concentrations before power plant operation. In Figure 10.6, the correlation coefficients for AERMOD (Figure 10.6c) and ISCST3 (Figure 10.6d) were calculated as 0.756 and 0.754, respectively. The R^2 values were found to be comparable for both figures; however, AERMOD predicted the observed SO_2 concentrations marginally better than ISCST3.

TABLE 10.6
Predicted and Observed Values of SO₂

Receptor	Background Concentration (µg/m³) Before Power Plant Operation	Predicted GLCs of SO₂ (µg/m³)		Observed GLC After the Power Plant Operation (µg/m³)
		AERMOD	ISCST3	
1	12.82	51.51	46.21	45.51
2	11.25	62.46	52.9	51.23
3	13.86	41.94	35.94	34.34
4	27.28	56.56	51.94	37.09
5	23.44	50.04	43.57	27.40
6	15.25	24.17	23.49	23.30
7	24.92	32.87	33.05	29.42
8	5.95	14.06	11.09	8.87

10.6.3 VALIDATION OF SPM MODEL PERFORMANCE

Table 10.7 displays the predicted and observed SPM concentrations at all receptors. Similar to the SO₂ results, predicted SPM values under both models were close at most receptors, after being combined with background SPM concentrations. Table 10.4 shows that the predicted SPM concentrations were considerably lower than background and observed concentrations at each receptor location. The higher background and observed SPM concentrations at all receptor locations prior to and after power plant operation are due to the concentration effect attributable to vehicular emissions, domestic heating, and open burning. These are known culprits contributing to ambient air SPM concentrations, but were not

TABLE 10.7
Predicted and Observed Values of SPM

Receptor	Background Concentration (µg/m³) Before the Power Plant Operation	Predicted GLCs of SPM (µg/m³)		Observed GLC After the Power Plant Operation (µg/m³)
		AERMOD	ISCST3	
1	109.95	112.59	112.92	127.35
2	93.14	96.96	96.06	143.44
3	118.37	120.71	120.27	157.56
4	218.88	220.71	220.37	239.38
5	170.49	171.92	171.32	177.45
6	97.21	97.98	97.83	102.48
7	177.78	178.26	178.39	187.46
8	89.39	90.06	89.78	94.57

considered in the model specifications. The results show that cumulative background levels and predicted concentrations were lower than monitored concentrations at all receptors. The low predicted SPM values are a result of low SPM emission rates from the power plant and could be the reason for the lower cumulative predicted concentrations. Figure 10.6 compares predicted and observed SPM concentrations at all the receptors. The correlation coefficients for AERMOD (Figure 10.6e) and ISCST3 (Figure 10.6f) were 0.860 and 0.855, respectively. Similar to the findings of SO_2 and NO_2, the R^2 values were comparable in both cases; however, when compared with observed data, the AERMOD model predicted SO_2 concentrations more accurately than the ISCST3 model. Overall, it can be speculated that when predicting the dispersion of pollutants, the AERMOD model is more accurate, significant, and valid than ISCST3.

10.7 CONCLUSION

The chapter addresses air quality estimation models using observed data and comparing this to the outcome of two commonly used USEPA models, AERMOD and ISCST3. The study area and period concern sensitive and highly populated areas around the supercritical technology-based thermal power plant. Inventory data from the thermal power plant stacks and meteorological readings were compiled and processed in both models to predict the (24-h) daily average ground-level concentrations of common GLCs, SO_2, NO_2, and SPM; these were used to form contour maps for the study zone. The predicted GLCs from both models were analyzed, compared with each other, and then observed data. Although, the spatial dispersion of the pollutants in the contour maps showed that AERMOD and ISCST3 generate a similar map pattern under similar conditions, the GLCs were found to be different for each pollutant at each receptor. Additionally, the AERMOD predictions were more accurate and the contour maps were more characteristic of the observed data, when compared to the ISCST3 projections. The GLC levels predicted by both models were superimposed over the background concentration levels from before the power plant commenced operation, then the cumulative concentrations were compared with observed concentrations after power plant operation began. Overall, the AERMOD results were more valid and accurate in terms of correlation coefficient. The predicted and observed values appeared closer to the trend line when compared to the ISCST3 results.

REFERENCES

Baumbach, G. (1996). Air Quality Control: Formation and Sources, Dispersion, Characteristics and Impact of Air Pollutants: Measuring Methods Techniques for Reduction of Emissions and Regulations for Air Quality Control (pp 15–78). Spring-Verlag, New York.

Central Electricity Authority (CEA) Annual Report 2017–18. Ministry of Power, Government of India (pp 74-93). Access to report: http://www.cea.nic.in/annualreports.html.

Cohen, A.J., Anderson, H.R., Ostro, B., Pandey, K.D., Krzyzanowski, M., Künzli, N., Gutschmidt, K., Pope, C.A. III, Romieu, I., Samet, J.M., Smith, K.R. (2004). Urban Air Pollution. Comparative Quantification of Health Risks: Global and Regional Burden of Disease Attributable to Selected Major Risk Factors. World Health Organization, Geneva, Switzerland. pp 1153–1433.

Cooper, C. D., & Alley, F. C. (2002). Air Pollution Control: A Design Approach, 3rd edition. Waveland Press, Inc, Long Grove, United State of America.

CPCB (2009). National ambient air quality standards, notification dated November 18, 2009. Central Pollution Control Board, Ministry of Environment and Forests, and Climate Change, Government of India, New Delhi. Access: https://cpcb.nic.in/air-quality-standard/

Faulkner, W.B., Shaw, B.W., Grosch, T. (2012). Sensitivity of two dispersion models (AERMOD and ISCST3) to input parameters for a rural ground-level area source. *Journal of the Air & Waste Management Association*, 58(10), 1288–1296.

Holmes, N. S., & Morawska, L. (2006). A review of dispersion modelling and its application to the dispersion of particles: An overview of different dispersion models available. *Atmospheric environment*, 40(30), 5902–5928.

Ministry of Environment of New Zealand (MENZ), (2004). Good Practice Guide for Atmospheric Dispersion Modelling (pp 8–21), Wellington, New Zealand. Access report at: http://www.mfe.govt.nz/publications/air/atmospheric-dispersion-modelling-jun04/html/page5.html

Mishra, U.C. (2004). Environmental impact of coal industry and thermal power plants in India. *Journal of Environmental Radioactivity*, 72(1–2), 35–40.

Pokale, W.K. (2012). Effects of thermal power plant on environment. *Scientific Reviews and Chemical Communications*, 2(3), 212–215.

Silverman, K.C., Tell, J.G., Edward, S.V., Zeyuan, Q. (2012). Comparison of the industrial source complex and AERMOD dispersion models: Case study for Human Health Risk Assessment. *Journal of the Air & Waste Management Association*, 57(12), 1439–1446.

U.S. Environmental Protection Agency (1995). Industrial Source Complex (ISC3) Dispersion Model User's Guide. EPA 454/B 95 003a (vol. I) and EPA 454/B 95 003b (vol. II). U.S. Environmental Protection Agency, Research Triangle Park, NC.

U.S. Environmental Protection Agency (1999). PCRAMMET User's EPA- 454B-96-001 Research Triangle Park, North Carolina, Office of Air Quality Planning and Standards Emission, Monitoring and Analysis Division.

U.S. Environmental Protection Agency (2003). AERMOD: Latest Features and Evaluation Results EPA-454/R-03-003, Research Triangle Park, North Carolina, Office of Air Quality Planning and Standards Emission, Monitoring and Analysis Division.

U.S. Environmental Protection Agency (2004). AERMOD: Description of Model Formulation. U.S. Environmental Protection Agency, Research Triangle Park, NC.

U.S. Environmental Protection Agency (2012). Addendum User's Guide for the AERMOD Meteorological Preprocessor—AERMET [Addendum to EPA- 454/B-03-002 (dated November 2004)]. U.S. Environmental Protection Agency, Research Triangle Park, NC.

WHO (2002). World Health Report 2002: Reducing Risk, Promoting Healthy Life. World Health Organization, Geneva, Switzerland.

11 Spatiotemporal Dispersion of Particles in the Vicinity of an Opencast Mine in India

Aditya Kumar Patra
Department of Mining Engineering, Indian Institute
of Technology Kharagpur, Kharagpur, India

Satya Prakash Sahu
Mahanadi Coalfields Limited,
Coal India Limited, Kolkata, India

CONTENTS

11.1 INTRODUCTION

Particulate matter (PM), also commonly known as dust, is the primary air pollutant that affects humans, plants, animals and structures. PM will negatively affect the health of humans and cause asthma (Pless-Mulloli et al., 2000; Banks et al., 1998), bronchitis, cardiovascular diseases (Hendryx, 2009; Chen et al., 1990), Parkinson's and Alzheimer's disease (Buzea et al., 2007). The location where PM is deposited in the human respiratory tract will determine how the health is affected. Particles that are harmful to human health are PM_{10} (particulate matter having aerodynamic diameter 10 µm or less) and the finer fraction, $PM_{2.5}$ (particulate matter having aerodynamic diameter 2.5 µm or less) because $PM_{2.5}$ can reach the

DOI: 10.1201/9781003140382-11

alveolar region of lungs and cause more severe health damages (Gautam et al., 2016). Mining of coal is undertaken through two methods: the underground method for deep deposits and the opencast method for shallow deposits close to the surface. In comparison to underground mines, surface mines usually produce more due to the large-scale of the operations and use of high-capacity machines. Opencast mining emits PM of different sizes directly to the atmosphere and deteriorates human health and the surrounding environment (Kumar et al., 2014; Patra et al., 2016). Pandey et al. (2014) assessed $PM_{2.5}$ and PM_{10} in four monitoring sites. Each site varies in direction and distance from the Ena colliery in Jharia Coal Fields of India. The concentrations of PM_{10} and $PM_{2.5}$ during winter were 2–2.7 times and 1.6–1.9 times higher than the limits (100 μg m^{-3} for PM_{10} and 60 μg m^{-3} for $PM_{2.5}$; 24-h time weighted average), set by NAAQS of India (CPCB, 2009). In a community exposure study in Virginia, USA, Aneja et al. (2012) reported that PM_{10} of coal dust generated by hauling coal in trucks during surface mining operations significantly exceeded the safe limit (150 μg m^{-3}; 24-h time weighted average, NAAQS, USA). Another study of coal mining in Zongultak city, Turkey revealed that mining activities generated more inhalable coarse PM ($PM_{2.5}$–PM_{10}) than fine PM ($PM_{2.5}$), with average concentrations of 29.38, 23.85 and 53.23 μg m^{-3} for $PM_{2.5}$, $PM_{2.5-10}$ and PM_{10}, respectively (Tecer et al., 2008). This study revealed that relative humidity, cloudiness and low temperature were found to contribute to the increase of PM episodic events. Several other related studies have also been assessing the PM level in terms of mining locality in the last two decades (Pandey et al., 2014; Aneja et al., 2012; Kakosimos et al., 2011; Tecer et al., 2008; Ghose and Majee, 2001).

Understanding the impact PM pollution in the mining local environment can determine the types of mitigation measures required to constrain the transport of pollutants from the mine. Few studies have been carried out to explore how far PM can travel while still significantly affecting air quality. This chapter presents findings from several studies and has two broad objectives: (a) to measure quantitatively the temporal and spatial variation of mass concentrations of PM_{10}, $PM_{2.5}$ and PM_1 escaping an active pit and (b) to assess the role of meteorology and the distance PM can travel from mine areas.

11.2   STUDY SITE AND METHODOLOGY

The study was conducted in a large opencast coal project (OCP), located in the eastern part of India. The project produced 10 million tonnes of coal per annum. The surface RL (RL stands for "Reduced Level") was 273.41 m and the pit bottom RL was 189.24 m. Mining operations were carried out using surface miners, pay loaders and tippers. The pit layout is shown in Figure 11.1.

PM concentration was measured using aerosol spectrometer (Model 1.108, Grimm, GRIMM Aerosol Technik GmbH & Co. KG, Germany) (Grimm, 2010). The instrument used light scattering technology to measure particle concentration in 15 channels with the size varying from 0.3 to 20 μm. A portable weather

FIGURE 11.1   Layout of study mine (Google maps).

station (Spectrum Technologies, Inc., Model WatchDog 2000) was used to measure local meteorological parameters (Figure 11.2). It recorded wind direction,
wind speed, humidity, pressure, temperature, dew point, solar radiation and rain
fall (Spectrum, 2010). Measurements were taken at 24 locations which consisted
of 4 sections with 6 locations along each section. The sampling was collected in
the predominant downwind direction (Figure 11.3). The study was carried out
during November 2016–January 2017.

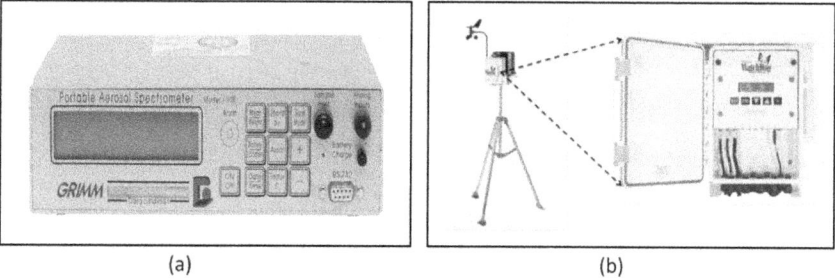

(a)   (b)

FIGURE 11.2   (a) Aerosol spectrometer (Grimm, 2010) and (b) weather station (Spectrum,
2010).

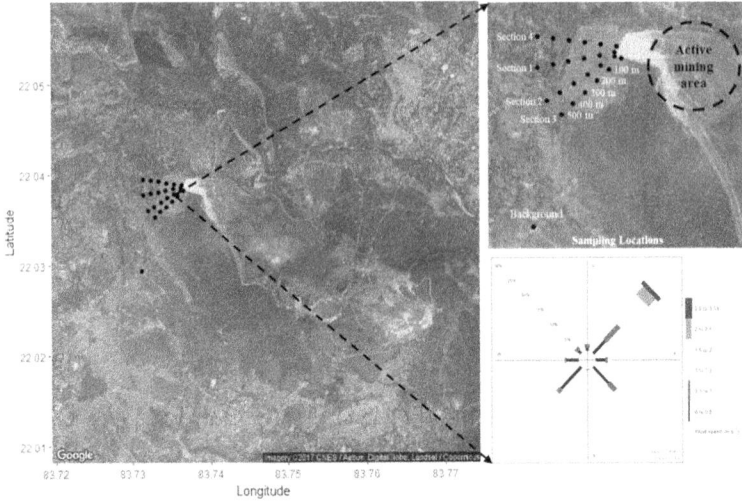

FIGURE 11.3   Sampling locations.

11.3   RESULTS AND DISCUSSION

11.3.1   Local Meteorology

Typical meteorological parameters at the study site are given in Figure 11.4. RH varied in the range 27.1–92.8%. The high relative humidity at the start of the days between 6 and 7 am was primarily contributed by the fog in the winter months.

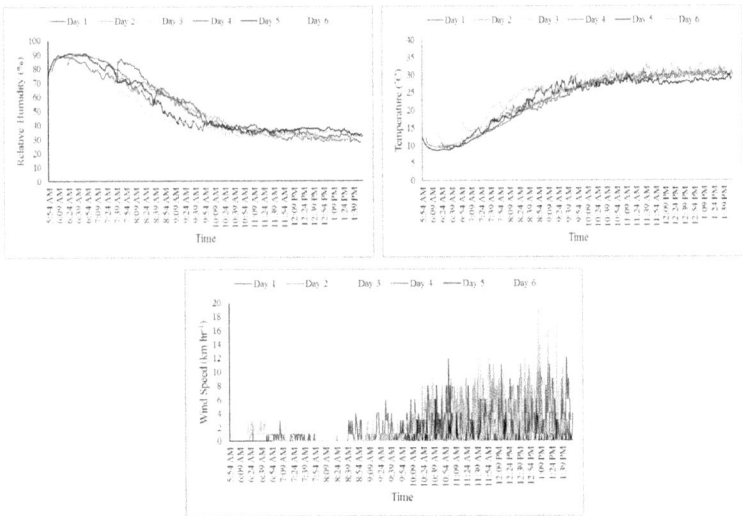

FIGURE 11.4   RH, temperature and wind speed at study site (Section 1; 25.11.2016–30.11.2016).

The temperature varied from 9.1 to 33.6°C. As expected, with the increase of temperature, the RH would decrease. During the period of the study, wind speed varied from 0 to 19 km h^{-1} (Figure 11.4).

11.3.2 PM Concentration

The mean ranges of PM concentrations were – PM$_{10}$: 281.08–509.06 μg m^{-3}, PM$_{2.5}$: 151.30–205.83 μg m^{-3} and PM$_1$: 118.89–160.58 μg m^{-3}. The PM$_{10}$ concentration was therefore 2.81–5.00 times over the limit set by the National Ambient Air Quality Standard (NAAQS) of India (CPCB, 2009) (100 μg m^{-3}; 24-hr average). A box plot of PM concentrations of Section 1 is shown in Figure 11.5. The frequency distribution plot shows that up to 84% of the PM$_{10}$ concentrations lies between 0 and 500 μg m^{-3}, up to 85% of the PM$_{2.5}$ concentrations lies within 300 μg m^{-3} and up to 80% of the PM$_1$ concentrations lies within 200 μg m^{-3} (Figures 11.6 and 11.7).

Average PM concentration between the start of the experiment and 8:00 am was low at each location due to the low temperature and the low wind speed in the morning. It began to increase 5–10 times between 8:00 am and 10:30 am because the higher temperature and the higher wind speed would assist the mobility of PM in the air. A typical temporal concentration profile is shown in Figure 11.8.

11.3.3 Contribution of Mining to PM Level at its Surroundings

The proportions of different particle sizes in the airborne PM in mine surroundings varied in distance. In general, the proportion of coarse particles (PM$_{2.5-10}$) would decrease in distance and fine particles (PM$_{1-2.5}$ and PM$_1$) would increase in distance. In Section 1, at pit boundary, the average contribution of PM$_{2.5-10}$, PM$_{1-2.5}$ and PM$_1$ were 73%, 9% and 18%, respectively. It changed to 58%, 11% and 31%, respectively, at the distance of 500 m (Figure 11.9). The background PM levels were 55%, 11% and 34% for PM$_{2.5-10}$, PM$_{1-2.5}$ and PM$_1$, respectively (Figure 11.10). These findings indicated that mining would generate more coarse particles than fine particles (Tecer et al., 2008). With the increase of the distance from the mine, the coarse fractions in air would gradually decrease and fine fractions would increase. This was mainly because coarse particles proportionately had a higher settling rate than fine particles.

11.3.4 Role of Meteorology and Distance on PM Concentration

The effect size statistics (partial eta squared) was estimated using GLM to determine the PM concentration variability affected by the parameters (RH, wind speed and distance). GLM results are summarized in Tables 11.1–11.3.

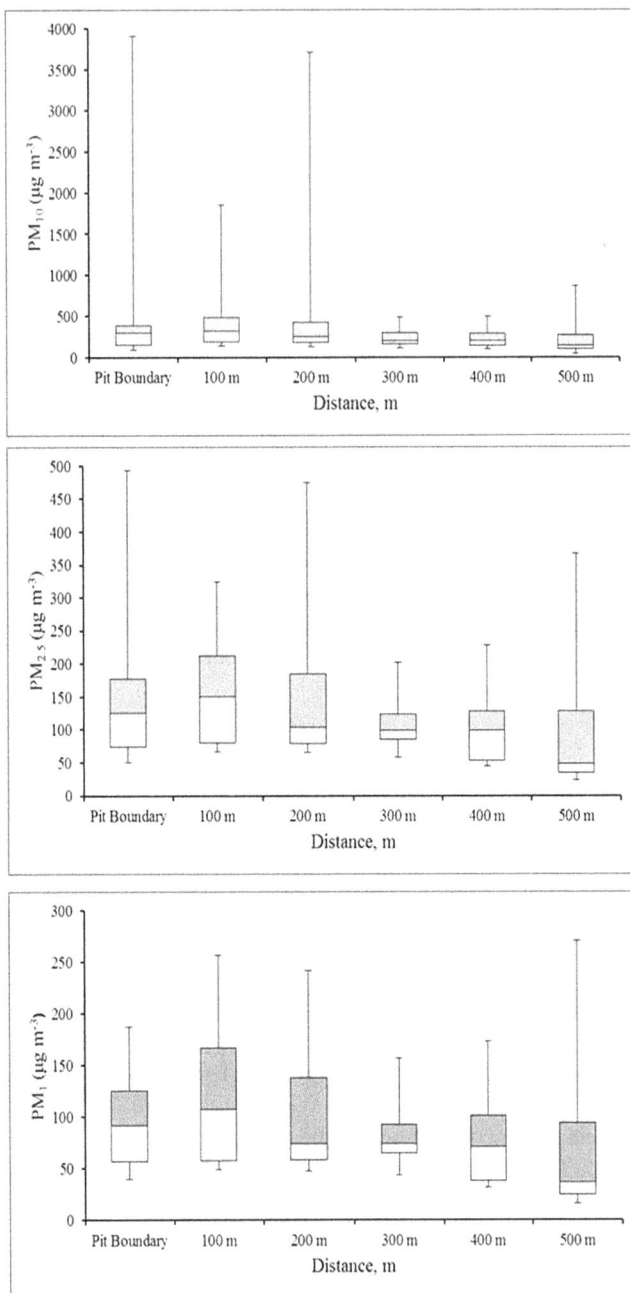

FIGURE 11.5   Box plot of PM$_{10}$, PM$_{2.5}$ and PM$_1$ along Section 1.

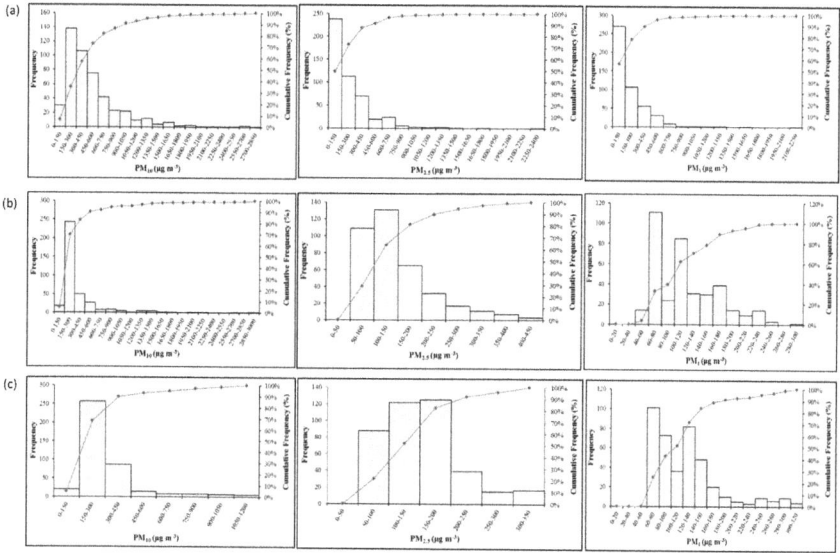

FIGURE 11.6 Frequency distributions of PM concentrations for Section 2 (a) pit boundary, (b) 100 m and (c) 200 m.

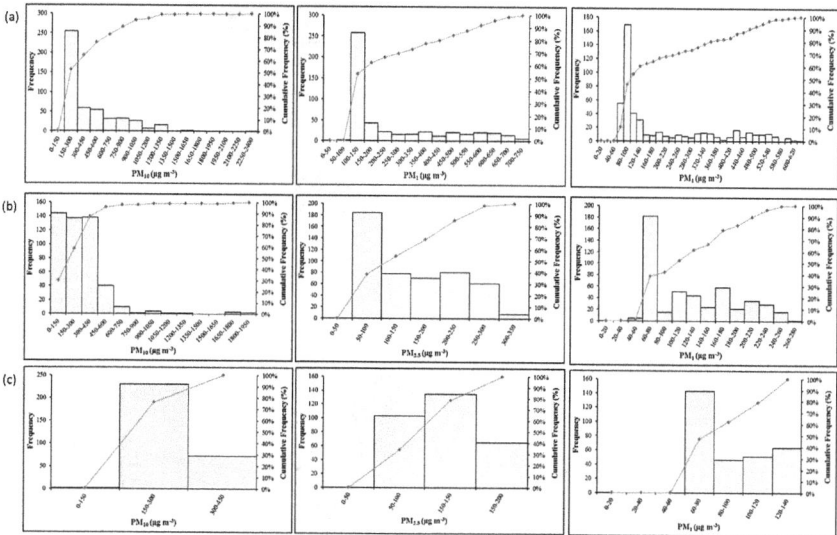

FIGURE 11.7 Frequency distributions of PM concentrations for Section 2 (a) 300 m, (b) 400 m and (c) 500 m.

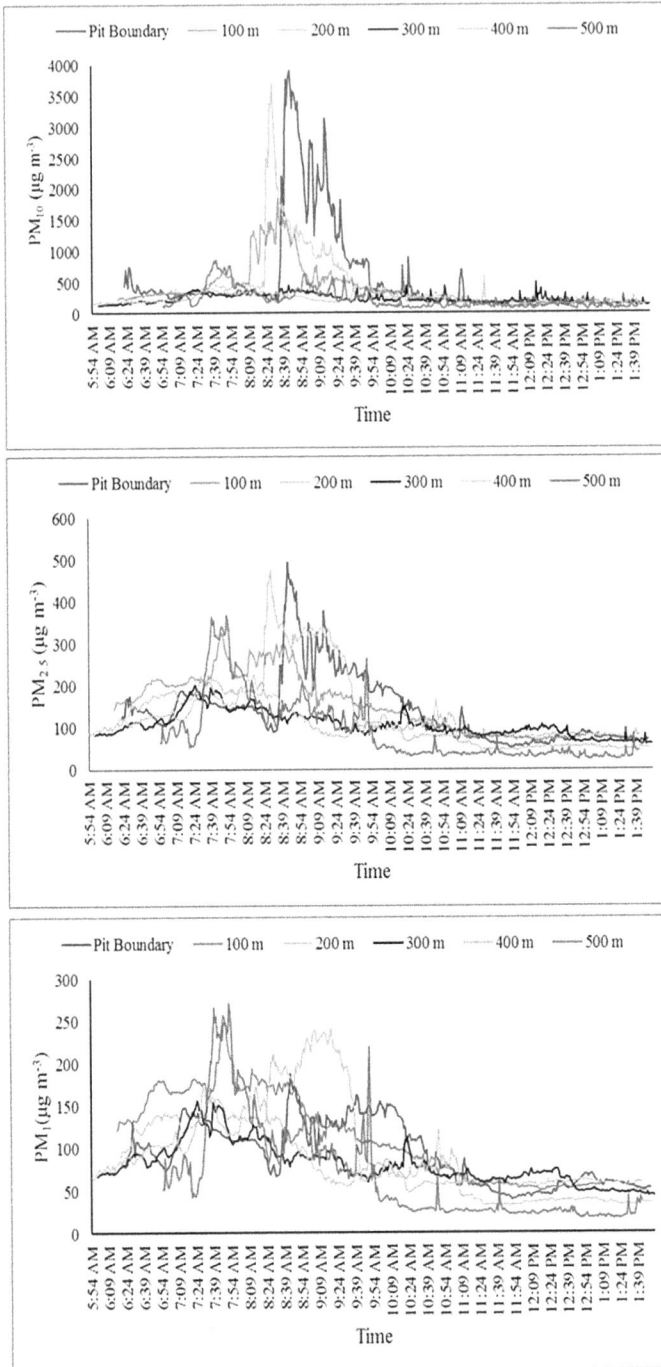

FIGURE 11.8 Temporal variations of PM concentrations along Section 1.

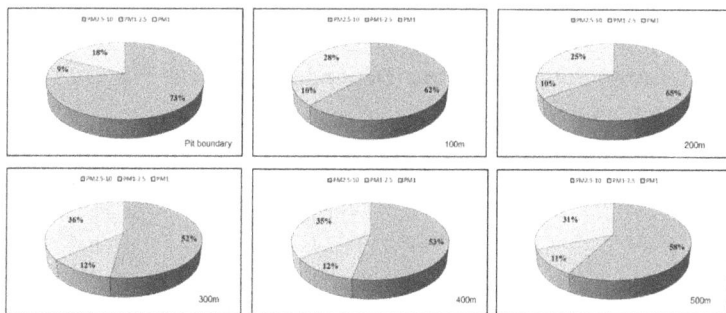

FIGURE 11.9 Proportions of $PM_{2.5-10}$, $PM_{1-2.5}$ and PM_1 (Section 1).

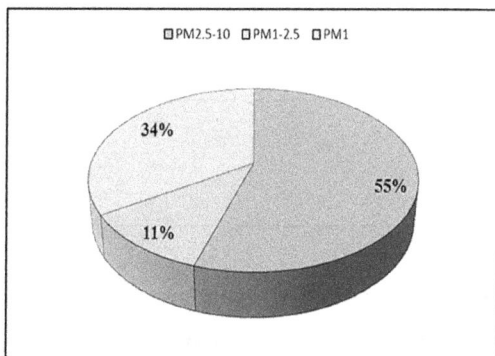

FIGURE 11.10 Proportions of $PM_{2.5-10}$, $PM_{1-2.5}$ and PM_1 at background location.

TABLE 11.1
Tests of between-Subjects Effects (Dependent Variable: PM_{10})

Source	Type III Sum of Squares	df	Mean Square	F	Sig.	Partial Eta Squared
Corrected Model	15.221[a]	7	2.174	14.714	0.000	0.464
Intercept	7.286	1	7.286	49.306	0.000	0.293
RH	4.185	1	4.185	28.319	0.000	0.192
WS	2.333	1	2.333	15.787	0.000	0.117
PMi[b]	1.355	1	1.355	9.171	0.003	0.072
D	4.614	4	1.154	7.806	0.000	0.208
Error	17.586	119	0.148			
Total	4202.241	127				
Corrected Total	32.807	126				

[a] $R^2 = 0.464$ (Adjusted $R^2 = 0.432$).
[b] PMi = PM concentration at the pit limit.

TABLE 11.2
Tests of between-Subjects Effects (Dependent Variable: PM$_{2.5}$)

Source	Type III Sum of Squares	df	Mean Square	F	Sig.	Partial Eta Squared
Corrected Model	19.653[a]	7	2.808	17.001	0.000	0.500
Intercept	71.904	1	71.904	435.428	0.000	0.785
RH	6.495	1	6.495	39.333	0.000	0.248
WS	1.646	1	1.646	9.965	0.002	0.077
PMi	1.489	1	1.489	9.019	0.003	0.070
D	3.006	4	0.752	4.551	0.002	0.133
Error	19.651	119	0.165			
Total	3241.160	127				
Corrected Total	39.304	126				

[a] $R^2 = 0.500$ (Adjusted $R^2 = 0.471$).

TABLE 11.3
Tests of between-Subjects Effects (Dependent Variable: PM$_1$)

Source	Type III Sum of Squares	df	Mean Square	F	Sig.	Partial Eta Squared
Corrected Model	23.849[a]	7	3.407	20.381	0.000	0.545
Intercept	66.849	1	66.849	399.908	0.000	0.771
RH	8.547	1	8.547	51.128	0.000	0.301
WS	1.396	1	1.396	8.353	0.005	0.066
PMi	2.480	1	2.480	14.838	0.000	0.111
D	3.278	4	0.819	4.902	0.001	0.141
Error	19.892	119	0.167			
Total	2909.938	127				
Corrected Total	43.741	126				

[a] $R^2 = 0.545$ (Adjusted $R^2 = 0.518$).

RH was the dominating factor for the variability of concentration of finer particles (PM$_{2.5}$ and PM$_1$) and distance was the predominant parameter for PM$_{10}$. Distance accounted for as high as 20.8% of the variation of the PM concentration, and RH accounted for 14.1–24.8% of the variability. The range of variability that the GLM accounted for PM concentration was 46–55%.

11.4   CONCLUSION

The study particularly revealed that surface coal mine is a significant source of respirable particle matter in the air around the mining surroundings. The following conclusions are drawn from the study.

a. The contribution of the mine to the PM level in its surroundings spreads beyond the distance of 500 m from the mine.

b. Surface mining operations involving coal cutting by surface miners would increase the pollution level in the vicinity of the mine up to 2.74, 1.34 and 2.77 times in the background concentration of $PM_{2.5-10}$, $PM_{1-2.5}$ and PM_1, respectively, at pit boundary. During peak production period, the concentrations rise was 5–15 times the average during non-active phase mining operations.

c. Over the distance of 500 m from the mine, the average proportion of $PM_{2.5-10}$ would decrease from 73% to 27%, and PM_1 would increase from 18% to 61%. However, $PM_{1-2.5}$ showed very little variation over the distance (8–12%).

This study provided a better understanding of the spatial distribution of PM concentrations around a typically highly-mechanized surface mining site and the influence of PM on local air quality, which was rarely found in present literatures. It therefore would require the implementation of more effective dust control used by the mine operators in order to protect the health of the people living around the mine. Moreover, these people were also depending on the mine for their livelihood. We believe that more studies aiming to evaluate how the mining would influence local pollution level at different mining sites should be carried out to gain further understanding of the air pollution caused by PM.

REFERENCES

Aneja, V.P., Isherwood, A., Morgan, P., 2012. Characterization of particulate matter (PM_{10}) related to surface coal mining operations in Appalachia. Atmos. Environ. 54, 496–501.

Banks, D. E., Wang, M. L., Lapp, N. L., 1998. Respiratory health effects of opencast coalmining: a cross sectional study of current workers. Occup. Environ. Med. 55(4), 287.

Buzea, C., Pacheco, I.I., Robbie, K., 2007. Nanomaterials and nanoparticles: sources and toxicity. Biointerphases. 2(4), MR17–MR71.

Chen, S.Y., Hayes, R.B., Liang, S.R., Li, Q.G., Stewart, P.A., Blair, A., 1990. Mortality experience of hematite mine workers in China. Br. J. Ind. Med. 47(3), 175–181.

CPCB, 2009. http://cpcb.nic.in/National_Ambient_Air_Quality_Standards.php (accessed on September 28, 2019).

Gautam, S., Kumar, P., Patra, A.K., 2016. Occupational exposure to particulate matter in three Indian opencast mines. Air Qual. Atmos. Health 9(2), 143–158.

Ghose, M.K., Majee, S.R., 2001. Air pollution due to opencast coal mining and its control in Indian context. J. Sci. Ind. Res. 60, 786–797.

Grimm, 2010. Operational manual of Portable Laser Aerosol spectrometer and dust monitor (Model 1.108/1.109). GRIMM Aerosol Technik GmbH & Co. KG, Ainring, Germany.

Hendryx, M., 2009. Mortality from heart, respiratory, and kidney disease in coal mining areas of Appalachia. Int. Arch. Occup. Environ. Health. 82(2), 243–249.

Kakosimos, K.E., Assael, M.J., Lioumbas, J.S., Spiridis, A.S., 2011. Atmospheric dispersion modelling of the fugitive particulate matter from overburden dumps with numerical and integral models. Atmos. Pollut. Res. 2(1), 24–33.

Kumar, P., Morawska, L., Birmili, W., Paasonen, P., Hu, M., Kulmala, M., Britter, R., 2014. Ultrafine particles in cities. Environ. Int. 66, 1–10.

Pandey, B., Agrawal, M., Singh, S., 2014. Assessment of air pollution around coal mining area: emphasizing on spatial distributions, seasonal variations and heavy metals using cluster and principal component analysis. Atmos. Pollut. Res. 5(1), 79–86.

Patra, A.K., Gautam, S., Kumar, P., 2016. Emissions and human health impact of particulate matter from surface mining operation—a review. Environ. Technol. Innovat. 5, 233–249.

Pless-Mulloli, T., Howel, D., King, A., Stone, I., Merefield, J., Bessell, J., Darnell, R., 2000. Living near opencast coal mining sites and children's respiratory health. Occup. Environ. Med. 57(3), 145–151.

Spectrum, 2010. Operational manual of Watchdog 2000 series Portable Weather Station. Spectrum Technologies, Inc., USA. https://www.specmeters.com/

Tecer, L.H., Süren, P., Alagha, O., Karaca, F., Tuncel, G., 2008. Effect of meteorological parameters on fine and coarse particulate matter mass concentration in a coal-mining area in Zonguldak, Turkey. J. Air Waste Manag. Assoc. 58(4), 543–552.

12 An Overview to Quantitative Risk Assessment Methodologies

Amarpreet Singh Arora
School of Chemical Engineering, Yeungnam
University, Gyeongsan, South Korea

Rahil Changotra and Himadri Rajput
School of Energy and Environment, Thapar Institute
of Engineering and Technology, Patiala, India

CONTENTS

12.1 INTRODUCTION

Creating a safe working environment is the key to the proper functioning in any industry (Summers, 2003). It is crucial to adopt methodologies that ensure safety and help avoid undesirable effects of serious accidents for either people or the environment. In recent years, complying with new laws and regulations, along with increased significance of the human factor, were the two main reasons why companies were compelled to address security issues (Grassi et al., 2009). Qi et al. (2012) reported that even though process safety had increased with time, accidental events still happened repeatedly for two reasons. Firstly, the organizations failed to achieve and execute best practices without delay. Secondly, they lacked the efficiency required to boost enhancements in the corporate environment. Accidental risks associated with an industry will always exist, but the organizations must make efforts to minimize the probability of such occurrences by implementing a specific study related to hazards identification (HAZID). Risk assessments study the knowledge of evaluating the uncertainties associated with a system, the possible threats and hazards, along with their causes and consequences. The acquired knowledge provides a basis for comparing available options with tolerable and acceptable alternatives. Risk assessments thus are instrumental in detecting flaws and enhancing safety in complex technical systems. Quantitative risk assessment (QRA) is a systematic technique for analyzing the risks from hazardous events, primarily employed by NASA and the Nuclear Regulatory Commission in the aerospace and nuclear industries. Since then, it has been applied in various fields, mainly in the chemical industries (Apostolakis, 2004). QRA involves prediction of the frequency and magnitude of effects linked to a hazard. Then information regarding these aspects is converted into numerical values – typically the risk of fatality. The development of QRA for the chemical industry is based on research aimed at understanding individual accidents as they relate to hazardous materials, such as toxic releases, fires, and explosions. Risk assessment and consequence analysis models were proposed by the researchers. However, in recent years, multi-hazard disasters have been increasingly recognized to have more serious effects and social influences than individual accidents in the chemical industry (Khan et al., 2015; Khakzad et al., 2018). Further stringent safety requirements have been set by the authorities, such as the European SEVESO Directive for the chemical industry (European Commission, 2020).

12.2   QRA PROCESS

In a broad sense, QRA includes four key processes/stages: hazard identification, hazard evaluation, exposure evaluation, and risk estimation. Figure 12.1 is a flowchart of each stage.

12.2.1   Hazard Identification

The first step in the QRA process is to identify the hazards in a facility by conducting a HAZID – a hazard study. The results of this study work as a starting point for identifying the possible scenarios in the QRA. Qualitative ranking based on the likely consequences and expected frequency of the identified hazards are constructed so that the significant hazards could be covered in QRA. QRA encompasses all parts of the plant that uses a hazardous material either flammable or toxic, or both. This procedure generally leads to hundreds of different scenarios, which can then be simplified for analysis by splitting the facility into segments (referred as isolatable sections), each of which contain related materials that share similar processing aspects such as temperature and pressure.

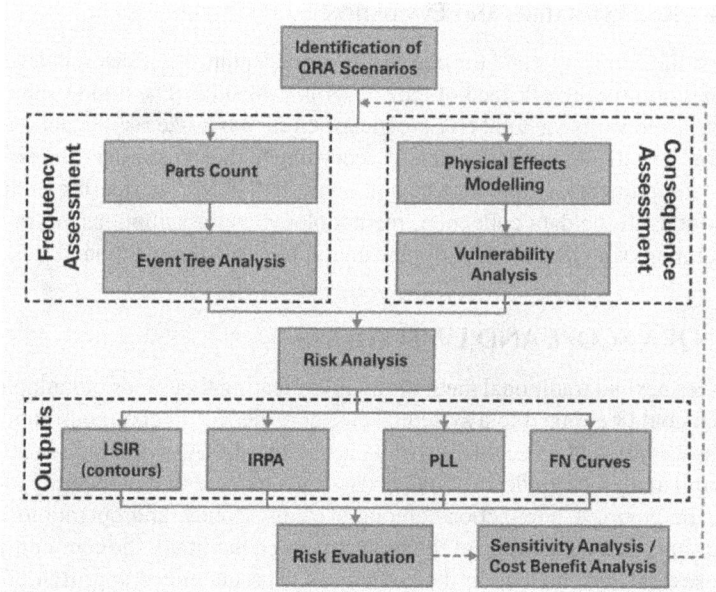

FIGURE 12.1 QRA process flow diagram. (Adapted from © Risktec Solutions Ltd. 2018.)

12.2.2 HAZARD EVALUATION (FREQUENCY AND CONSEQUENCE ASSESSMENT)

A range of different models are available for estimating the consequences of various events, such as gas explosion, gas dispersion, resulting fires, etc. An appropriate criterion is used to determine the susceptibility of people to these physical effects by estimating the probabilistic rate of fatality. This consequence assessment can also detect and identify probable escalated situations that may lead to additional critical consequences. A QRA may also contemplate the impact on the environment, the asset itself, and the reputation of the company.

12.2.3 EXPOSURE EVALUATION/RISK ANALYSIS

Numerical values of a risk analysis can be obtained by integrating the consequences and frequencies in the QRA model. Spreadsheets are used for conducting the offshore QRA whereas available commercial software is typically used for the onshore QRA. Besides the process hazards, other hazards such as occupational hazards, aircraft impact, ship collision, personnel transport, and natural hazards are referred to as non-process hazards, which are also ought to be analyzed. Each of the model has its own specialized procedure for risk analysis. Risks associated with all possible outcomes are calculated, summed, and reported in an easy-to-use form. Offshore risk is typically stated in the form of potential loss of life (PLL) and individual risk per annum (IRPA), whereas onshore risk is usually estimated in FN-curves, IRPA, and location-specific individual risk (LSIR) for societal risk considerations.

12.2.4 RISK ESTIMATION AND EVALUATION

Based on the company's and the country's risk acceptance criteria, risk levels are compared and the significance of each level is evaluated. The highly significant contributors towards the total risk are identified to maximize risk reduction. Risk reduction options can further be ranked according to their cost-effectiveness using a QRA complimentary tool, cost-benefit analysis (CBA). Considering numerous uncertainties in the data collection, methodology, and assumptions, an in-depth interpretation of a QRA study will play a vital role in risk reduction.

12.3 QRA SCOPE AND LIMITATIONS

QRA goes beyond traditional methods by investigating thousands of multiple failures that could be linked to a system. Thus, considering an enormous number of possible scenarios of an accident will be necessary. QRA investigation enhances the overall analysis significantly, and it also leads to increased probability of identifying the complex inter-actions among systems, events, and operations. QRA offers a shared understanding of the problem, and it facilitates the communication among several stakeholders in the companies. The integrated approach of QRA is derived from diverse fields relating to engineering, sciences, behavioral and social sciences. QRA creates an overall picture of the knowns and the unknowns by considering the uncertainties involved, and it provides valuable revelations for the overall decision-making process.

It also helps improve risk management by finding the main scenarios linked with accidents in order to promote improved resource allocation that eliminates insignificant risk scenarios, this risk reduction is referred to as risk-based decision-making.

Despite the QRA process has a broad scope and advantages, there are a few limitations that cannot be neglected. First, the level of accuracy in QRA does not really exist. QRA can be sometimes misleading as it may divert attention from preventative or precautionary measures. There is paucity of reliable data in certain specialized areas. It is frequently dwarfed by hardware issues and the rigid management of human factors. Theoretical modelling in QRA may not reflect the actual operations in some cases.

12.4 QRA MODELS AND SOFTWARE

There are numerous models and software available for QRA. The choice of the software depends upon the following key aspects.

- Scope of the model – detailed modeling requirements, needs, and processes. Software's capability can meet the requirements without overwhelming functionality.
- Transparency and Repeatability – data sets, methods, and rule-based scenarios can be easily viewed and traced.

- Cost – CBA, such as the cost of license, the training, time, and the external consultants.
- Integration – ease of integrating the processes into the company's management system.

Transparency and flexibility in methods, data sets, and rules is one of the most essential traits to be considered when selecting a model/software.

Larger organizations having multiple amenities will want a flexible and a more robust methodology than the spreadsheets. They employ an alternative cost-effective option by developing their own bespoke model using Microsoft.NET and/or ActiveX technology (Lewis, 2005). The direct development cost is comparable to a multi-user license, but it provides far greater flexibility to the organization because it grants a clearer understanding of the technical and user needs when modeling specific organizational risk issues.

But there are key specific advantages and disadvantages of using an integrated model of QRA and the spreadsheet. Table 12.1 below has illustrated the details.

TABLE 12.1
Advantages and Disadvantages of Integrated QRA and Spreadsheet Models

Integrated QRA Models		Spreadsheet Models	
Pros	**Cons**	**Pros**	**Cons**
Multiple model integration in a common computing environment	Complexity of use and understanding – requires customer training and familiarity	Relatively easy to understand	Prone to errors by the analyst
Models validated against experiment	Modifications in software are not possible – no flexibility and control	Lower user training requirements and user friendly	Specific to analyst and difficult to update by others without errors
Software quality and technical support from supplier	Black box approach involving hidden assumptions and calculation methods	Transparency in calculations and assumptions	Macro programming can be difficult to check
Easily available for immediate implementation	High initial and ongoing costs (licenses)	Flexibility of calculation-possible to develop detailed spreadsheet model to the required level of precision	More time consuming to demonstrate validation
Recognized and generally accepted within the industry		Low initial cost but labor intensive	False perception of less sophistication

12.5 QUANTITATIVE RISK ASSESSMENT USING MACHINE LEARNING

Machine learning and Artificial Intelligence are powerful tools that have been widely used for prediction purposes. Large data sets are collected and analyzed in order to identify the relationships or patterns. An integrated model can be generated, and the most effective out-of-box predictions can be provided. Creating such model will need to run model variables on subsampled data for identifying the key predictors, and then performs a model testing on several different data sub-sets. These steps can be repeated on numerous occasions. The model has the capability to "learn" from the data by itself and enhances its predictive power and performance. This great power of computational modelling can aid in the risk assessment where a model can learn from the past experiences to minimize the occurrence of uncalled events. Careful integration of artificial intelligence and machine learning to meet the organization needs, type, information requirements will contribute immensely to bring the agility and efficiency to the organizational structure (Gao, 2015). Recent advances are being made to use the machine learning algorithms for QRA. Paltrinieri et al. (2019) used deep neural network (DNN) for risk assessment linked to a drive-off situation of an Oil & Gas drilling rig.

The results indicated that using the DNN model could produce better accuracy, in comparison to a multiple linear regression (MLR) model in terms of accurate dynamic assessment and the flexibility in dealing with unexpected events. However, appropriate metrics and criteria should always be considered while evaluating the limitations associated with any model so that the model would enhance safety-related decision-making. Figure 12.1 provides the overall framework where the machine learning and artificial intelligence can be incorporated into the overall system of QRA proposed in the study of Žigiene et al. (2019), focusing on the commercial risk management for small and medium-sized enterprises (SMEs). This flow diagram can help and add a step to develop AI based QRA approach (Figure 12.2).

12.6 CONCLUSIONS

QRA has been around for over four decades, preceded by the nuclear and petrochemical industries (onshore), and it will soon be followed by the offshore and rail industries. The level of complexity, differences, and details of QRA in each industry varies according to the level of understanding of the critical risk concerns relevant in that industry. QRA cannot be considered as a panacea or a solution to all problems, but all the industries do agree that it is vital in making risk-informed decisions. Thus, lives can be saved, economic losses will be reduced, the environment can be safeguarded, and the repute of the associated organization will be preserved. This chapter represents an effort to point out a probable new conceptual foundation for the assessment of risks, but further research is needed for developing the next level of methods based on AI and machine learning. It hopes to identify and analyze the uncertainties that extends beyond the predictive

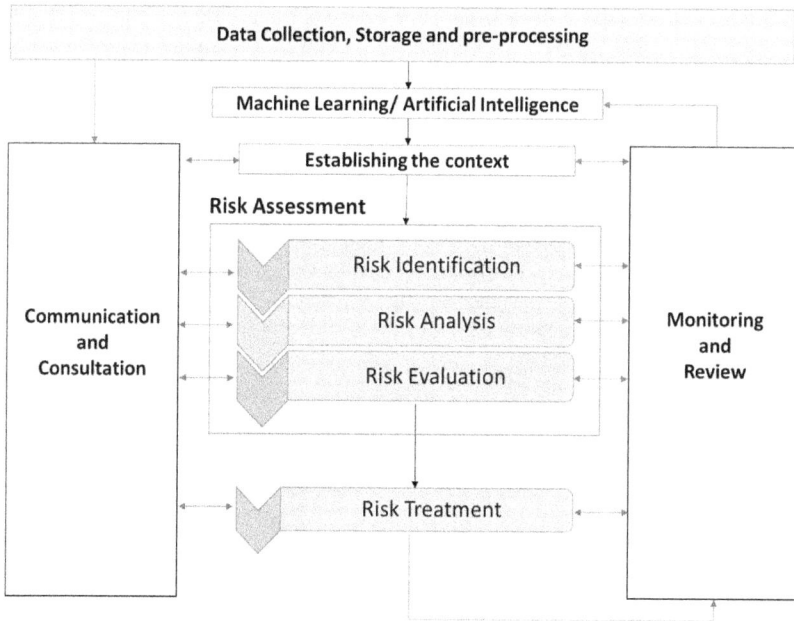

FIGURE 12.2 Machine Learning or AI-based decision making in risk assessment and management. (Modified from Žigiene et al. (2019) with reference to ISO 31000.)

capabilities of the deep learning tools. The purpose of the chapter is not to provide a complete solution but to point out key issues and to outline some possible directions for the future research in QRA methodologies.

REFERENCES

Apostolakis, G.E., 2004. How useful is quantitative risk assessment? Risk Anal. 24, 515–520, http://dx.doi.org/10.1111/j.0272-4332.2004.00455.x.

European Commission, 2020. The Seveso Directive-Technological Disaster Risk Reduction (accessed 30 January 2020) https://ec.europa.eu/environment/ seveso/.

Gao, L., 2015. Long-term contracting: the role of private information in dynamic supply risk management. Prod. Oper. Manag. 24, 1570–1579, doi:10.1111/poms.12347

Grassi, A., Gamberini, R., Mora, C., Rimini, B., 2009. A fuzzy multi-attribute model for risk evaluating in workplaces. Saf. Sci. 47, 707–716, http://dx.doi.org/10.1016/j.ssci.2008.10.002

Khan, F.I., Rathnayaka, S., Ahmed, S., 2015. Methods and models in process safety and risk management: past, present and future. Process Saf. Environ. Prot. 98, 116–147, http://dx.doi.org/10.1016/j.psep.2015.07.005

Khakzad, N., Landucci, G., Cozzani, V., Reniers, G., Pasman, H., 2018. Cost-effective fire protection of chemical plants against domino effects. Reliab. Eng. Syst. Safe. 169, 412–421, http://dx.doi.org/10.1016/j.ress.2017.09.007

Lewis S. 2005. An overview of the leading software tools for QRA, American Society of Safety Engineers –Middle East Chapter (161), 7th Professional Development Conference & Exhibition, Kingdom of Bahrain.

Paltrinieri, N., Comfort, L., Reniers, G., 2019. Learning about risk: machine learning for risk assessment. Saf. Sci. 118, 475–486, https://doi.org/10.1016/j.ssci.2019.06.001.

Risktec Solutions Ltd. 2018. Risktec Essentials: Quantitative Risk Assessment (QRA) An introduction to the quantitative assessment of risks associated with high hazard facilities (accessed 28 September 2020) http://risktec.tuv.com.

Summers, A.E., 2003. Introduction to layers of protection analysis. J. Hazard. Mater. 104, 163–168, http://dx.doi.org/10.1016/S0304-3894(03)00242-5.

Qi, R., Prem, K.P., Ng, D., Rana, M.A., Yun, G., Sam Mannan, M., 2012. Challenges and needs for process safety in the new millennium. Process Saf. Environ. Prot. 90, 91–100, http://dx.doi.org/10.1016/j.psep.2011.08.002.

Žigienė, G., Rybakovas, E., Alzbutas, R., 2019. Artificial intelligence based commercial risk management framework for SMEs. Sustainability. 11, 4501, https://doi.org/10.3390/su11164501.

13 Understanding Spatio-Temporal Variability of Groundwater Level Changes in India Using Hydrogeological and GIS Techniques

Suneel Kumar Joshi
Research and Development Centre, Geo Climate
Risk Solutions, Visakhapatnam, India

Sneha Gautam
Department of Civil Engineering, Karunya Institute
of Technology and Sciences, Coimbatore, India

CONTENTS

13.1 INTRODUCTION

Globally, groundwater is the worlds' most considerable freshwater resource that is vulnerable to unsustainable abstraction (Wada et al., 2010; Taylor et al., 2013). Groundwater depletion is a global issue, the magnitude of which was poorly known until recently (Konikow and Kendy, 2005; Joshi et al., 2021a). Currently,

DOI: 10.1201/9781003140382-13

the global groundwater abstraction rate is about 1500 km³/year (Döll et al., 2012), and the global recharge rate is approximately 12600 km³/year (Doll and Fiedler, 2008). Recently, the Gravity Recovery and Climate Experiment (GRACE) satellite mission is being widely used to monitor large-scale groundwater storage changes of aquifer systems (Richey et al., 2015). However, high-resolution monitoring data for groundwater levels remain indispensable because groundwater depletion can be highly localized in many aquifer systems (Scanlon et al., 2012; Joshi et al., 2021a & b; Shekhar et al., 2020).

Groundwater is essential for food security and drinking water, besides rendering a significant service toward supporting ecosystems and human health (Gleeson et al., 2016). Aquifers are the primary source of available drinking water and irrigation in many places. Because of this, groundwater is a significant contributing factor to the climate change scenario in view of its relative stability in quantity and quality (Saha et al., 2018; Shekhar et al., 2020). Groundwater draft in India is approximately 243 km³ and meets more than 50% of urban water demand, about 62% of irrigation demand, and about 85% of rural drinking needs (Saha et al., 2018). As a result, groundwater resources have declined rapidly in many parts of India (Rodell et al., 2009; Tiwari et al., 2009; Chen et al., 2014; Joshi et al., 2014 & 2021a; Panda and Wahr, 2016; Bhanja et al., 2017; Sinha et al., 2019; Shekhar et al., 2020; van Dijk et al., 2020).

This study uses earlier published historical groundwater observation well records and GRACE data to examine the spatio-temporal variability of groundwater levels and trends in groundwater storage. The historical groundwater observation wells records can be helpful to validate GRACE-based estimates. No existing literature was discovered on this topic, making it the first systematic study to investigate groundwater level changes and their connection to various hydrometeorological parameters, for example, recharge rate, and Groundwater-Surface water (GW-SW) interaction.

13.2 DATA AND METHODS

The in-situ measurement of groundwater observation wells occurs four times each year in January, May, August, and November from >15000 wells, and are organized by the Central Groundwater Board (CGWB), India. However, relatively few in situ measurement data are available in the public domain. Taking these datasets, it was possible to prepare spatio-temporal plots of groundwater level changes in India for the period 1996–2016. Then, using GRACE satellite data, spatial and temporal variations in groundwater storage were identified (e.g., Rodell et al., 2009; Scanlon et al., 2012).

13.3 GROUNDWATER RECHARGE

Precipitation is the primary groundwater recharge source in India (Joshi et al., 2018 & 2020; Swarnkar et al., 2021). Annual recharge through rainfall is about 342 km³ in India (Kumar et al., 2005). The recharge from precipitation shows seasonality across the country (Joshi et al., 2020 & 2021b; Swarnkar et al., 2021); for example, monsoonal precipitation comprises a significant recharge in shallow aquifers. The water table is lower just before the onset of the Indian Summer Monsoon (ISM), whereas the shallow water table can be seen during the ISM, particularly during August. During the post-monsoon period, groundwater is essential for water resource utilization strategies and planning purposes. During the non-monsoon season, about 90% of groundwater is consumed by irrigation (Shekhar et al., 2018; Swarnkar et al., 2021). Canals also play a vital role in groundwater recharge through return irrigation flow and canal leakage (Joshi et al., 2014, 2018, & 2020), this contribution to annual recharge is approximately 89 km^3 (Kumar et al., 2005).

13.4 GROUNDWATER DEPLETION

In the north of India, there has been a rapid fall in groundwater levels observed through GRACE-based estimates (Tiwari et al., 2009; Chen et al., 2014; Long et al., 2016). Rodell et al. (2009), assessed groundwater volume using GRACE data from 2002 to 2008, and suggested a net groundwater storage loss of about 109 km^3, at a depletion rate of 4.0±1.0 cm/year for Punjab, Haryana, Rajasthan, and Delhi regions including Chandigarh. Tiwari et al. (2009) estimated the total groundwater storage loss of 54±9 km^3 for north India between 2002 and 2008. Chen et al. (2014) estimated groundwater storage loss at a rate of 2.4±0.59 cm/year for northwest India, between 2003 and 2013. Long et al. (2016) estimated the groundwater decline rate to be 3.1±0.1 cm/year for northwest India between 2003 and 2010 (Joshi et al., 2021a). Panda and Wahr (2016) estimated groundwater depletion rate (approximately 2.1 cm/year) for the Ganga River basin in north India. MacDonald et al. (2016) also estimated groundwater storage decline (at a rate of 1.4±0.5 km^3/year) for Haryana, and 2.6±0.9 km^3/year for Punjab between 2002 and 2012, using in situ measurements of groundwater observation well records. They also assessed groundwater contamination and water levels within the top 200 m bgl of the aquifers, producing high-resolution estimates of groundwater storage change. Further, Joshi et al. (2021a & b) have used high-resolution in situ measurement of observations wells record from CGWB and State Government Agencies of Punjab and Haryana, and prepared high resolution spatial and temporal pattern of groundwater level changes. Joshi et al. (2021a) also identified several controlling factors of groundwater. This result provided a more detailed view of groundwater changes across the Indo-Gangetic basin than the GRACE data (MacDonald et al., 2016; Joshi et al., 2021a). In contrast, in southern India, groundwater levels are rising (Panda and Wahr, 2016; Asoka et al., 2017; Bhanja et al., 2017). These studies provide regional estimates based on a specific region. Therefore, they cannot be extrapolated more generally to understand India's groundwater scenario at a high-resolution scale as groundwater depletion is highly variable in most Indian aquifers (Saha et al., 2018). Based on the GRACE and historical groundwater observation well records, it can be observed that groundwater storage has been undergoing a continuous depletion during the last four decades in northern India (Joshi et al., 2014 & 2021a; Shekhar et al., 2020; van Dijk et al., 2020).

13.5 GROUNDWATER OCCURRENCE

India's geographical area is over 32.87 lakh km^2, divided into three geomorphic divisions, the Indo-Ganga-Brahmaputra Plain, Peninsular India, and Extra Peninsular India. Thick unconsolidated deposits of Quaternary age underlie the Indo-Ganga-Brahmputra Plain, and these hold some of the highest potential aquifer systems (Joshi et al., 2018 & 2020). This aquifer is characterized by its multilayered structure (van Dijk et al., 2016), regional and sub-regional; furthermore, it is reported to run several hundreds of meters below the groundwater level. This aquifer recharges through local and regional flow systems in the region (Joshi et al., 2018 & 2021a). The groundwater is stored under various conditions, ranging from phreatic to semi-confined, at shallow (aquifer-I) and deep (aquifer-II) depths. At the same time some parts are confined at an even deeper (aquifer-III) level (Table 13.1). The groundwater in levels I and II gets recharged by rainfall,

TABLE 13.1
Aquifer System in India (CGWB, 2009; Joshi et al., 2021a, b)

Aquifer Group I	• Aquifer group comprises coarse sand and gravel.
	• Sand layers are generally thick and are separated by a small, thin clay layer.
	• The depth range of the aquifer extends up to ~167 m bgl from the surface.
	• Transmissivity value varies from ~800 to 5210 m^2/day, with an average value of 2200 m^2/day.
	• Hydraulic conductivity ranges from ~8.75 to 47.10 m/day with an average value of 24 m/day and
	• Storativity value varies from ~2.10 to 24, with an average value of 12.
Aquifer Group II	• Aquifer group comprises alternating sequences of thin sand and clay layers.
	• Sand layers are thick and separated by thin and discontinuous clay layers.
	• The depth ranges of the aquifer are from ~150 to 283 m bgl.
	• Transmissivity value varies from ~350 to 1050 m^2/day, with an average value of 700 m^2/day.
	• Hydraulic conductivity value ranges from ~3.95 to 10.70 m/day, with an average value of 7.2 m/day.
	• Storativity value varies from ~5.60 × 10^{-4} to 1.70 × 10^{-3} with an average value of 1.0 × 10^{-3}.
Aquifer Group III	• Aquifer group comprises thin sand layers alternated with thicker clay layers.
	• This aquifer behaves like a confined aquifer in the maximum part of the study area.
	• The depth ranges of the aquifer are from ~197 to 346 m bgl.
	• Transmissivity value varies from ~345 to 830 m^2/day, with an average value of 525 m^2/day.
	• Hydraulic conductivity value ranges from ~3.50 to 10.70 m/day with an average value of 7.1 m/day
	• Storativity value varies from ~6.60 × 10^{-4} to 2.40 × 10^{-4} with an average value of 4.5 × 10^{-4}.

but aquifer-III may contain older water, even up to several thousand years old (Saha et al., 2011; Joshi et al., 2018 & 2020).

Extra Peninsular India and Peninsular India are predominantly represented by hilly and undulating topography, interspersed with alluvial deposits from rivers. These two units are mainly characterized by hard work aquifers, where groundwater occurs within the top weathered and underlying fractures and joints. The fractured aquifer is highly heterogeneous and primarily local to sub-regions. In these areas, the recharge is mainly through the local recharge. The groundwater remains in the weathered zone's phreatic conditions, whereas groundwater remains semi-confined to confined in the fractures below. The groundwater in hard rock terrain is comparably younger, indicating recharge on a seasonal scale (CGWB, 2017).

13.6   AQUIFER SYSTEM IN INDIA

Alluvial aquifers usually consist of various grains of varying sizes, shapes and proportions materials including sand, silt, gravel, and clay deposits, which act as lenses and layers of varying thicknesses (Kresic and Mikszewski, 2012). When sand and gravels dominate, with finer fractions forming thin interbeds and lenses, the aquifer may be considered one continuum, providing water to a pumping well through its entire screen. India's aquifer system shows some heterogeneity and stratification (van Dijk et al., 2016; Lapworth et al., 2017). Aquifer heterogeneity refers to the spatial variation in an aquifer system's hydraulic, transport, and geochemical properties. The heterogeneity may be within or between beds (or both). Aquifer heterogeneity is particularly crucial at contaminated sites where dissolved contaminants may move faster through layers or more permeable porous media, creating convoluted preferential pathways intersecting a well at discrete intervals. Detecting such paths, although challenging, is often the key for successful groundwater remediation, though it may not be of much importance when quantifying groundwater flow rates for water supply considerations.

Alluvial aquifer's areal extent and thickness depend on the parent stream size and the aquifer location, relative to the drainage area. Aquifers in floodplains of smaller streams and higher upstream areas are of limited scale, rarely exceeding 10 m in thickness (Kresic and Mikszewski, 2012). On the other hand, alluvial aquifers developed in major river floodplains are among the most prolific and widely used for water supplies worldwide. In addition to thick, extensive sand and gravel deposits, they are typically directly hydraulic connected to the river, providing abundant recharge potential.

The aquifer system transmits, stores, and yields a significant amount of water to the wells. The distribution of all these properties in the study area depends on the sediment depositional environment in the basin-fill aquifer. The aquifer system is interconnected and hydraulically continuous across the study area. It has been categorized by the CGWB using the hydrogeological, hydrogeochemical and geomorphological settings for northwestern India (Table 13.1).

13.7 GROUNDWATER REGIME

India's water table varies considerably in response to geologic, terrain, climatic conditions, and anthropogenic interventions (Joshi et al., 2021a & b). Figure 13.1 shows a simplified version of the groundwater level map that shows pre-monsoon water table data for 2016 (CGWB, 2017).

From the southeast to the northwest of India, four broad zones were found (Figure 13.1). The trans-Himalayan foothill zone and the East Coast have a nearly continuous band of <5 m water table, apart from a few isolated patches primarily concentrated in India's eastern half and coastal Maharashtra. About 5–10 m below ground level (bgl) occurs in large parts of India, except in northwest Rajasthan and Punjab (CGWB, 2017). More in-depth 10–20 m bgl follows this, mostly in small pockets concentrated in the country's western region. The deeper levels are recorded in Rajasthan's arid region and the central part of Punjab, where levels vary from 20 to 40 mbgl and deeper. Apart from this, deeper levels are observed

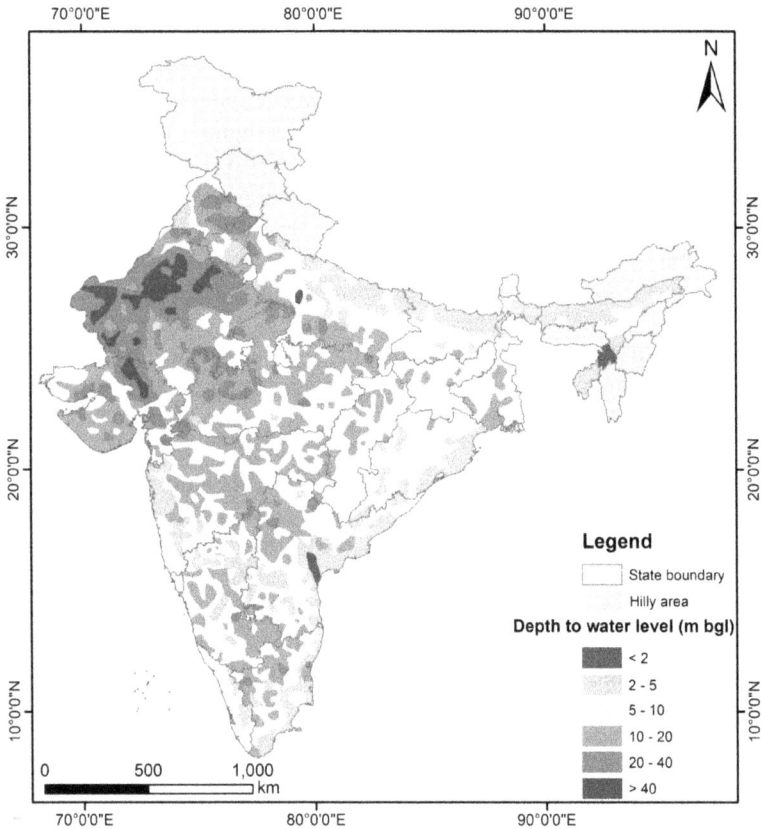

FIGURE 13.1 Depth to water level map (pre-monsoon 2016) prepared using historical groundwater observation well records in India. (From CGWB, 2017)

in Gujarat, Haryana, Telangana, and the Bundelkhand region of Uttar Pradesh and Madhya Pradesh. The perennial shallow groundwater levels (2–5 mbgl) are mostly the regional discharge zones, either along the foothills of the Himalayas or along the East Coast. Similarly, the deepest groundwater level zones are the negligible groundwater recharge areas in the more arid parts of India.

13.8 GROUNDWATER ABSTRACTION

Precipitation is the primary recharge source for groundwater system in India (Joshi et al., 2018 & 2020; Swarnkar et al., 2021). The ratio of groundwater recharge to abstraction on an annual scale provides the groundwater development stage. The groundwater development stage is considered in India as an indcator of dynamic water resource stress. The latest estimation of dynamic groundwater resources (CGWB, 2017) indicates the country's overall groundwater development stage at approximately 62% (Figure 13.2). The assessment unit,

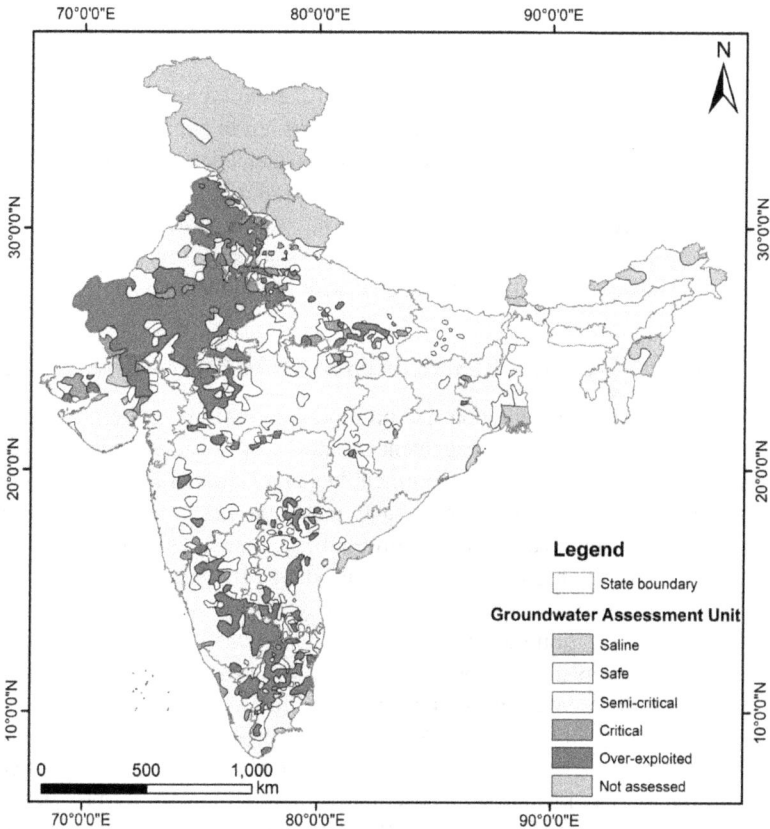

FIGURE 13.2 Groundwater assessment units in India. (From CGWB, 2017; Sidhu et al., 2020.)

that is, blocks, are categorized based on their stage of groundwater development as – "safe" (groundwater development stage <70%), "semi-critical" (ground-water development stage 70–90%), "critical" (groundwater development stage 90–100%), and "over-exploited" (groundwater development stage >100%). The groundwater development stage is not uniform throughout the country due to differing aquifer potentials, recharge capabilities, and the extent of groundwater exploitation. Currently 1071 units of the country are classified as over-exploited, and another 914 are classifed as significant, where exploitation levels exceed 70% (CGWB, 2017; Joshi et al., 2021a & b). All these units are primarily concentrated in the northwestern parts, as well as deep in Peninsular India. These areas are also characterized by deeper groundwater levels (Figure 13.1). Groundwater over-abstraction has resulted in declining water levels, dwindling well yields, and dug wells drying ups. Additionally, there is an overall increase in background salinity levels in groundwater. This is apart from the known ambient higher EC values in India's arid regions and coastal tracts. This has prompted researchers to conclude that higher abstraction deteriorates ambient groundwater quality (Mukherjee et al., 2015). Shifts to drinking water extracted from deeper zones may have trig-gered certain pollutants to enter the aquifer-based drinking water cycle (Saha, 2009; Mukherjee et al., 2015).

India's groundwater resource management is at a crossroads due to the acceler-ated pace and unevenness of abstraction. This has led to a significant decline in groundwater levels, expansion of waterlogged areas, and deteriorating water qual-ity, resulting in a considerable reduction in the availability of fresh and potable water. Strategies for better groundwater governance and management need to be developed through broad consultation among various stakeholders, academia, researchers, and state departments that deal with groundwater.

13.9   GROUNDWATER RESOURCES MANAGEMENT STRATEGY

Significant groundwater depletion has prompted central and state governments to launch a series of water management programs on local and regional levels (Sinha et al., 2019). The Indian government began a National Project on Aquifer Management (NAQUIM) to identify and map aquifers at a high-resolution, quantifying the available district-level groundwater resources across India (Saha et al., 2018). The Indian government also introduced the Pradhan Mantri Krishi Sichayee Yojana (PMKSY) program for five years from 2015/16 to 2019, to enhance India's irrigation capabilities. This scheme was helpful for the agricul-ture sector, and the primary motivation was "More crop per drop". The proper use of water for different purposes such as agriculture, industrial, and household, could be monitored using the PMKSY scheme as it amalgamated other concurrent government schemes, such as the On-Farm Water Management (OFWM) com-ponent of the National Mission on Sustainable Agriculture (NMSA); Integrated Watershed Management Programme (IWMP), initiated by the Department of Land Resources; and the Accelerated Irrigation Benefit Programme (AIBP) undertaken by the Ministry of Water Resources, River Development, and Ganga

Rejuvenation. This scheme's primary objectives were to provide optimized solutions for different sectors such as irrigation supply, water resources, distribution networks, and improvements to water use efficiency. The rapid drawdown of the aquifer system in northwestern India indicates that the management of alluvial aquifers in India now needs a serious rethink.

Sustainability of groundwater resources can be achieved by either increasing the supply of surface water, or reducing demand for groundwater. Under the given conditions, the former is implausible; therefore, the demand side needs to be addressed through (a) improvements in irrigation efficiency, (b) diversification of crop types to more minor water-hungry variants, or (c) addressing farm economics through energy pricing. It is also important to emphasize that depletion is quite diverse, and this region's substantial control of geomorphic setting on groundwater depletion is more than evident. The combination of geomorphic and stratigraphic mapping is made possible by both the wealth of data held by the CGWB and state groundwater boards, and the advent of new technological developments in the field of terrain mapping – this is a promising but relatively under-explored approach. Understanding geomorphic/stratigraphic controls of groundwater depletion can help determine the most appropriate management strategies, such as planning artificial recharge strategies and advice on crop management. The depletion of groundwater resources is also detrimental to rivers draining in any region. GW-SW interaction functions in two ways, for example, the rivers recharge the groundwater system during the monsoon period, when the water level in rivers is higher than the groundwater table, and during non-monsoonal flows (base flows) are generally maintained by the groundwater system's influx. In semi-arid parts of Haryana and Rajasthan, the monsoonal flows last for a short period, and it is the groundwater-derived baseflow that keeps the rivers flowing for most parts of the year. Excessive groundwater abstraction in NW India has disturbed this delicate balance over the last several decades. Since 2002, annual rainfalls have been insufficient to overcome the declining groundwater levels, and as such the levels are now unable to return to the pre-2001 state. As a result, rivers flowing in this region have also dried up, and their rejuvenation will depend strongly on groundwater resource development.

The severity of the problem demands that some regulatory mechanisms are put into place that will arrest the over-exploitation of groundwater across the region. The government of India has made some efforts in this direction by circulating the Model Ground Water Control & Regulation Bill to all states as early as 1970; this bill was subsequently revised in 1992, 1996, 2005 and then again redrafted in 2016. It is incredibly heartening to note that more than 15 states across the country have already made some groundwater legislation provisions based on the Model Bill circulated in 2016. Some more laudable efforts include the Central Ground Water Authority (CGWA) constitution by the Supreme Court of India to regulate indiscriminate boring and withdrawal of underground water. The CGWA was formed in 1997, and since then, several areas have been "notified" where no new wells are permitted, except for small diameter drinking wells. More recently in 2011, the Planning Commission came up with a draft Model Bill for the conservation, and protection of groundwater which is still under government consideration.

In addition to these regulations, public awareness about the groundwater crisis has also increased, and community efforts to save this valuable resource have multiplied. In several states such as Maharashtra, Madhya Pradesh, Gujrat, and Sikkim, the farmers have come together to manage this valuable resource by developing protocols for pumping water, sequencing of water use, etc., establishing distance norms between wells and tube wells (Saha et al., 2018) and these actions have already begun yielding positive results. However, these efforts have to be upscaled to cover the entire country, considering its varied hydrogeology and water stress levels, which requires massive support from governmental agencies. Ironically, the state and central groundwater departments have shrunk to an alarming size in terms of workforce and resources available, which is counterproductive to determining long-term solutions to this crisis.

REFERENCES

Asoka, A., Gleeson, T., Wada, Y. and Mishra, V. (2017) Relative contribution of monsoon precipitation and pumping to changes in groundwater storage in India. Nature Geoscience 10(2), 109–117.

Bhanja, S.N., Rodell, M., Li, B., Saha, D. and Mukherjee, A. (2017) Spatio-temporal variability of groundwater storage in India. Journal of Hydrology 544, 428–437.

CGWB (2009) Methodology for assessment of development potential of deeper aquifers, Central Groundwater Board, Ministry of Water Resources, River Development & Ganga Rejuvenation, Government of India, Faridabad.

CGWB (2017) Dynamic groundwater resources of India, Central Groundwater Board, Ministry of Water Resources, River Development & Ganga Rejuvenation, Government of India, Faridabad.

Chen, J., Li, J., Zhang, Z. and Ni, S. (2014) Long-term groundwater variations in Northwest India from satellite gravity measurements. Global and Planetary Change 116, 130–138.

Döll, P., Hoffmann-Dobrev, H., Portmann, F.T., Siebert, S., Eicker, A., Rodell, M., Strassberg, G. and Scanlon, B.R. (2012) Impact of water withdrawals from groundwater and surface water on continental water storage variations. Journal of Geodynamics 59, 143–156.

Döll, P. and Fiedler, K. (2008) Global-scale modeling of groundwater recharge. Hydrology and Earth System Sciences 12, 863–885.

Gleeson, T., Befus, K.M., Jasechko, S., Luijendijk, E. and Cardenas, M.B. (2016) The global volume and distribution of modern groundwater. Nature Geoscience 9(2), 161–167.

Hora, T., Srinivasan, V. and Basu, N.B. (2019) The groundwater recovery paradox in South India. Geophysical Research Letters 46(16), 9602–9611.

Jia, X., Hou, D., Wang, L., O'Connor, D. and Luo, J. (2020) The development of groundwater research in the past 40 years: A burgeoning trend in groundwater depletion and sustainable management. Journal of Hydrology 587, 125006.

Joshi, S.K., Rai, S.P., Sinha, R., Gupta, S., Densmore, A.L., Rawat, Y.S. and Shekhar, S. (2018) Tracing groundwater recharge sources in the northwestern Indian alluvial aquifer using water isotopes ($\delta^{18}O$, $\delta^{2}H$ and ^{3}H). Journal of Hydrology 559, 835–847.

Joshi, S.K., Rai, S.P., Sinha, R., Gupta, S., Shekhar, S., Rawat, Y.S., Kumar, M., Mason, P.J., Densmore, A.L. and Singh, A. (2014) Spatio-temporal Variations in Groundwater Levels in Northwest India and Implications for Future Groundwater Management. AGUFM 2014, H41G-0903.

Joshi, S.K., Gupta, S., Sinha, R., Densmore, A.L., Rai, S.P., Shekhar, S., Mason, P.J. and van Dijk, W.M. (2021a) Strongly heterogeneous patterns of groundwater depletion in northwestern India. Journal of Hydrology 598, 126492.

Joshi, S.K., Swarnkar, S. and Shukla, S. (2020) Variability in snow/ice melt, surface runoff and groundwater to Sutlej river runoff in the western Himalayan region. Geological Society of America, USA, Abstracts with Programs (Vol. 52, No. 6). doi: 10.1130/abs/2020AM-355211

Joshi, S.K., Tiwari, A., Kumar, S., Saxena, R., Khobragade, S.D. and Tripathi, S.K. (2021b). Groundwater recharge quantification using multiproxy approaches in the agrarian region of Bundelkhand, Central India. Groundwater for Sustainable Development 13, 100564.

Kresic, N. and Mikszewski, A. (2012) Hydrogeological conceptual site models: data analysis and visualization. CRC Press, Boca Raton, FL, 33487–2742

Konikow, L.F. and Kendy, E. (2005) Groundwater depletion: A global problem. Hydrogeology Journal 13(1), 317–320.

Kumar, R., Singh, R. and Sharma, K. (2005) Water resources of India. Current Science 89(5), 794–811.

Lapworth, D.J., Krishan, G., MacDonald, A.M. and Rao, M.S. (2017) Groundwater quality in the alluvial aquifer system of northwest India: New evidence of the extent of anthropogenic and geogenic contamination. Science of the total Environment 599, 1433–1444.

Long, D., Chen, X., Scanlon, B.R., Wada, Y., Hong, Y., Singh, V.P., Chen, Y., Wang, C., Han, Z. and Yang, W. (2016) Have GRACE satellites overestimated groundwater depletion in the Northwest India Aquifer?. Scientific Reports 6, 24398.

MacDonald, A.M., Bonsor, H.C., Ahmed, K.M., Burgess, W.G., Basharat, M., Calow, R.C., Dixit, A., Foster, S.S.D., Gopal, K., Lapworth, D.J. and Lark, R.M. (2016) Groundwater quality and depletion in the Indo-Gangetic Basin mapped from in situ observations. Nature Geoscience 9(10), 762–766.

Mishra, V., Asoka, A., Vatta, K. and Lall, U. (2018) Groundwater depletion and associated CO_2 emissions in India. Earth's Future 6(12), 1672–1681.

Mukherjee, A., Gupta, A., Ray, R.K. and Tewari, D. (2017) Aquifer response to recharge–discharge phenomenon: Inference from well hydrographs for genetic classification. Applied Water Science 7(2), 801–812.

Mukherjee, A., Saha, D., Harvey, C.F., Taylor, R.G., Ahmed, K.M. and Bhanja, S.N. (2015) Groundwater systems of the Indian sub-continent. Journal of Hydrology: Regional Studies 4, 1–14.

Panda, D.K. and Wahr, J. (2016) Spatiotemporal evolution of water storage changes in India from the updated GRACE-derived gravity records. Water Resources Research 52(1), 135–149.

Richey, A.S., Thomas, B.F., Lo, M.H., Reager, J.T., Famiglietti, J.S., Voss, K., Swenson, S. and Rodell, M. (2015) Quantifying renewable groundwater stress with GRACE. Water Resources Research 51(7), 5217–5238.

Rodell, M., Velicogna, I. and Famiglietti, J.S. (2009) Satellite-based estimates of groundwater depletion in India. Nature 460(7258), 999–1002.

Saha, D. (2009) Arsenic groundwater contamination in parts of middle Ganga plain, Bihar. Current Science 97(6), 753–755.

Saha, D., Marwaha, S. and Mukherjee, A. (2018) Clean and sustainable groundwater in India. Springer, Singapore, pp. 1–11.

Saha, D., Sinha, U. and Dwivedi, S. (2011) Characterization of recharge processes in shallow and deeper aquifers using isotopic signatures and geochemical behavior of groundwater in an arsenic-enriched part of the Ganga Plain. Applied Geochemistry 26(4), 432–443.

Scanlon, B.R., Longuevergne, L. and Long, D. (2012) Ground referencing GRACE satellite estimates of groundwater storage changes in the California Central Valley, USA. Water Resources Research, 48(4), 1–9.

Shekhar, S., Kumar, S., Sinha, R., Gupta, S., Densmore, A., Rai, S., Kumar, M., Singh, A., Van Dijk, W. and Joshi, S. (2018) Clean and sustainable groundwater in India. Springer, Singapore, pp. 117–124.

Shekhar, S., Kumar, S., Densmore, A.L., van Dijk, W.M., Sinha, R., Kumar, M., Joshi, S.K., Rai, S.P. and Kumar, D. (2020) Modelling water levels of northwestern India in response to improved irrigation use efficiency. Scientific Reports 10(1), 1–15.

Sidhu, B.S., Kandlikar, M. and Ramankutty, N. (2020) Power tariffs for groundwater irrigation in India: A comparative analysis of the environmental, equity, and economic tradeoffs. World Development 128, 104836.

Sinha, R., Joshi, S.K. and Kumar, S. (2019) Green revolution in northwest India. Geography and You 19(24), 12–19.

Swarnkar, S., Prakash, S., Joshi, S.K. and Sinha, R. (2021). Spatio-temporal rainfall trends in the Ganga River basin over the last century: understanding feedback and hydrological impacts. Hydrological Sciences Journal. https://doi.org/10.1080/02626667.2021.1976783

Taylor, R.G., Scanlon, B., Döll, P., Rodell, M., Van Beek, R., Wada, Y., Longuevergne, L., Leblanc, M., Famiglietti, J.S. and Edmunds, M. (2013) Ground water and climate change. Nature Climate Change 3(4), 322–329.

Tiwari, V.M., Wahr, J. and Swenson, S. (2009) Dwindling groundwater resources in northern India, from satellite gravity observations. Geophysical Research Letters 36(18), 1–5.

Van Dijk, W., Densmore, A., Singh, A., Gupta, S., Sinha, R., Mason, P., Joshi, S., Nayak, N., Kumar, M. and Shekhar, S. (2016) Linking the morphology of fluvial fan systems to aquifer stratigraphy in the Sutlej-Yamuna plain of northwest India. Journal of Geophysical Research: Earth Surface 121(2), 201–222.

van Dijk, W.M., Densmore, A.L., Jackson, C.R., Mackay, J.D., Joshi, S.K., Sinha, R., Shekhar, S. and Gupta, S. (2020) Spatial variation of groundwater response to multiple drivers in a depleting alluvial aquifer system, northwestern India. Progress in Physical Geography: Earth and Environment 44(1), 94–119.

Wada, Y., Van Beek, L.P., Van Kempen, C.M., Reckman, J.W., Vasak, S. and Bierkens, M.F. (2010) Global depletion of groundwater resources. Geophysical Research Letters 37(20), 1–5.

14 Electrocoagulation Influencing Parameters Investigation on Reactive Dyes in Textile Wastewater

A Simple Optimization Method

Mohit Garg
Department of Chemical Engineering, Indian
Institute of Technology Roorkee, Roorkee, India

Sarbani Ghosh
Department of Chemical Engineering, Birla
Institute of Technology and Science (BITS),
Pilani Campus, Vidyavihar, India

Amit Kumar
Department of Chemical Engineering,
Institute of Technology, Nirma University,
Ahmedabad, India

Vikram Chopra and Indra Deo Mall
Department of Chemical Engineering, Indian
Institute of Technology Roorkee, Roorkee, India

Sneha Gautam
Department of Civil Engineering,
Karunya Institute of Technology and Sciences,
Coimbatore, India

DOI: 10.1201/9781003140382-14

CONTENTS

14.1 INTRODUCTION

The dyeing and finishing of fabrics is a major source of wastewater generation in textile industries (Zaroual et al., 2005, 2006). The discharged dyes are harmful to humans and aquatic animals if not treated adequately (Bell and Buckley, 2003). The effluent from cotton textile dyeing operations consists mainly of reactive dyes of which a maximum of 20–50% may be released into waterways (Soares et al., 2004). Due to their low degree of fixation, these dyes are released in huge quantities and are major contributors to water pollution. These effluents from the dyeing process contain not only aesthetic pollutants but may also interfere with biological processes due to low light penetration in the receiving bodies of water (Lang, 1991). Apart from these negative impacts, the reactive dye bath effluents for dyeing cotton yarn are highly basic in nature and contain high color, chemical oxygen demand (COD), dissolved solids, and a huge amount of chlorides and sulfates. Therefore, their treatment prior to disposal is of major concern to the environment. Traditionally, various techniques such as coagulation, adsorption, Fenton's reagent, and photochemical degradation have been discussed from time to time by different authors (Harrelkasa et al., 2009; Joo et al., 2007; Mahmoodi et al., 2010; Kalra et al., 2011; Bilgi and Demir, 2005). Electrocoagulation, on the other hand, is an electrochemical technique which utilizes sacrificial anodes that corrode to release ions in the solution which then act as active coagulant precursors (Cenkin and Belevtsev, 1985). Insoluble metallic hydroxide flocs are produced due to the electrostatic interaction between the coagulants and the dyes, which removes pollutants (Mollah et al., 2001; Rajeshwar et al., 1994). EC is preferred over regular coagulation in color removal from wastewater because of a decrease in the discharge of chemicals into the waste stream and a lower chemical pollution in the decolorized effluent (Hashim et al., 2019, 2020; Vepsäläinen and Sillanpää, 2020). The addition of excessive amounts of coagulants can be avoided using electrocoagulation. Photodegradation experiments are regarded as highly efficient, but the process is costly.

Over the years, the EC technique for a single dye or mixture has been discussed by various researchers (Ashtoukhy and Amin, 2010; Durango-Usuga et al., 2010; Aoudj et al., 2010 Mollaha et al. 2010; Daneshvar et al., 2006). This method has a prominent effect on their removal. The reactive dye bath, however, consists of a mixture with a minimum of at least two dyes along with auxiliary chemicals such as NaCl, Na_2CO_3, etc. (Kabdasl et al., 2007). This dye bath effluent containing real reactive dyes and auxiliary chemicals follows different behavior. Recently, a lot of work has been done on simulated dye bath effluent (Arslan-Alaton et al., 2008; Vijayaraghavan et al., 2008; Kabdasl et al., 2009) using various techniques. The simulated dye bath effluent is relatively simple to treat compared to undiluted effluent due to the low value of color, COD, and all the other auxiliary chemicals (Vepsäläinen and Sillanpää, 2020).

In the present study, experiments were performed to remove reactive dyes from undiluted dye bath effluent in a textile industry. The removal of color and COD from the wastewater was evaluated using the electrocoagulation method. Various parameters (i.e., current density, initial pH, spacing between electrodes, and contact time of the solution) were varied in order to optimize the operating parameters for maximum removal efficiency.

14.2   EXPERIMENTAL METHODOLOGY

Ten samples of waste generated from dye baths in a textile mill were collected. The waste contained used up dyes, NaOH, inorganic salts, detergents, as well as dispersing, fixing and softening agents. The effluent was dark red in color. The collected samples of dye bath effluent were stored in an incubator at 4°C before use in the experimental study. The characterization of wastewater was done according to standard procedures previously described (Cleceri et al., 1998). The characteristics of the reactive dye bath effluent before and after treatment have been tabled in Table 14.1.

TABLE 14.1
Characterization of Textile Wastewater Before and After Treatment

Parameter	Before Treatment	After Treatment
pH	11.20	9.6
Color	7000 Hazen	787 Hazen
COD	1780 mg/l	780 mg/l
TOC	198.6 mg/l	114.5 mg/l
TDS	35000 mg/l	18400 mg/l
Turbidity	184 NTU	185 NTU
Conductivity	102.9 mS	76.4 mS
Chlorides	8600 mg/l	5800 mg/l
Sulfate	2160 mg/l	1296 mg/l
Nitrate	245.56 mg/l	60.3 mg/l

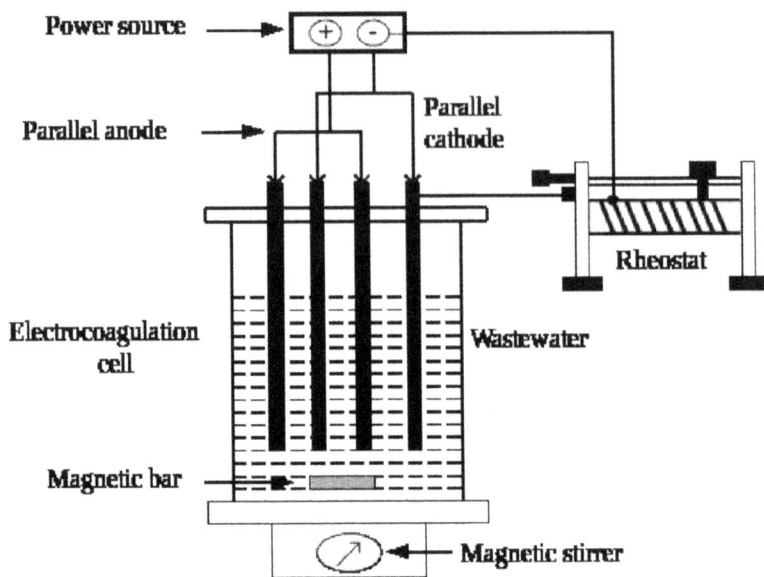

FIGURE 14.1 Schematic diagram of laboratory scale batch experimental setup.

A schematic diagram of the experimental setup is shown in Figure 14.1. It consisted of a 500 ml beaker, a pair of Cathode and Anode aluminum sheets with dimensions of 1 mm and 80 mm × 65 mm. Each electrode pair was fixed in parallel configuration and spaced 10 mm apart.

The submerged surface area of each electrode was kept at 52 cm². To control the current, a Galvanostat with a power supply of 12 Volt DC with variable current was connected to the electrodes. In each experimental run, 10 ml samples were taken in the reactor. Samples were withdrawn at different intervals of time in sec (300) and collected in the centrifuge bottles. After completion of the batch, the experimental study samples were centrifuged at 1000 rpm for 5 min until all of the suspended matters were removed. The samples were then diluted for measurement of color and COD. The scum and sludge were collected and dried. The electrodes were cleaned and the loss in weight was measured. If the loss in weight of the electrodes exceeded 10% of the initial values, a replacement was made (Arslan-Alaton et al., 2007).

14.3 ANALYTICAL MEASUREMENTS

Experiments were conducted under various conditions of pH, electrode spacing, time, and stirring rate. The pH of the wastewater varied according to the addition of 0.1N NaOH or 0.1N H_2SO_4 solutions. In each experimental run, the color and COD of the samples were measured to calculate the removal efficiency. The following relationship was used to calculate the percentage of COD removal:

$$\text{Percent COD removal} = \frac{(COD_0 - COD_t)}{COD_0} \times 100$$

where COD_0 is the COD (mg/l) at the beginning of the experiment and COD_t is the COD after time t (min).

Similarly, the decolorization efficiency was calculated as follows:

$$\text{Percent color removal} = \frac{(\text{initial color} - \text{final color})}{\text{initial color}} \times 100$$

The current density (CD) can be calculated as follows:

$$CD = \frac{I(A)}{2 \times S_{electrode}(m^2)} \frac{I(A)}{2 \times S_{electrode}(m^2)}$$

where I and S refer to the current and surface area of the electrode, respectively.

After completion of the EC batch study, the electrodes were removed from the experimental setup and the anode consumption was determined. The solution left in the batch reactor was stirred again, and the samples were analyzed for aluminum content using Atomic Adsorption Spectroscopy (AAS). The electrode and residue morphologies were obtained using SEM and EDX analysis. To compare the surface texture, a LEO 435 VP scanning electron microscope was used to obtain SEM images of the aluminum electrode before and after use, as well as of the dried residue. A PerkinElmer (Pyris Diamond) thermogravimetric analyzer was also used to obtain Thermogravimetric analysis (TGA) of the residues of sludge and scum. The TGA scans were recorded separately from 25 to 1000°C using a scanning rate of 10°C /min in a nitrogen atmosphere.

14.4　RESULTS AND DISCUSSION

14.4.1　THE REACTION MECHANISM AT ELECTRODES

The results showed that the initial pH had a predominant effect on the performance of the EC process (Daneshvar et al., 2006). The following reaction occurred at the aluminum electrodes:

- At anode:

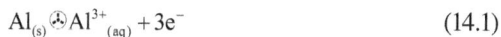

$$Al_{(s)} \rightarrow Al^{3+}_{(aq)} + 3e^- \tag{14.1}$$

$$2H_2O \rightarrow 4H^+_{(aq)} + O_{2(g)} + 4e^- \tag{14.2}$$

- At cathode:

$$2H_2O + 4e^- \oplus H_{2(g)} + 2OH^- \qquad (14.3)$$

The electrolytic dissolution of the electrodes produced Aluminum ions (Al^{3+}) which reacted with the water to produce various aluminum hydroxides and protons in the solution according to the following reaction:

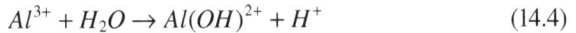

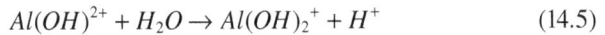

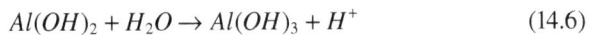

$$Al^{3+} + H_2O \rightarrow Al(OH)^{2+} + H^+ \qquad (14.4)$$

$$Al(OH)^{2+} + H_2O \rightarrow Al(OH)_2^+ + H^+ \qquad (14.5)$$

$$Al(OH)_2 + H_2O \rightarrow Al(OH)_3 + H^+ \qquad (14.6)$$

This reaction can undergo further hydrolysis to produce other polynuclear materials such as $Al_2(OH)_2^{4+}$, $Al_3(OH)_4^{5+}$, and $Al_6(OH)_{15}^{3+}$(Kobya et al., 2006; Benefield et al., 1982; Mouedhen et al., 2008), but the final product of this complex process leads to $Al(OH)_3$ precipitates. The aggregates formed were generated via mechanisms of hydrolysis followed by the precipitation of aluminum into colloids. Furthermore, the effects of different parameters on COD and color removal were studied.

14.4.2 EFFECT OF CONTACT TIME

The effect of electrolysis time at a uniform current density of 60 mA/cm² and initial pH of 11.2 was studied. Coagulation started at the anode while the charge ions neutralized the particles present at the anode. It was observed that the complete reaction took from 0 to 150 min. With an increase in time, the concentration of ions and their hydroxide flocs increased. Thus, an increase in time leads to an increase in the removal efficiency of color and COD. Figures 14.2 and 14.3 show

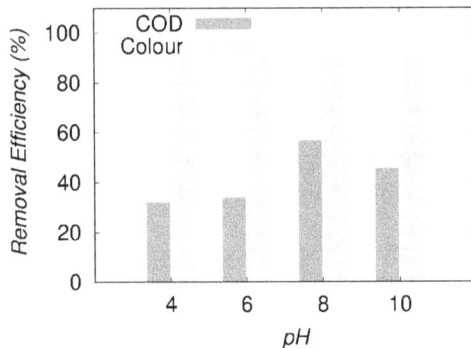

FIGURE 14.2 Effect of initial pH on COD and color removal in EC process (current density (j) = 60 mA/cm², contact time (t) = 150 min, and inter electrode distance (z) = 1 cm).

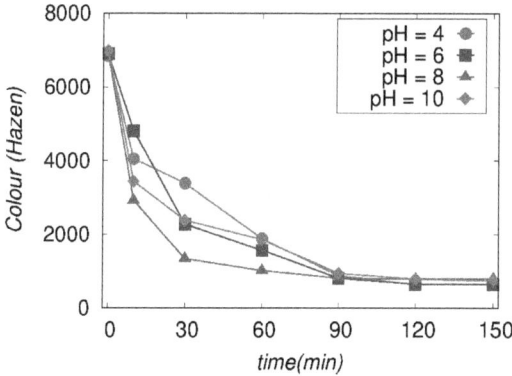

FIGURE 14.3 Effect of initial pH on color removal with time in EC process (current density (j) = 60 mA/cm², contact time (t) = 150 min, and inter electrode distance (z) = 1 cm).

the trends for color and COD with respect to time at various initial pH values, respectively. Figure 14.3 depicts the rate of removal of color and shows how it decreases with time and is completed within 90 sec for all pH values and remains constant thereafter, whereas COD is achieved in 120 min (as shown in Figure 14.4).

14.4.3   THE pH

In order to study pH influence, the initial pH of the wastewater was adjusted to between 4 and 10. The removal efficiency in terms of color and COD was found to increase with an increase in pH up to 8. This increase of COD at higher pH levels may be due to the oxidation process. Figures 14.3 and 14.4 show the color and COD removal percentage against time for different pH values.

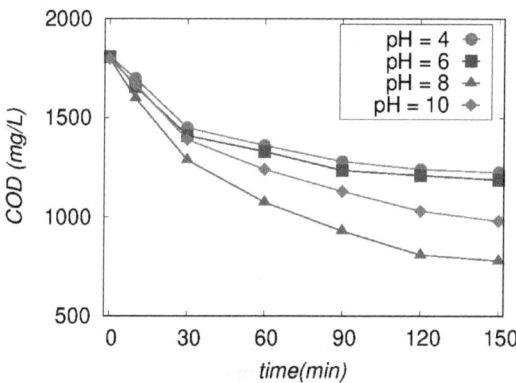

FIGURE 14.4 Effect of initial pH on COD and color removal in EC process (current density (j) = 60 mA/cm², contact time (t) = 150 min, and inter electrode distance (z) = 1 cm).

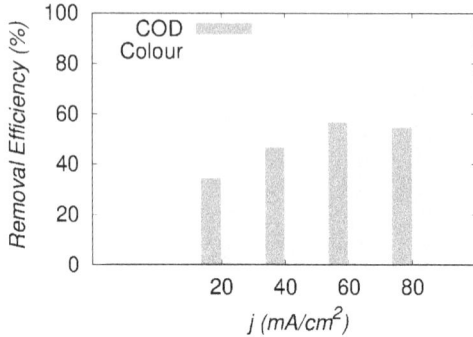

FIGURE 14.5 Effect of current density on COD and color removal efficiency in EC process (initial pH = 8, contact time (t) = 150 min, and inter electrode distance (z) = 1 cm).

14.4.4 CURRENT DENSITY

To observe the effect of current density on color and COD abatement rates, current density was varied from 20 to 80 mA/cm². We observed that the dissolution rate of aluminum electrodes in the solution increased with an increase in current density. Figure 14.5 shows the COD and color removal efficiency at different current densities.

Figure 14.5 shows that as the current density increases, the removal efficiency, in terms of both COD and color, increases because of the faster degradation of the electrodes which produces ions in the solution. An increase in current density from 20 to 80 mA/cm² increased the COD removal rates from 36% to 56% and the color removal efficiency from 80% to 90%. Beyond a current density of 60 mA/cm², the efficiency of the EC process did not change significantly. Taking into consideration the power involved in the EC process for higher current density and the higher costs of the process, a value of 60 mA/cm² was considered optimal in this study. Figures 14.6 and 14.7 show the color and COD removal over time with various current densities applied.

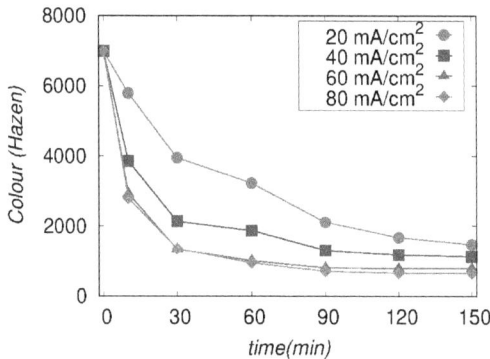

FIGURE 14.6 Influence of current density on color removal with time in EC process (initial pH = 8, contact time (t) = 150 min, and inter electrode distance (z) = 1 cm).

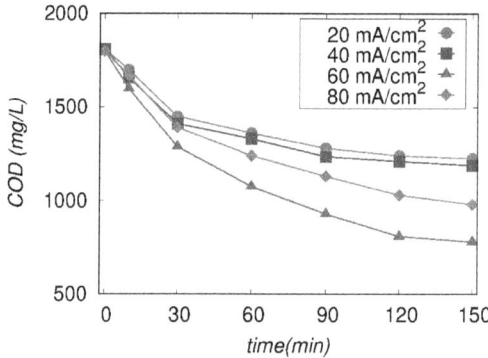

FIGURE 14.7   Influence of current density on color removal with time in EC process (initial pH = 8, contact time (t) = 150 min, and inter electrode distance (z) = 1 cm).

14.4.5   INTER-ELECTRODE SPACING

Inter-electrode spacing also influences the rate of COD and color abatement. Several authors (Aoudj et al., 2010) have studied the effect of electrode spacing on removal efficiency and articulated that both the removal efficiency and COD depend on the nature of the pollutant, the electrode setup, and other conditions. In the present study, inter-electrode spacing varied from 0.5 to 2 cm. It was observed that the maximum removal efficiencies were obtained at small distances and that the color and COD values first increased when the electrode spacing increased from 0.5 to 1 cm and then decreased up to 2 cm. Figure 14.8 shows the COD and color removal efficiency at different inter-electrode spacing.

The effect of inter-electrode spacing over time is shown in Figures 14.9 and 14.10. It was noted that as the spacing between the electrodes increased, electrical current decreased for a constant voltage supply. This is due to the fact that less interaction occurred between the dye ions and the hydroxyl ions generated within the solution.

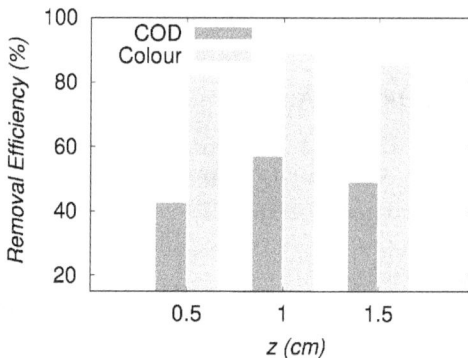

FIGURE 14.8   Effect of inter-electrode distance on color and COD removal in EC process (initial pH = 8, contact time (t) = 150 min, and current density (j) = 60 mA/cm^2).

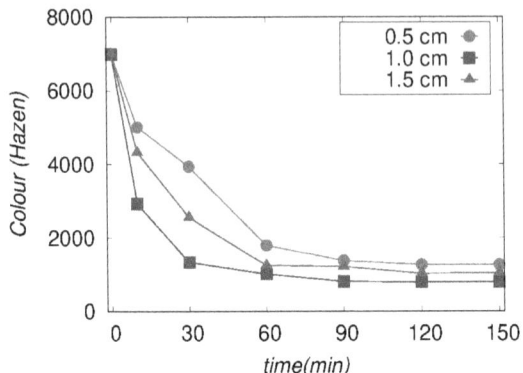

FIGURE 14.9   Effect of current density on color removal with time in EC process with time (initial pH = 8, contact time (t) = 150 min, and current density (j) = 60 mA/cm²).

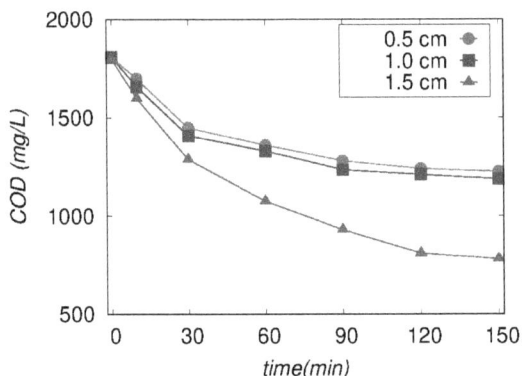

FIGURE 14.10   Effect of current density on COD removal with time in EC process with time (initial pH = 8, contact time (t) = 150 min, and current density (j) = 60 mA/cm²).

14.4.6   CHARACTERIZATION OF ALUMINUM ELECTRODE, SCUM, AND SLUDGE

An SEM analysis using a LEO 435 VP scanning electron microscope was performed to study the surface morphology of the anode, scum, and sludge generated in the electrochemical treatment process. SEM images of the aluminum electrode prior to and after the EC process, and dried sludge and scum produced after the EC treatment of the dye bath effluent were analyzed. Figure 14.11 shows an image of the fresh aluminum electrode. It is noted that the fresh aluminum electrode is uniform with nano-sized crystals.

Figure 14.12 (a) shows the image of the electrode before cleaning. It is clear that the uncleaned electrode has material attached to it which is basically the salt

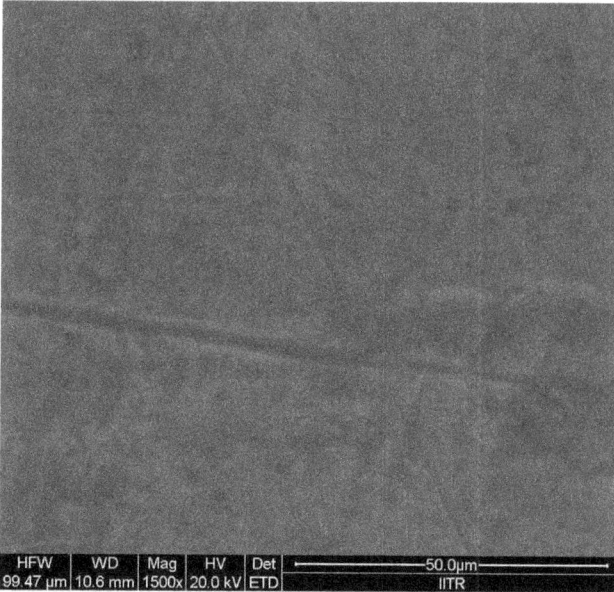

FIGURE 14.11   SEM image of fresh aluminum electrode.

and organics present due to the dyes and other inorganic substances used in various operations in textile industries.

Similarly, Figure 14.12 (b) shows the image of the electrode after cleaning. It shows the presence of dents which are due to the consumption of aluminum electrodes during the process. Due to these dents, electrocoagulation activity is reduced because the active sites get blocked for further precipitation. So, periodic cleaning of the electrode with 10% NaCl is necessary.

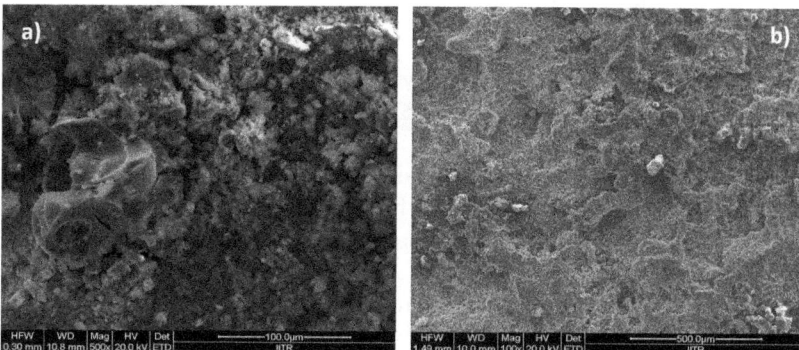

FIGURE 14.12   SEM images of uncleaned and cleaned electrodes after electrocoagulation process.

FIGURE 14.13   SEM images of scum and sludge produced after the electrocoagulation process.

Figure 14.13 shows the images of scum and sludge. The scum produced during the EC process is porous in nature due to hydrogen evolution. The sludge mainly contains aluminum and sodium which can be clearly observed in the image.

Furthermore, an EDX analyzer was used to determine the elemental compositions that the scum and sludge produced. It showed the presence of 7.01% and 14.33% carbon, 18.49% and 10.08% oxygen, 18.09% and 1.86% aluminum, 14.62% and 26.35% sodium, 40.66% and 42.86% chlorine, and 1.13% and 4.52% sulfur in sludge and scum, respectively. The sludge and scum contained huge amounts of sodium and chlorine. This was due to the presence of sodium chloride in the dye bath effluent. We noticed that under optimum conditions, out of a total of 3.78 gm of consumed aluminum electrode, 84.2% could be attributed to sludge and the rest to scum and wastewater.

The treated dye bath effluent contained 24 mg/l of aluminum. EPA regulations do not consider aluminum to be a potential hazard in primary drinking water. However, the national secondary drinking water guidelines recommend 0.2 mg/l concentration. Hence, further treatment is required to remove COD, aluminum, etc.

The disposal of residue generated during the process is of great concern because if the residue generated cannot be disposed of or used in some way, further environmental problems could be created.

Therefore, to study the temperature at which scum and sludge are produced and can be disposed of, TGA and differential thermal analysis (DTA) were performed. The rate of heating was fixed at 10°C/min. The TG curve produced on heating was further divided into various oxidation regions. The total weight loss of 43.55% and 48.38% were noted for sludge and scum, respectively. Figures 14.14 and 14.15 show TGA and DTA curves of the sludge and scum.

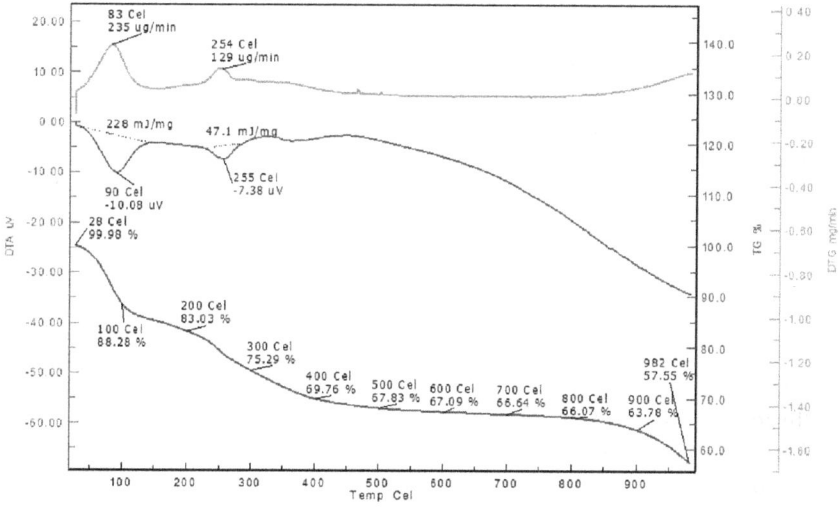

FIGURE 14.14   TG, DTA, and DTG analysis of sludge produced during EC process.

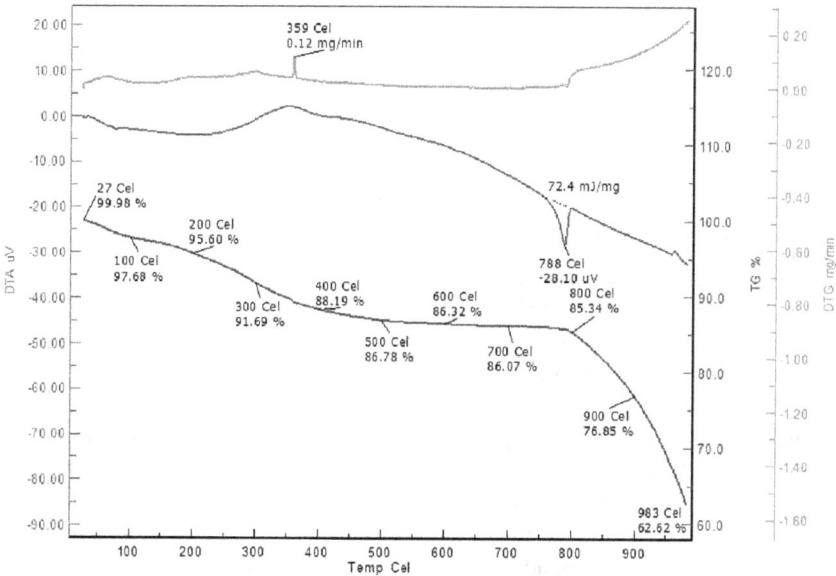

FIGURE 14.15   TG, DTA, and DTG analysis of scum produced during EC.

14.5 COST ANALYSIS

The cost of electrical energy as well as the aluminum electrodes consumed during the process were considered for the cost analysis. The EC process was limited by electrical consumption. The electrical energy consumed per liter of sample treated during EC is calculated as follows:

$$\text{Electrical Energy (EE)} \frac{U \times I \times t}{V}$$

where U is the voltage (Volts), I is the applied electrical current (Ampere), and t is the time of reaction (sec).

To convert this value into kWh/l, multiply this factor by 3.6×10^{-6}.

Similarly, the electrical energy per liter of sample treated in the EC process was found to be 1.65 kWh/m³ of wastewater treated.

The following calculated values were obtained:

Electrical energy consumed = 1.65 kWh/m³ of wastewater treated
Electrical energy price in India = $0.069/kWh
C_{energy} = $2.61 per meter cube of wastewater treated.
Aluminum electrode consumed = 7.5kg/m³ of wastewater treated
Aluminum price in the India = $1/kg
$C_{electrode}$ = $7.5 per meter cube of wastewater treated
Total cost of EC process = $C_{energy} + C_{electrode}$ = $10.11 per cubic meter of wastewater treated.

14.6 CONCLUSION

From this study, we can observe that electrocoagulation is an effective technique for the removal of color and COD from a single dye solution where initial color and COD are very low. This is not true in the case of real wastewater because it also contains a lot of organic and inorganic substances. In this study, the color and COD removal were found to be 90% and 56%, respectively. SEM was used to create images of scum and sludge produced during the process. EDAX was used to determine the amount of aluminum present in the scum and sludge, respectively. The electrical energy consumed per cubic meter of the sample treated was found to be 1.65 kWh/m³ of wastewater treated. The removal efficiency in terms of color and COD was found to be 90% and 56%, respectively. The electrical energy consumed during the process was 1.65 kWh/m³ of wastewater treated. Further treatment of this solution is necessary to keep various constituents within the environmental norms. For this reason, further studies regarding electrocoagulation coupled with other techniques like adsorption and advanced oxidation processes are needed.

REFERENCES

Aoudj S., Khelifa A., Drouiche N., Hecinia M., Hamitouche H., Electrocoagulation process applied to wastewater containing dyes from textile industry, Chemical Engineering and Processing 49 (2010) 1176–1182.

Arslan-Alaton I., Kabdas I., Hanbaba L.D., Kuybu F., Electrocoagulation of a real reactive dyebath effluent using aluminum and stainless steel electrodes, Journal of Hazardous Materials 150 (2008) 166–173.

Ashtoukhy El E.S.Z., Amin N.K., Removal of acid green dye 50 from wastewater by anodic oxidation and electrocoagulation—A comparative study, Journal of Hazardous Materials 179 (2010) 113–119.

Bell J., Buckley C.A., Treatment of a textile dye in the anaerobic baffled reactor, Water SA 29 (2003) 129–134.

Benefield L.D., Judkins J.F., Weand B.L., Process Chemistry for Water and Wastewater Treatment, Prentice-Hall Inc., New Jersey, 1982.

Bilgi Saliha, Demir Cevdet, Identification of photooxidation degradation products of C.I. Reactive Orange 16 dye by gas chromatographyemass spectrometry, Dyes and Pigments 66 (2005) 69–76.

Cenkin V.E., Belevtsev A.N., Electrochemical treatment of industrial wastewater, Effluent Water Treatment Journal 25 (1985) 243–247.

Cleceri L.S., Greenberg A.E., Eaton A.D., Standard Methods for the Examination of Water and Wastewater, 20th edition, American Public Health Association, Washington DC, 1998.

Daneshvar N., Oladegaragoze A., Djafarzadeh N., Decolorization of basic dye solutions by electrocoagulation: An investigation of the effect of operational parameters, Journal of Hazardous Materials B129 (2006) 116–122.

Durango-Usuga P., Guzmán-Duque F., Mosteo R., Vazquez M.V., Gustavo Peˉnuela, Torres-Palma R.A., Experimental design approach applied to the elimination of crystal violet in water by electrocoagulation with Fe or Al electrodes, Journal of Hazardous Materials 179 (2010) 120–126.

Harrelkasa Farida, Azizib Abdelaziz, Yaacoubib Abdelrani, Benhammoua Ahmed, Ponsc Marie Noelle, Treatment of textile dye effluents using coagulation–flocculation coupled with membrane processes or adsorption on powdered activated carbon, Desalination 235 (2009) 330–339.

Hashim Khalid S., Al Khaddar Rafid, Jasim Nisreen, Shaw Andy, Phipps David, Kot P., Pedrola Montserrat Ortoneda, Alattabi Ali W., Abdulredha Muhammad, Alawsh Reham, Electrocoagulation as a green technology for phosphate removal from river water, Separation and Purification Technology 210 (2019) 135–144.

Hashim K. S., Ali S. S. M., AlRifaie J. K., Kot P., Shaw A., Al Khaddar R., Idowu, I., Gkantou, M., *Escherichia coli* inactivation using a hybrid ultrasonic–electrocoagulation reactor. Chemosphere 247 (2020) 125868.

Joo Duk Jong, Shin Won Sik, Choi Jeong-Hak, Choi Sang June, Kim Myung-Chul, Han Myung Ho, Ha Tae Wook, Kim Young-Hun, Decolorization of reactive dyes using inorganic coagulants and synthetic polymer, Dyes and Pigments 73 (2007) 59–64.

Kabdasl I., Vardar B., Arslan-Alaton I., Tunay O., Effect of dye auxiliaries on color and COD removal from simulated reactive dyebath effluent by electrocoagulation, Chemical Engineering Journal 148(1) (2009) 89–96.

Kalra Shashank Singh, Mohan Satyam, Sinha Alok, Singh Gurdeep, Advanced Oxidation Processes for Treatment of Textile and Dye Wastewater: A Review, 2011 2nd International Conference on Environmental Science and Development IPCBEE vol. 4 (2011).

Kobya M., Hiz H., Senturk E., Aydiner C., Demirbas E., Treatment of potato chips manufacturing wastewater by electrocoagulation, Desalination 190 (2006) 201–211.

Lang I.G., The impact of effluent regulations on the dyeing industry, Review of Progress in Coloration 21 (1991) 56–71.

Mahmoodi N.M., Arami M., Bahrami H., Khorramfar S., Novel biosorbent (Canolahull): Surface characterization and dye removal ability at different cationic dye concentrations, Desalination 264 (2010) 134–142.

Mollaha M.Y.A., Gomes J.A., Das KK, Cocke DL, Electrochemical treatment of Orange II dye solution—Use of aluminum sacrificial electrodes and floc characterization, Journal of Hazardous Materials 174 (2010) 851–858

Mollah M.Y.A., Schennach R., Parga J.P., Cocke D.P., Electrocoagulation (EC) science and applications, Journal of Hazardous Materials B84 (2001) 29–41.

Mouedhen G., Feki M., Wery M.P., Ayedi H.F., Behavior of aluminum electrodes in electrocoagulation process, Journal of Hazardous Materials 150 (2008) 124–135.

Rajeshwar K., Ibanez J.B., Swain G.M., Electrochemistry and the environment, Journal of Applied Electrochemistry 24 (1994) 1077–1091.

Soares G.M.B., Miranda T., Campos A.M.F.O., Hrdina R., Fereira M.C., Amorim M.T.P., Current situation and future perspectives for textile effluent decolourisation. Textile Institute World Conference, 83, Shanghai, China (2004) 8 p.

Vijayaraghavan K., Lee Min Woo, Yun Yeoung-Sang, A new approach to study the decolorization of complex reactive dye bath effluent by biosorption technique, Bioresource Technology 99 (2008) 5778–5785.

Zaroual Z, Azzi M, Saib N., Chainet E, Contribution to the study of electrocoagulation mechanism in basic textile effluent, Journal of Hazardous Materials B131 (2006) 73–78.

Zaroual Z., Azzi M., Saib N., Karhat Y., Zertoubi M., Treatment of tannery effluent by an electrocoagulation process. Journal-American Leather Chemists Association 100 (2005) 16–21.

Vepsäläinen Mikko, and Sillanpää Mika, "Electrocoagulation in the treatment of industrial waters and wastewaters," Advanced Water Treatment (2020) 1–78.

15 Evaluation of Historic Trends on Hydro-Meteorological Variables in the Cauvery Basin, India

Dhasarathan R, Brema Jayanarayanan,
and Sneha Gautam
Department of Civil Engineering, Karunya Institute
of Technology and Sciences, Coimbatore, India

CONTENTS

15.1 INTRODUCTION

Water is an important natural resource that is essential for sustaining life on earth. The availability of water, however, is decreasing due to changes in the patterns of land use/land cover (LU/LC), climate, and reduced storage areas. Additionally, changes in regional rainfall patterns, i.e., changes in frequency and intensity of rainfall, have reduced the temporal availability of regional water when required

(Gautam et al. 2021, Kalpana et al. 2020). The IPCC (2007), in its fourth assessment report, projected a decrease in winter (December to February) precipitation over the Indian subcontinent implying lesser storage and greater water stress during the lean period. Thus, India is predicted to be severely impacted by climate change. Watersheds are receptive to the dynamics of land use that are induced by human activities. Changes in land cover are predicted to have a significant effect on climatological parameters, river flows, and sediment yields (Welde, 2017).

Runoff and geomorphic impacts may become more severe as the transition progresses, and this would be compounded by more extreme and frequent impacts as well as cumulative minor impacts (Clifton, 2018). A wide variety of scenarios used in recent national climate projections have helped to quantify the consequences of climate change on hydrology in Germany's major river basins (Hattermann, 2014). The results also indicated that hydrology is very vulnerable to climate change, and the effects of an overall precipitation increase can even be overcompensated by a rise in evapotranspiration.

Jin (2018) evaluated the changes in precipitation, temperature, and runoff in the region of the Yellow River over the past six decades using the Mann – Kendall and Spearman correlation tests. Hydrological changes can increase the risk of flooding, while reduced flows will significantly impact inland navigation during the summer and reduce the supply of water for industry and agriculture (Middelkoop, 2001). Sensitivity analysis provides help in understanding the hydrological response to possible climate change fluctuations (Singh, 2015). The Soil and Water Assessment Tool (SWAT) can be used to determine potential shifts in sub-basin river flow rates and water balance under the climate projections predicted (Singh, 2015). The simulated daily runoff is substantiated by relatively high Nash-Sutcliffe simulation coefficients and low root mean square errors for the calibration and validation periods, respectively (Khare, 2014). Land development and the changing environment are expected to alter the hydrology of the tropical forests significantly (Farinosi, 2019). The SWAT results show the consistency between observed and simulated flow (Ayivi, 2018). Compared with La-Nina years, the SWAT analysis showed that the risk of failure in rice crop productivity is low during El Niño years (Bhuvaneswari and Geethalakshmi, 2013). Rainfall intensity analysis shows more frequent and higher level floods in the wet regions and more intense droughts in the arid areas (Abbaspour, 2009).

There is a high correlation between the efficiency of reservoirs and the measurements reflecting the rainfall patterns during the winter and summer seasons, as well as the average cumulative precipitation total which evaluates the variation of rainfall during the year (Afzal, 2015); 60% of the precipitation contributed to the runoff toward Chilika Lake and the efficiency of the analysis resulted in a Nash-Sutcliffe coefficient of 0.88 and a RSME of 54.5 mm (Santra, 2013). SWAT results can help in decision-making regarding prevention and mitigation measures, and this research helps us to better face the demands of potentially greater flood risks and is thus of significant social and environmental value (Qihui Chen, 2019). Land use change is an especially important issue considering global dynamics and their response to the hydrologic characteristics of soil and water management in a catchment (Welde, 2017).

SWAT is a continuous time model which operates at basin scale on a daily basis. Its aim is to forecast the long-term impacts of planning and scheduling of agricultural practices within one year. It can also aid in determining management strategies and alternate management strategies for environmental efficiency (Food and Agriculture Organization of the United Nations, 2019). A study conducted by Bhuvaneswari and Geethalakshmi (2011) studied the effects of climate change on agriculture in the Cauvery basin. The study suggested the use of adaptation techniques that included the rice augmentation program, the use of temperature responsive cultivars, and the use of organic manures and bio-fertilizers to conserve water and improve rice production under tropical temperatures.

Limitations of the above-referenced studies advocate for a systematic investigation into future scenarios in order to better understand the water sources and hydrodynamics of the Cauvery River. In the present study, the sensitivity of the Cauvery basin is investigated for its hydrologic response to potential changes in climate variability using the SWAT model. This effort is built on an integrated approach to watershed modeling and climate change. The grid-based time series of temperature, precipitation, humidity, wind speed, and solar radiance datasets are generated from the IMD (Indian Meteorological Department) and are used in four scenarios.

15.2   STUDY AREA

The Cauvery River basin (CRB), with a surface area of 81155 km^2, covers 24% of the geographical area of the country. CRB is situated in the three southern states of Tamil Nadu, Kerala, and Karnataka and originates from Talakaveri, Coorg District, in Karnataka and ends in the Nagapattinam District of Tamil Nadu, stretching 805 km. The whole basin obtains an average annual rainfall of 956 mm, and over 4.4 million people are employed in the agricultural sector (Bhuvaneswari and Geethalakshmi, 2011). Agriculture is based on the southwest monsoon period. The water requirement for agriculture in the study area is mainly based on the Cauvery River. CRB is a vital river basin in India in the field of food security and agriculture in southern India (Bhuvaneswari and Geethalakshmi, 2013).

The basin is divided into three parts: the Delta region in western Tamil Nadu, the highlands of Mysore, and the Western Ghats area. The soil types found in the basin area are black soil, alluvial soil, red soil, and laterites. However, red soil occupies the largest area of the river basin. The delta region of the study area is covered with Alluvial soil. The hydrology report by the National Institute of Hydrology, Roorkee, states that the surface water potential of the basin has dropped to 66.88 km^3 and the groundwater potential has dropped to 16.46 km^3.

The average discharge from the 805km Cauvery River is 680 m^3/sec and leads into the Bay of Bengal. The primary utilization of Cauvery is for electricity generation, irrigation for agricultural activities, and for water supply for domestic consumption. The major reservoirs of the Cauvery basin are the Mettur Dam which has a 2644 Mm^3 capacity and the Krishna Raja Sagara Dam which has a 1368 Mm^3 capacity and contributes to major activities. It was observed that inflow into the reservoirs was 58% short of total inflow capacity in 2003. The districts of Thanjavur, Nagapattinam, Thiruvarur, Pudukottai, and Cuddalore have been declared protected Agricultural zones in the Cauvery delta region.

The basin experiences a tropical climate which features the monsoon rain flourishing the agriculture and meeting the domestic and commercial needs. The average precipitation of the Cauvery basin during the study period ranged from 1000 mm to 1500 mm with a summer rainfall of 13.8%, winter rainfall of 1.4%, monsoon rainfall of 60.6%, and post-monsoon rainfall of 24.2%. The highest rainfall is received in the southwest basin during the northeast monsoon. The average maximum temperature in the study area is 44°C and the average minimum temperature is 18°C, respectively. The Cauvery basin region situated in Tamil Nadu mostly receives rain during the months of October to December. In Karnataka, the rainfall is extremely high, which floods the farmlands and affects agricultural activities in the region. The crops widely cultivated in the study area are Sugarcane, Paddy, and Ragi. The kharif crops are cultivated between the months of May and July and are based on the southwest monsoon in the basin.

15.3 WEATHER DATABASE

The weather data for the watershed modeling was collected on a daily basis from the IMD and included rainfall data in mm, temperature in °C, wind speed in km/hr, relative humidity in % and solar radiance. The collected data was formatted for SWAT files. The GRD data (Gridded data) from the IMD was converted to SWAT files to import the weather database into the ArcSWAT workspace. GRD is widely used in image processing for grid formats to represent data in a grid pattern. The years 1970–1980 were taken as the warm-up period for the ArcSWAT simulation. A sample dataset from the meteorological data is shown in Table 15.1.

TABLE 15.1
Sample Meteorological Data from 1990 for Thanjavur District

Parameter	Precipitation mm/Day	Relative Humidity (%)	Temperature (C)	Wind Speed (m/s)	Specific Humidity
January	84.19	71.11	25.33	6.07	0.014613
February	2.01	66.52	28.82	5.82	0.016467
March	16.99	62.85	31.5	4.94	0.017665
April	9.74	59.51	33.56	4.97	0.018695
May	81.87	64.5	31.64	7.03	0.019005
June	7.2	59.36	31.26	7.55	0.017804
July	40.02	62.06	30.29	7.06	0.017222
August	47.54	60.4	30.39	7.02	0.017083
September	171.81	69.63	29.45	4.58	0.018111
October	183.37	77.44	28.38	4.9	0.018524
November	62.12	81.02	26.93	5.31	0.017945
December	13.6	75.65	26.76	7.04	0.016802
Annual	720.47	67.51	29.52	6.03	0.017496

15.4   SWAT MODELLING

15.4.1   Watershed Delineation

The Digital Elevation Model for the study area was obtained from ASTER (Advanced Spaceborne Thermal Emission and Reflection Radiometer) by NASA (United States National Aeronautics and Space Administration) which was collected in a resolution of 30 m for an area of 82000 km^2 covering the CRB area. The elevation raster was masked for the study area using a study area boundary. The elevation of the study area was found to vary from 0 m to 2629 m. The highest point of elevation was found in the Nilgiris District, a southwestern region of the river basin. The delta region in the study area is 3% of the total river basin located in the Nagapattinam District.

15.4.2   Land Use/Soil/Slope Definition

15.4.2.1   Land Use

The type of land in the study area is described in the model using Land use/Land cover extracted with respect to the basin area (Figure 15.1). The dataset was classified based on the International Geosphere-Biosphere Programme (IGBP) scheme.

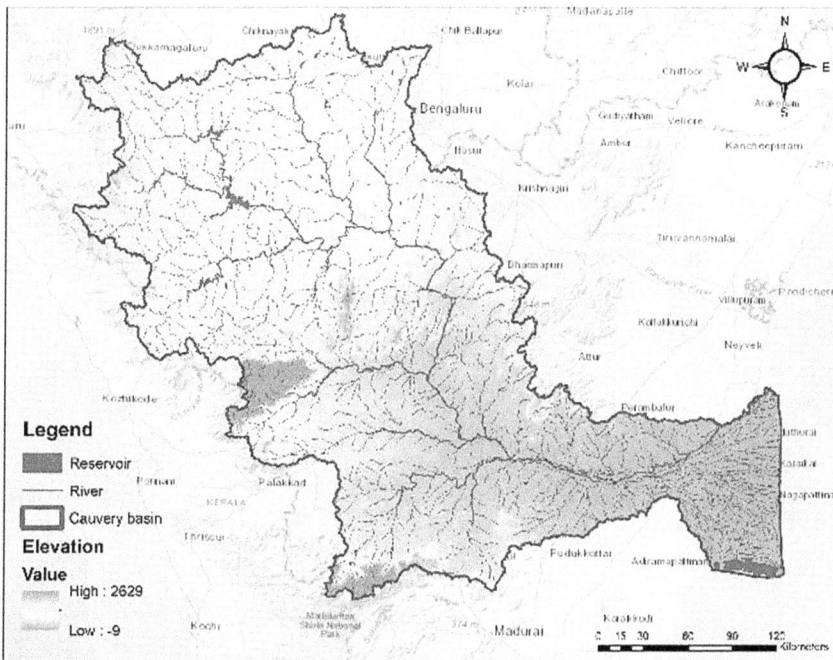

FIGURE 15.1   Study area.

TABLE 15.2

Land Use Categories

Type	SWAT Class	Area (ha)	Coverage (%)
Forest-Deciduous	FRSD	9188.144	11.91
Agricultural Land-Generic	AGRL	23860.63	30.93
Residential	URBN	22369.49	29
Forest-Mixed	FRST	3841.381	4.98
Water	WATR	1213.288	1.57
Plantains	PLAN	13344.76	17.3
Forest-Evergreen	FRSE	3320.09	4.3
Wetlands-Forested	WETF	8.044256	0.01

The Land use dataset was obtained with 100-m resolution from a Linear Imaging
Self-Scanning Sensor—3 (LISS-III). Furthermore, the land use dataset was classified based on LULC USGS classes in ArcSWAT. Agricultural land in the study
area was found to occupy the largest share, followed by residential or urban land
at 30.93% and 29%, respectively. The land use categories of the Cauvery basin are
identified as given in Table 15.2 and Figure 15.2.

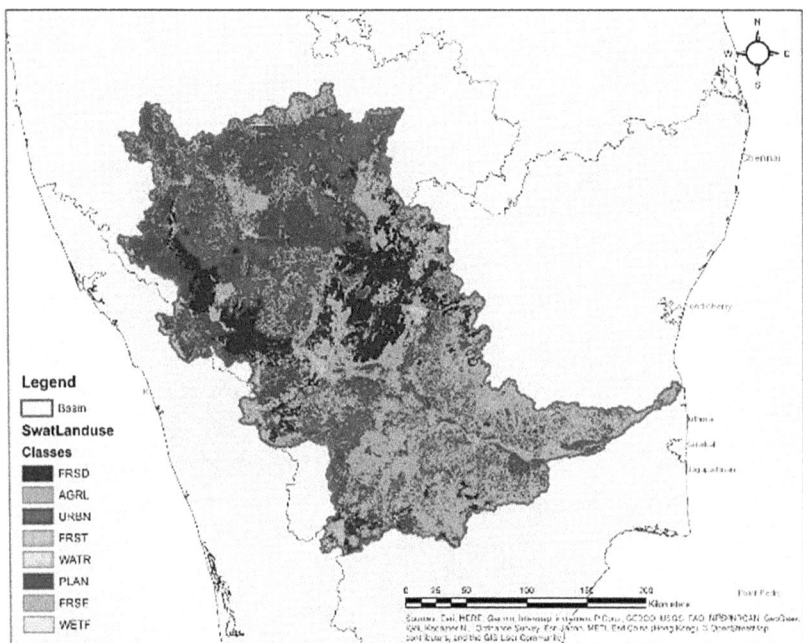

FIGURE 15.2   SWAT land use classification.

TABLE 15.3
Soil Classification

HSG	Soil Texture Class	Runoff Potential	Area (km²)	Coverage (%)
D	Sandy clay	High	43731	55.65
C	Clay loam	Moderately high	30809	36.14
C	Silty loam	High (Unless drained)	3888	3.86
D	Silt	High (Unless drained)	5901	4.35

15.4.3 SOIL

The soil dataset for the study area was obtained using global hydrologic soil groups (HSG) with 250-m resolution and a geographical resolution of 1/480 decimal degrees (Table 15.3).

The HSG classification was based on the runoff potential for each type of soil in the study area. Among the four classes from Figure 15.5, Sandy clay has moderate runoff potential with <50% sand and 20–40% clay, Clay loam has high runoff potential with <50% sand and 40% clay, and Silty loam and Silt have high runoff potential unless the water is drained with <50% sand and 40% clay.

15.4.4 SLOPE

It is of vital importance to assess the nutrient, water, and sediment transport through a watershed. For the HRU analysis, a DEM was used to create a slope map for the study area; 81% of the study area is covered by a slope of 0 to 10% of the total elevation. The slope was reclassified as shown in Table 15.4. The statistical details of the slope were generated. The minimum and maximum elevations of the study area were 0 and 2629 m. The mean elevation was 621.95 m with a Standard deviation of 376.46 m.

TABLE 15.4
Slope Classification

Slope Class	Area (km²)	Coverage (%)
0–10	62540.16	81.07
10–50	12991.10	16.84
50–100	1581.99	2.05
100–200	32.14	0.04
200–9999	0.42	0

15.5 RESULTS AND DISCUSSION

The results from the SWAT simulation in the study area stated the impact of climate and land use change on the meteorological variables. In order to perform a comparative analysis, the study period was split into four decadal periods.

15.5.1 Scenarios and Period of Study

The simulation time was divided into four groups to differentiate between climate change and land use change. The land use change and climate dataset for the decadal period were classified and simulated distinctively. The Scenario details are given in Table 15.5.

15.5.2 Trend Analysis of History Data

The Meteorological data for the ArcSWAT simulation was obtained from the IMD. The parameters considered for the analysis were Precipitation (mm), Temperature (°C), Wind speed (km/hr), Solar radiation (cal/sq. cm), and Relative humidity (%). Figure 15.3 displays the average precipitation from the various monitoring stations in the study area.

TABLE 15.5
Scenario and Period Classification

Scenario	Period
S1	1980–1990
S2	1991–2000
S3	2001–2010
S4	2011–2019

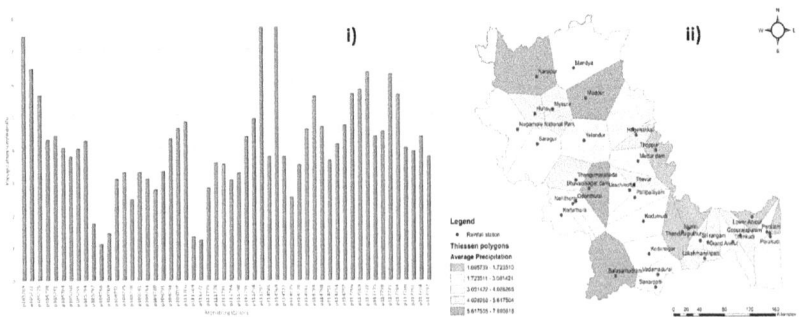

FIGURE 15.3 (i) Average precipitation from rain gauge stations, (ii) Theissen polygons from average precipitation.

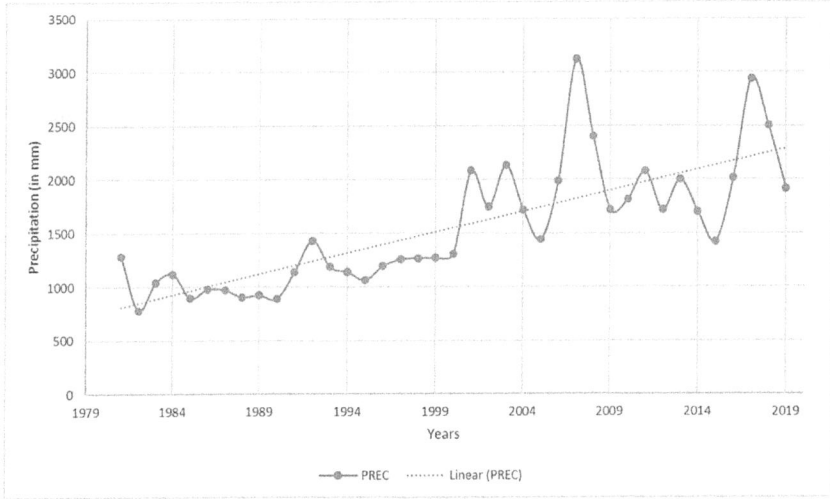

FIGURE 15.4   Average precipitation in years.

In Figure 15.4, the average precipitation trend graph is plotted for the years 1980–2019, as obtained from the SWAT simulation. During 2007, the study area experienced the highest rainfall. However, there was an abrupt change in precipitation after the year 2000 which showed a sudden rise and fall. The year 1981 had the lowest rainfall during the study period.

From Figure 15.5 and Table 15.6, it can be observed that there is an increasing trend in rainfall. Using the rainfall data, Theissen polygons were constructed

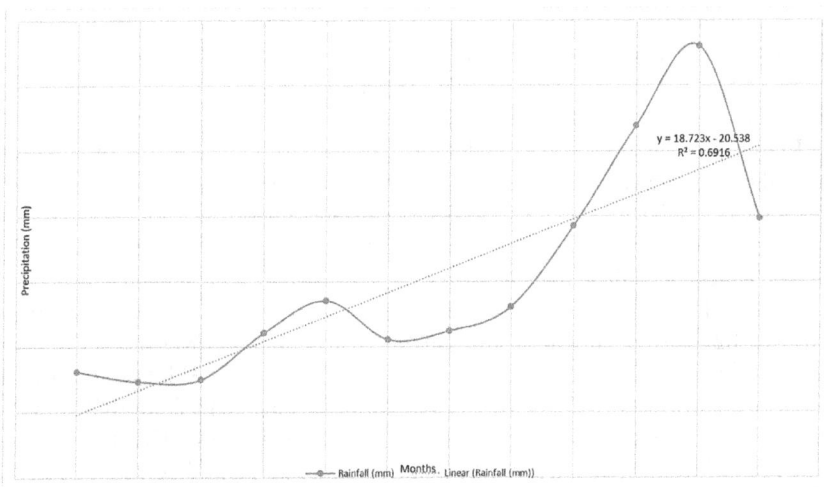

FIGURE 15.5   Monthly average precipitation in the study area during the period 1981–2019.

TABLE 15.6

Results from Mann-Kendall Analysis

Month	Rainfall (mm)	Variance (S)	Test Statistics (Z)	Significant Level (%)
January	31.37	91	1.18	<90
February	23.43	92	1.19	<90
March	25.08	22	0.27	<90
April	60.76	25	0.31	<90
May	85.37	142	1.92	<90
June	55.36	0	0	<90
July	62.2	−74	−0.95	<90
August	80.55	155	2.01	95
September	142.3	−212	−2.76	<90
October	219.03	132	1.71	<90
November	280.17	163	2.12	95
December	148.31	75	0.97	<90
Annual	**1213.93**	**154**	**2**	**<90**

which classified the basin area into 38 polygons according to the average precipitation, which revealed that the region surrounding the Krishna Raja Sagara Dam and the Bhavani Sagar Dam received high rainfall of 5.6 to 7.69 mm/day, while the main catchment areas of the basin area received an average precipitation of 4 to 5.6 mm/day.

The Mann-Kendall analysis was conducted to understand the trend of precipitation in the study area. The analysis showed that there was a linear decrease in precipitation from January to July, and a sudden increase from August to December with high rainfall during September.

Taking precipitation into consideration, the study area received the highest rainfall during the years 2007 and 2016, and average precipitation was highest in Mayiladuthurai, Tamil Nadu and Mysore, Karnataka. However, the precipitation change was abrupt with a sudden rise and fall after the year 2000. The year 1981 had the lowest rainfall in the study period.

Figures 15.6 and 15.7 display the annual average precipitation received from the 82 locations in CRB. The states of Karnataka and Kerala lie between 11.36 N and 13.06 N latitude and between 75.19 E and 77.88 E longitude. Tamil Nadu stretches from 11.77 N to 10.22 N latitude and 78.86 E to 76.77 E longitude, which extends from Coimbatore up to Nagapattinam. The delta region of the river basin extends from 11.72 N to 10.17 N and from 79.87 E to 78.95 E. The average distribution of precipitation and evapotranspiration are shown

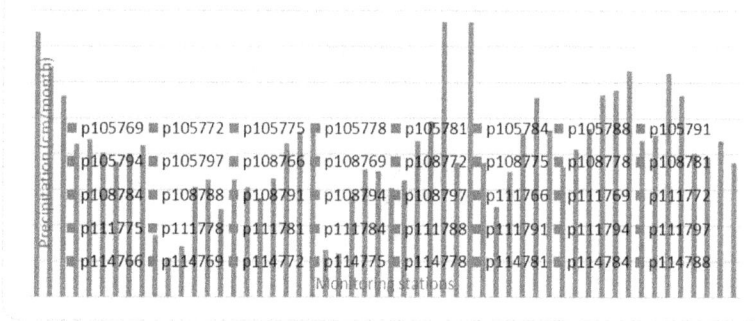

FIGURE 15.6   Annual average precipitation in study area during the period 1981–2019 – station wise.

in Tables 15.7–15.10 and Figures 15.8–15.12. These graphs help to identify the behavior of evapotranspiration with respect to the geographical and meteorological characteristics of the study area. The precipitation in the study area was found to be linear during the years 1980–2000, whereas the years 2001–2019 experienced a discrete variability with exceedingly high rainfall during the years 2007 and 2017.

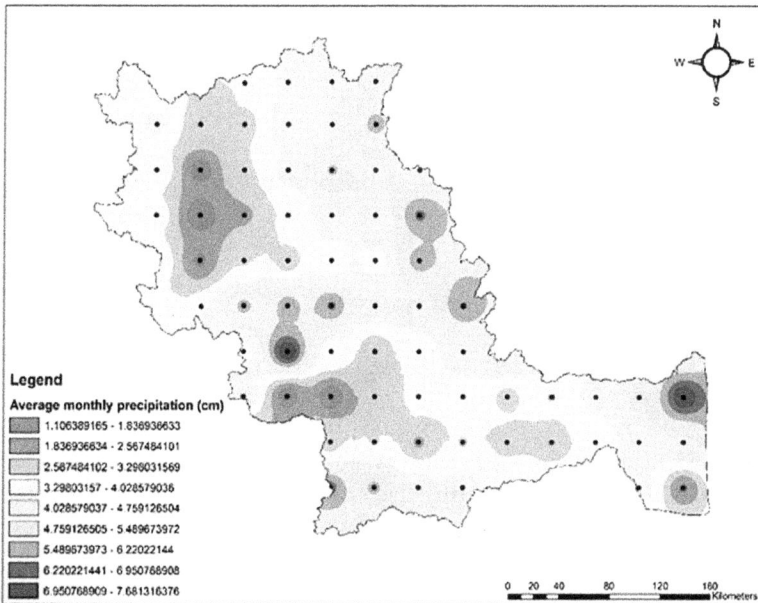

FIGURE 15.7   Annual average precipitation in study area during the period 1981–2019 – interpolated distribution.

TABLE 15.7
SWAT Output for the Years 1981–1990

Year	Annual Average Precipitation (mm)	ET (mm)
1981	1282.93	288.17
1982	783.87	251.57
1983	1041.33	259.71
1984	1120.51	252.31
1985	896.95	248.28
1986	980.79	269.1
1987	972.48	263.43
1988	907.16	290.05
1989	923.65	281.87
1990	889.67	237.78

TABLE 15.8
SWAT Output for the Years 1991–2000

Year	Annual Average Precipitation (mm)	ET (mm)
1991	1138.86	290.9
1992	1428.85	295.69
1993	1184.21	282.99
1994	1136.86	261.92
1995	1063.37	294.36
1996	1191.52	309.45
1997	1254.94	323.45
1998	1266.53	293.23
1999	1270.55	288.75
2000	1304.25	310.15

Figure 15.13 shows that the average evapotranspiration occurring in the study area represents a linear rate during the first two decades of the study period, whereas the years 2001–2019 experienced a periodic rise and fall in the rate of evapotranspiration.

The SWAT-CUP calibration process was implemented for the years 1993–2005. The calibration process gave a Percent bias (PBIAS) of 14% and a Nash-Sutcliffe (NS) coefficient of 0.77. The validation process implemented for the years 2006–2019 returned a Percent bias (PBIAS) of 11% and a Nash-Sutcliffe (NS) coefficient of 0.81 (Table 15.11).

TABLE 15.9
SWAT Output for the Years 2001–2010

Year	Precipitation (mm)	ET (mm)
2001	2080.79	474.32
2002	1744.86	412.54
2003	2129.51	416.83
2004	1715.58	364.86
2005	1444.67	332.63
2006	1987.05	476.14
2007	2937.07	430.54
2008	2399.21	463.32
2009	1716.84	419.33
2010	1817.87	389.4

TABLE 15.10
SWAT Output for the Years 2011–2019

Year	Precipitation (mm)	ET (mm)
2011	2073.78	463.51
2012	1716.44	401.37
2013	1997.69	430.81
2014	1698.97	376.77
2015	1424.17	332.74
2016	2013.17	477.98
2017	2937.07	430.76
2018	2501.42	441.66
2019	1908.42	417.77

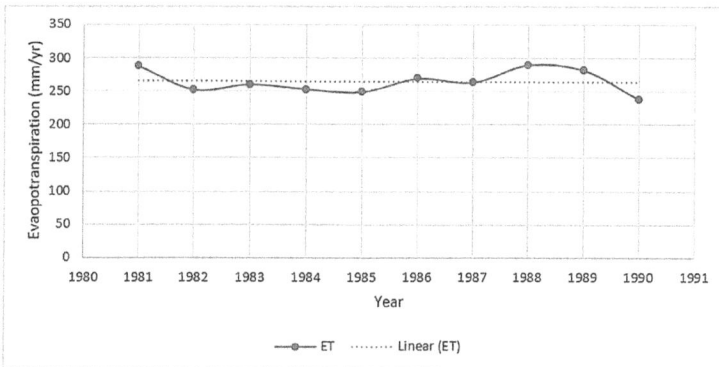

FIGURE 15.8 Evapotranspiration simulated from SWAT for the years 1981–1990.

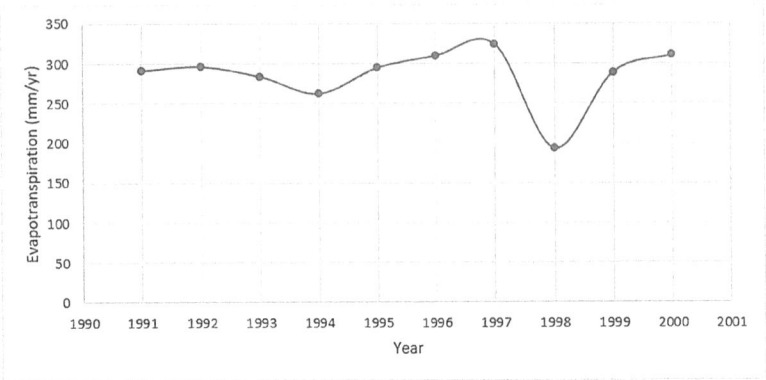

FIGURE 15.9   Evapotranspiration simulated from SWAT for the years 1991–2000.

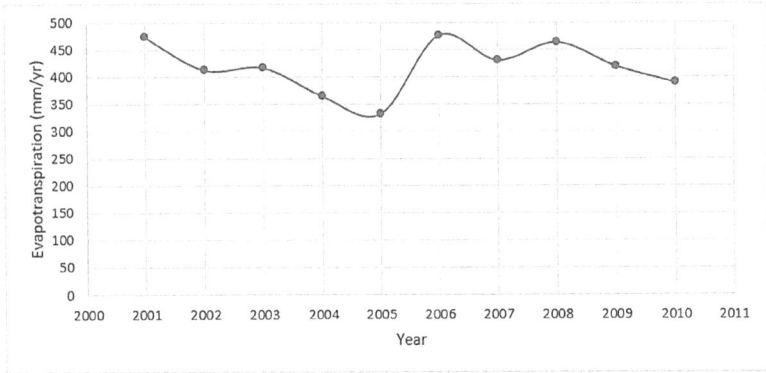

FIGURE 15.10   Evapotranspiration simulated from SWAT for the years 2001–2010.

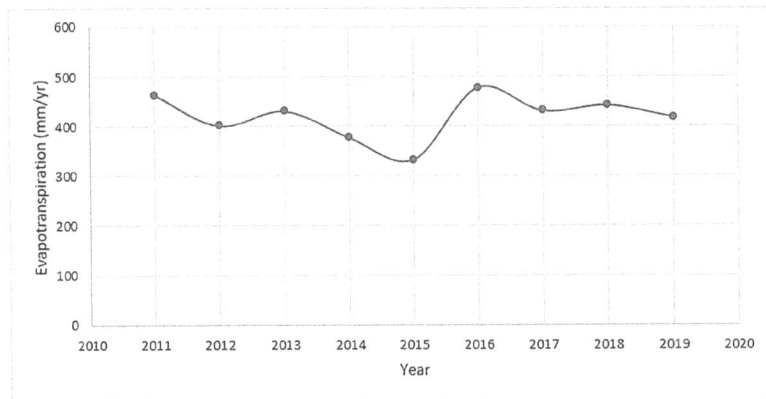

FIGURE 15.11   Evapotranspiration simulated from SWAT for the years 2011–2019.

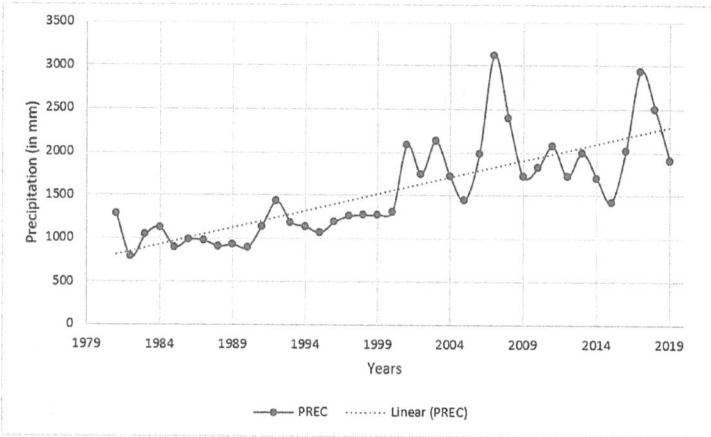

FIGURE 15.12 Temporal variation of average precipitation simulated from SWAT in the study area for the period 1981–2019.

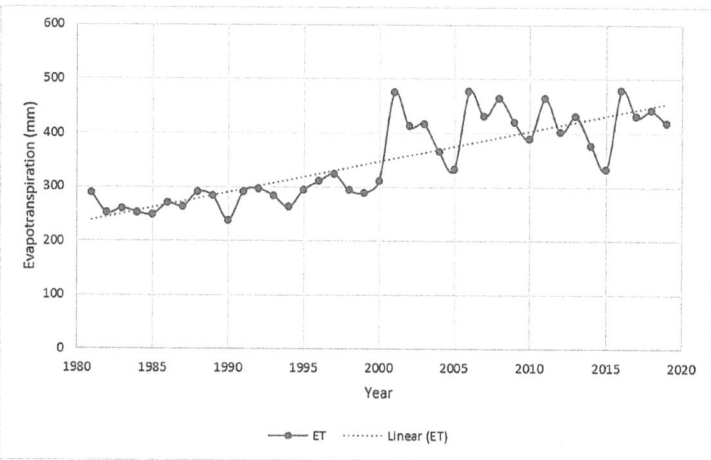

FIGURE 15.13 Temporal variation of average evapotranspiration simulated from SWAT in the study area for the period 1981–2019.

TABLE 15.11

Evaluation Indicators from Calibration and Validation in SWAT-CUP

Evaluation Indicators	Calibration Period (1993–2005)	Verification Period (2005–2019)
NS	14	11
PBIAS (%)	0.77	0.81

15.6   CONCLUSION

In this study, the variation and trends of hydro-meteorological parameters and land use change were investigated to study the impact of climate change in the hydrological components of CRB. The simulation implemented through scenarios paves the way to better understand the trend of land use and climate change impacts in the study area. The conclusions are as follows:

1. The meteorological parameters in the first two decades had regularity, whereas the following two decades displayed dissimilarity with a sudden rise and fall in the hydrological parameters.
2. The SWAT modeling resulted in a good performance of the evaluation indicators for the years 1993–2005 and 2006–2019 obtaining PBIAS (Percentage Bias) of 14% and 11% and a NS (Nash-Sutcliffe) efficiency of 0.77 and 0.81 from the SWAT statistics.
3. The study provides a facility to manage the available water resources and understand the impacts of various hydro-meteorological components and natural resources in the study area.

REFERENCES

Abbaspour, K.C. (2009). Assessing the impact of climate change on water resources in Iran. Water Resources Research, 45(10): 1–16.
Afzal, M. (2015). The impact of the variability and periodicity of rainfall on surface water supply systems in Scotland. Journal of Water and Climate Change, 7(2): 321–329.
Ayivi, F. (2018). Estimation of water balance and water yield in the Reedy Fork-Buffalo Creek Watershed in North Carolina using SWAT. International Soil and Water Conservation Research, 6(3): 203–213.
Bhuvaneswari, K., Geethalakshmi, V. (2011). Climate change impact assessment and adaptation strategies to sustain rice production in Cauvery basin of Tamil Nadu. Current Science, 101(3): 342–347.
Bhuvaneswari, K., Geethalakshmi, V. (2013). The impact of El Niño/Southern oscillation on hydrology and rice. Weather and Climate Extremes, 2: 39–47.
Clifton, C.F. (2018). Effects of climate change on hydrology and water resources in the Blue Mountains, Oregon, USA. Climate Services, 10: 9–19.
Farinosi, F. (2019). Future climate and land use change impacts on river flows in the Tapajos Basin in the Brazillian Amazon. Earth's Future, 7(8): 9–19.
Food and Agriculture Organization of the United Nations. (2019). Food and Agriculture Organization of the United Nations. Retrieved from Food and Agriculture Organization of the United Nations: http://www.fao.org/land-water/land/land-governance/land-resources-planning-toolbox/category/details/en/c/1111246/
Gautam S, Gautam AS, Singh K, James EJ, Brema J. (2021) Investigations on the relationship among lightning, aerosol concentration, and meteorological parameters with specific reference to the wet and hot humid tropical zone of the southern parts of India. Environmental Technology and Innovation. (doi.org/10.1016/j.eti.2021.101414).

Hattermann, F.F. (2014). Climate change impacts on hydrology and water resources in Germany. Meteorologische Zeitschrift, 2(2015): 201–211.

Jin, J. (2018). Impacts of climate change on hydrology in the Yellow River source region, China. Journal of Water & Climate Change, 11(3): 916–930.

Kalpana, P. Prathibha, S. Gopinathan, P. Subramani, T. Roy, P. D. Gautam, S. Brema, J. (2020). Spatio-temporal estimation of rainfall patterns in north and northwestern states of India between 1901 and 2015: change point detections and trend assessments. Arabian Journal of Geosciences. 13, 1116. https://doi.org/10.1007/s12517-020-06098-9.

Khare, D. (2014). Hydrological modelling of barinallah watershed using Arc-SWAT model. Earth & Environmental Sciences, 43(X–XII): 76–89.

Middelkoop, H. (2001). Impact of climate change on hydrological regimes and water resources management in the Rhine Basin, 49: 105–128. Climate change.

Qihui Chen, H.C. (2019). Impacts of climate change and land-use change on hydrological extremes in the Jinsha River Basin. Water, 11(7): 1–25.

Santra, P. (2013). Modelling runoff from an agricultural watershed of western catchment of Chilika lake through ArcSWAT. Journal of Hydro-Environment Research, 7(4): 261–269.

Singh, D. (2015). Assessment of impact of climate change on water resources. Arabian Journal of Geosciences, 8: 10625–10646.

Welde, K. (2017). Effect of land use land cover dynamics on hydrological response of watershed: Case study of Tekeze Dam watershed, northern Ethiopia. International Soil and Water Conservation Research, 5(1): 1–16.

16 Valuation of Environmental Externalities

A Tool for Sustainability

Hemant Bherwani
CSIR-National Environmental Engineering Research
Institute (CSIR-NEERI), Nagpur, India, and

Academy of Scientific and Innovative
Research (AcSIR), Ghaziabad, India

Alaka Das and Hariharan B
CSIR-National Environmental Engineering
Research Institute (CSIR-NEERI), Nagpur, India

Ankit Gupta and Rakesh Kumar
CSIR-National Environmental Engineering Research
Institute (CSIR-NEERI), Nagpur, India, and

Academy of Scientific and Innovative
Research (AcSIR), Ghaziabad, India

CONTENTS

DOI: 10.1201/9781003140382-16

16.1   INTRODUCTION

The depletion of natural resources, or ecocide (Higgins, 2012), in the name of development is a paradox that can occur when natural resources and their ecosystem services account for the economic growth of a nation. In this situation, natural resources are considered the real wealth affecting income, fiscal revenue, and poverty reduction (Organization for Economic Co-operation and Development [OECD], 2011). A country with an abundance of natural resources often falls victim to the "natural resource curse" and this leads to adverse ecosystem degradation in the name of economic development. The need for macro-economic policies is indispensable for long-term economic growth as well as for the conservation of natural resources (United Nations Development Programme, 2011). The valuation of natural resources and ecosystem services is essential in order for decision-makers to assess the natural capital. Environmental economics is a tool that measures the costs of natural resource degradation and the costs of damage due to an increase in pollution (Bherwani et al., 2020a).

Businesses involved in production/manufacturing, transportation; and agricultural and allied services significantly contribute to the overexploitation of natural resources. The loss of natural resources due to the adverse effects of pollution from man-made activities can be classified using cost-benefit analysis. The cost accumulated by the polluter for not taking any initiatives to curb the pollution is determined using cost-based equations, whereas the estimated damages to the ecosystem and human well-being are determined using benefits-based calculations. Regulatory instruments are utilized to analyze the relevant information about mitigating pollution through market-based or economic instruments (Marketable pollution permits & Pollution taxes) and non-market-based instruments (Command & Control). These environmental policy instruments can be utilized as tools to mitigate pollution by promoting a transition toward low-carbon technology. The data relating, the revenue collected by the European Union in the form of Environmental tax (European Commission, 2020) and Environmental Performance Index (EPI) report published in the year 2020 (where the UK formally left the EU in January, 2020) shows that the prior transition towards climate-resilient and low-carbon technologies by implementing pragmatic environmental legislations led the EU to be at the forefront of the move toward a sustainable future (World Economic Forum, 2020). The two crucial policies considered for the evaluation of the Environmental Performance Index of nations are Ecosystem vitality and Environmental health. Denmark holds the first position in EPI report and utilizes economic instruments effectively (pollution tax) (Oh & Sverndsen, 2015).

Recently Volkswagen, a prominent automobile manufacturer, was forced to pay compensation for installing a "defeat device" – software that manipulated the diesel emission test (Fitzgerald & Spencer, 2020). The assessment of pollution-related damages is now under significant consideration by most of the world's nations, and the environment has become a crucial part of sustainability agendas since the Paris Agreement (The United Nations [UN], 2015). Environmental Impact Assessment (EIA) and Life Cycle Assessment (LCA) are used as midpoint analysis methods to identify the causes of environmental impacts (Bherwani et al., 2019 and 2020a). The valuation of economic impacts can be accomplished by the step-by-step pathway assessment explained in Figure 16.1.

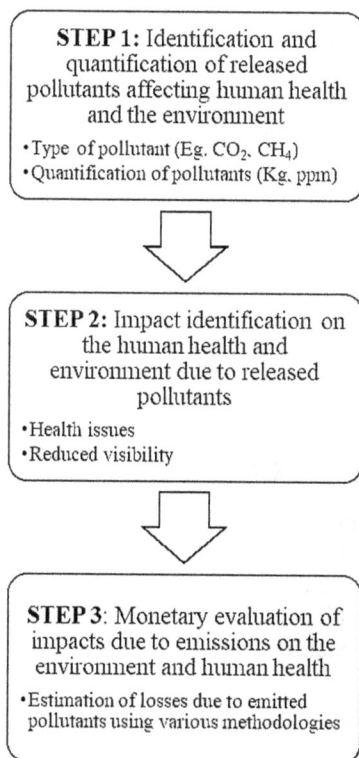

STEP 1: Identification and quantification of released pollutants affecting human health and the environment
- Type of pollutant (Eg. CO_2, CH_4)
- Quantification of pollutants (Kg, ppm)

STEP 2: Impact identification on the human health and environment due to released pollutants
- Health issues
- Reduced visibility

STEP 3: Monetary evaluation of impacts due to emissions on the environment and human health
- Estimation of losses due to emitted pollutants using various methodologies

FIGURE 16.1 Valuation of environmental damages.

The Economics of Ecosystem and Biodiversity (TEEB) project from the Millennium Ecosystem Assessment (MEA) framework is globally recognized for its economic valuation of ecosystems and their services. The core principles of TEEB are recognizing the importance of the ecosystem, defining the values of the ecosystem in terms of benefits/costs, and the valuation of the ecosystem (Millennium Ecosystem Assessment, 2005). The economics of evaluating natural ecosystems can be determined through various valuation methods (Pascual et al., 2010). The integration of socio-economic factors can mitigate adverse anthropogenic activities when the environmental external cost for energy is considered. The classification of driver-based impacts into three categories of pollutants (greenhouse gases, air pollution, and water pollution) which affect the environment are utilized for the monetary valuation of the damages caused to the environment (Bherwani et al., 2019 and 2020b).

16.2   GREENHOUSE GASES

Greenhouse gas emissions are one of the crucial contributors to environmental deterioration leading to potential impacts on the ecosystem. The extensive release of greenhouse gases into the atmosphere is due to inexorable anthropogenic activities in the last 150 years like urbanization, industrialization, and changes in land use for agricultural and other activities (United States Environmental Pollution Agency) (Intergovernmental Panel on Climate Change [IPCC], 2014). The adverse impacts of GHG emissions include a rise in temperature, the melting of glaciers, flash floods, extreme drought, changes in climate patterns, loss of biodiversity, crop yield reduction, etc. The classification of greenhouse gas contributors based on economic activities show that Electricity & heat production and Agriculture, forestry, and other allied activities contribute nearly 50% of emissions (United States Environmental Pollution Agency). China, the US, and Europe are the top three GHG contributors, adding 14 times the emissions contribution of the bottom 100 countries (Friedrich et al., 2013). Also, these three major contributors account for more than half of GHG emissions, whereas the bottom 100 nations contribute only 3.5% of total emissions.

16.2.1   QUANTIFICATION OF GHGs EMISSION

Various methods of quantification based on activities (such as energy, transportation, agricultural activities, etc.) exist in the form of formulae or monitoring instruments that allow calculations of GHG emissions and the methods include Quantification using continuous monitoring system, fuel analysis method and Life cycle assessment modelling.

16.2.2   MONETARY EVALUATION OF DAMAGES DUE TO GHGs

The social and environmental costs of climate change can be determined for financial decision-making using a tool called carbon pricing. The most common method for achieving the targets set for industries to reduce emissions provides value to emissions in order to accelerate low-carbon innovations (Bherwani, et al., 2019). Various approaches such as shadow price, hybrid carbon price, the social cost of carbon, and internal carbon fees are involved in determining carbon pricing.

16.2.3 SHADOW PRICE METHOD

The corporates assign the cost (hypothetical) of emissions according to various business activities to assess the associated risks. This aids in the identification of high-carbon emission activities and sets out investments in low-emission technologies for adhering to the regulations. The range of the Shadow price varies from $2 to $892 per ton depending on the business sectors (Center for Climate and Energy Solutions, n.d.). Infosys, a service sector company, assigned $11 per ton of CO_2, whereas the ESSAR group, from the production/manufacturing sector, assigned $15 per ton of CO_2 (Gajjar & Adhia, 2018).

16.2.4 INTERNAL CARBON FEE METHOD

The internal carbon fee method adopted by corporates involves assigning a cost to GHG emissions of various economic activities for prioritizing the high-emitter activities and shifting toward low-carbon technologies. The fee is internally developed by the organization and ranges from $5 to $20 per metric ton (Center for Climate and Energy Solutions, n.d.).

16.2.5 HYBRID CARBON PRICE METHOD

The organization incorporates all the aforementioned methods to meet the target emissions. A tractor manufacturing company from India, Mahindra & Mahindra, integrated hedonic pricing and the implicit price method for investments in eco-friendly initiatives. The company also fixed a shadow price of $10 per ton of CO_2 and is focusing on low-cost pollution mitigation measures (Bherwani et al., 2019).

16.2.6 THE SOCIAL COST OF CARBON

The social cost of carbon method is the general method used for calculating CO_2 equivalent emissions, irrespective of sectors. It provides a detailed estimation of climate change damages, including damages from changes in human health, net agricultural yield, changes in energy system costs (reduced costs for heating and increased costs for cooling), and property damage due to severe calamities. The social cost of carbon is a better method for analyzing climate change impacts on the environment, and it has been utilized by the US EPA (United States Environmental Pollution Agency, 2016).

$$\text{SCC} = \text{Quantity of } CO_2 \text{ emitted}(tCO_2) \times \text{Cost of } CO_2 \text{ per tonnes} \times \text{exchange rates}$$

(16.1)

16.3   AIR POLLUTION

Air pollution is one of the major causes of health risks as it leads to widespread morbidity and mortality each year (Nair et al., 2020). Pollutants that are emitted directly into the atmosphere are known as primary pollutants which includes particulate matter like $PM_{2.5}$, PM_{10}; oxides of nitrogen (NO_x); oxides of sulfur (SO_x) and carbon monoxide (CO). Those pollutants formed from various anthropogenic activities or due to the reaction of two or more primary pollutants in the atmosphere are known as secondary pollutants that encompasses ammonium (NH_4^+) and ozone (O_3). Negative impacts on human health, visibility, agriculture, tourism, etc., are mostly created due to primary and secondary air pollutants (Hyslop, 2009). There is a need to understand the level of impacts generated by these pollutants and monetizing these impacts/damages is one of the ways in which the scale of impact can be communicated.

16.3.1   IMPACT PATHWAYS FOR AIR POLLUTION

The monetary computation of the impacts can be understood by analyzing the characteristics and quantity of pollutants released into the ecosystem. The adverse outcomes caused by the impacts of emissions can be revealed by the impact pathway approach (Bherwani et al., 2020b and 2019).

16.3.2   QUANTIFICATION OF POLLUTANTS

The different methods used to quantify air pollutant concentration, aside from EIA and LCA reports, are used, as per CPCB guidelines 2013 (CPCB, 2013). Apart from ground measurements, modeling software like AERMOD and CALPUFF are also available to perceive information about the dispersion of air pollutants which incorporates various meteorological factors, like wind speed and wind direction, temperature, humidity, cloud cover, and precipitation.

16.3.3   VALUATION OF AIR POLLUTION DAMAGES

Once the above-mentioned outcomes have been determined, the monetary impacts of air pollution can be calculated, which is a driver of the impact. There are various approaches like cost of illness, value of statistical life, opportunity cost, and shadow cost which can be used for determining the monetary value. The various methods of calculation and their related results are explained below.

16.3.4   HEALTH IMPACTS (MORTALITY & MORBIDITY) VALUATION

Value of statistical life and Cost of illness method:
    The damage cost can be determined by comparing the pollutant concentration with the NAAQS standard and the health risk can be calculated using different

factors such as baseline incidence and relative risks (Ko & Hui, 2012; Maji et al., 2017). Evaluation of morbidity and mortality can be done using the Cost of Illness (COI) method and the Value for Statistical Life (VSL). VSL is used for mortality cases, which can be calculated on the basis of baseline incidences and the relative risk of a particular disease. COI includes the total costs incurred, such as medicinal cost, travel cost, hospital admission, and lost workdays (Maji et al., 2017).

The relative risk for the exposed category 'c' R_r (c) is calculated using Eq (16.2) where Ca is the ambient air concentration of pollutant 'K', Cw is the NAAQS permissible standard for the pollutant 'K', and R_r is the relative risk for the pollutant 'K'. The equation is shown below:

$$R_r(c) = 1 + (Ca - Cw)*(R_r - 1)/10 \qquad (16.2)$$

Population attributable risk (PAR) is calculated using Eq (16.3) where ρ(c) is the proportion of the population in the 'C' category being exposed to the pollutant.

$$PAR = \left(\Sigma[\{R_r(c) - 1\}]*\rho(c)\right)/\left(\Sigma[\{R_r(c) - 1\}]*\rho(c) + 1]\right) \qquad (16.3)$$

I_e is the rate attributed to exposure in the population and is calculated using Eq (16.4) where I_w is the Baseline incidence per 10^5 population.

$$I_e = I_w * PAR \qquad (16.4)$$

I_{Ne} is the estimated number of cases of mortality/morbidity and is calculated using Eq (16.5) where N is the total population.

$$I_{Ne} = I_e * N \qquad (16.5)$$

The most recent values of the Global Burden of Disease can also be used to understand the relative risk related to various diseases, and the Total health damage cost for estimated cases can be calculated by VSL/COI as suggested by Maji et al. (2017) according to the equation provided below:

$$VSL/COI_{study\ region} = VSL/COI_{Reference\ Study\ Region} *\left(Per\ capita\ income_{Study\ region}\right)/$$
$$\left(Per\ capita\ income_{Reference\ Study\ Region}\right)$$
$$(16.6)$$

16.3.5 MONETIZING DAMAGES DUE TO POOR VISIBILITY

In order to determine the damages due to poor visibility, the direct transfer function method and willingness to pay (WTP) method can be used. WTP is the direct survey of people to discuss their willingness to remove visibility constraints. Using the known parameters such as average income, pollution area, rainfall, population density, temperature, and air quality index, a statistical site-dependent equation can be correlated with the amount of willingness and used as the direct transfer function (Wang & Mullahy, 2006).

$$\begin{aligned} \text{Willingness to pay (INR)} = {} & A + B\left(\text{Population density}\right) + \\ & C\left(\text{Extent of pollution area}\right) + \\ & D\left(\text{Average annual temperature}\right) + \\ & E\left(\text{Average annual rainfall}\right) + \\ & F\left(\text{Air quality index}\right) \end{aligned} \qquad (16.7)$$

The air quality index is calculated as shown in Eq (16.8) for each of the pollutants and the maximum value is considered in estimating monetary loss.

$$Ip = \left[\left\{(IHI - ILO)/(BHI - BLO)\right\} * (Cp - BLO)\right] + ILO \qquad (16.8)$$

where, BHI is Breakpoint concentration greater or equal to given conc., BLO is Breakpoint concentration smaller or equal to given conc., IHI is AQI value corresponding to BHI, ILO is AQI value corresponding to BLO and finally AQI = Max (Ip) (where, p = 1, 2…n; denotes n pollutants).

16.3.6 MONETIZING DAMAGES DUE TO REDUCED AGRICULTURAL PRODUCTIVITY

The loss of agricultural yield in financial terms can be analyzed using the marginal cost method (Muller & Mendelsohn, 2007). This method analyzes the costs added due to the release of excess pollutants. In addition, a benefit transfer with marginal damage cost values can be integrated with appropriate uncertainties.

16.4 WATER POLLUTION

One of the main problems faced by developing countries is water pollution. The direct and indirect discharge of pollutants into bodies of water is caused by different economic activities such as production, transportation, and consumption in all the sectors which increases the problem of water pollution. This study includes methodology showing how the cost of water pollution can be assessed and identified. To evaluate the environmental impact of water pollution, the gathering of data should be done. This methodology considers three types of data that are explained in Figure 16.2 (Bherwani et al., 2019).

16.4.1 OVERVIEW OF IMPACT OF WATER POLLUTION

Studies have shown that most of India's surface water resources and groundwater reserves are contaminated with toxic, inorganic, organic, and biological pollutants. This greatly increases water pollution and has an adverse effect on the environment. Most of the industries in developing countries release an estimated

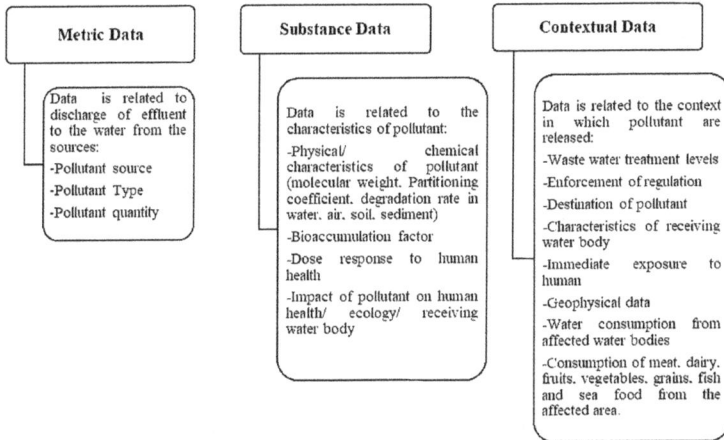

FIGURE 16.2   Types of data required for evaluation of monetary value of water pollution.

300–500 million tonnes of toxic pollutants into the water every year, and it was reported that approximately 0.64 million deaths have occurred in India due to water pollution (The United Nations, 2012).

16.4.2   MONETARY EVALUATION OF DAMAGES DUE TO WATER POLLUTION

Studies carried out by foreign authors as well as value transfer methods can be used to estimate environmental damages due to water pollution. The relationships between various activities, environmental impacts, and the resultant outcomes determined by impact pathway approach are used for monetary quantification of damages caused by water pollution (Bherwani et al., 2019).

16.4.3   DAMAGE TO HUMAN HEALTH

Damage to human health due to water pollution can be estimated using the Dose Response Method (Yongguan et al., 2001). In this method, a site survey for toxicity is carried out on those people affected to analyze the factor of dose-response to the heavy metal contamination in the water. The total health damage that is caused due to the pollutant can be calculated using the following formula:

Total health damage due to pollutant 'K' = Dose Response * discharge pollutant

concentration 'K' * Cost of

illness/DALY   (16.9)

where, DALY = Disability adjusted life year, which refers to the total working days lost due to illness, Cost of illness = Total cost (Hospital admission charges) incurred in treatment.

The damage cost for the death of a human being can be calculated by using the equation below in which the cost of one fatality depends on the cost of the dependent, the age of the victim, income, and the number of dependents.

[Assume life Expectancy = 80 years of human health]

$$
\begin{aligned}
\text{Cost of one fatality} &= \text{Victim's own loss in age} + \text{Cost of dependent's living needs} \\
&= [\text{Victim's income} \times (80 - \text{actual age})] + \\
&\quad [\text{Income} \times (\text{age of dependent*})]
\end{aligned} \tag{16.10}
$$

Age of dependent*: a) For dependent above 60 yrs.: 80 – age of dependent above 60 years, b) For dependent below 18 yrs.: 18 – age of dependent below 18 years.

The total cost related to the damage caused to humans due to water pollution can be found by adding the cost of fatality and the cost of human health as shown in the equation below:

$$
\begin{aligned}
\text{Economic loss of damage to human health} = &\big\{(\text{Economic loss of one fatality} \times \\
&\text{No. of fatalities}) + \text{Total health} \\
&\text{damage due to pollutant k}\big\}
\end{aligned} \tag{16.11}
$$

16.4.4 DAMAGE TO FISHERIES

The pollution of surface water hampers fish production and population because a certain period of time is required for the body of water to recover in order to retrieve its yield of fishes (Bherwani et al., 2019). The economic loss to a fishery can be determined using the equation below:

$$
\begin{aligned}
\text{Total monetary damage on fishes} = &\text{Daily gross income from fisheries} * \\
&\text{Total days for rejuvenation}
\end{aligned} \tag{16.12}
$$

In the above equation, the total days for rejuvenation is assumed. If fishing is banned or restricted in the area, the recovery cost of the water body surface is termed the control cost method and is expressed as total damage.

16.4.5 DAMAGE TO RECREATIONAL FUNCTION

Recreational activities such as swimming, angling, boating, etc., are hampered due to the pollution of water surfaces, and this affects the economic value as there is a loss of aesthetic value. The recreational damage cost can be determined by doing a willingness to pay (WTP) survey. In this survey, the factor considered is the total amount that people are willing to pay to restore the damaged recreational site, and this is referred to as the total monetary value. The control cost method can also be used to show the purpose of restoration. The cost related to recreational function (Yao et al., 2016) can be estimated using the following formulae (Table 16.1).

TABLE 16.1

Formulae and Nomenclature to Estimate Damage Associated with Recreational Area

Formulas	Nomenclature
$L_R = L_{SM} + L_{BT} + L_{AG}$	L_R: the loss of damage to recreation
	L_{SM}: the loss of swimming
	L_{BT}: the loss of boating
	L_{AG}: the loss of angling
	L_{LM}: loss of leisure means
$L_{SM} = P_{SM} \times N_{SM} \times d$	P_{SM}: the price of replacement of swimming per person (rs/cap/day)
	N_{SM}: the number of people swimming in the water per day (cap/day)
	d: duration of the pollution episode (day)
$L_{BT} = P_{BT} \times N_{BT} \times d$	P_{BT}: the price for replacement for boating (rs/cap/day)
	N_{BT}: the number of people boating in water per day
	d: duration of pollution episode (d)
$L_{AG} = P_{AG} \times N_{AG} \times d$	P_{AG}: the price for angling for boating (rs/cap/day)
	N_{AG}: the number of people angling in water per day
	d: duration of pollution episode (d)

Source: Bherwani et al. (2019).

16.4.6 DAMAGE TO ENVIRONMENTAL PROPERTY/OTHER DAMAGES

The Control Cost Method is used to analyze the costs incurred in operating the standard treatment technologies used to restore water bodies that become contaminated by taking into consideration all the total damages related to fisheries, recreational sites, environmental property, etc. In accordance with studies carried out by foreign authors, the damage cost values due to different pollutants like BOD_5, COD, TSS, heavy metals, etc., can be shifted to the study location with regard to uncertainties. Ground data can be used to do better calculations. The shadow prices of various physicochemical pollutants causing water pollution caused by some Indian industries are estimated and can be utilized for damage cost calculations (Murty & Kumar, 2004).

16.4.7 VALUE TRANSFER METHOD

Value transfer is the process of estimating the value of an ecosystem service that is of current policy interest (at a "policy site") by assigning an existing valuation estimate for a similar ecosystem elsewhere (at a "study site"). To determine the economic value of water pollution, the value transfer method can be used as it is less costly and time consuming. The key steps in the value transfer method's evaluation procedure are to first identify existing studies or values and determine whether they are transferable, and then to analyze the quality of the studies to be transferred and adjust the current value to better reflect the value for the

TABLE 16.2

Environmental Damages Cost Associated with Heavy Metal Above Permissible Limit

Pollutant	Cost of per kg Emission to Water (EURO, 1995)	Cost of per kg Emission to Water (INR, 2020)
Lead (Pb)	178	15,809
Cadmium (Cd)	622	55,242
Mercury (Hg)	1,022	90,767
Dioxins	62,824,889	5,579,688,118
Antimony (Sb)	121,366	10,778,920
Arsenic (As)	308	27,355
Barium (Ba)	31	2,753
Beryllium (Be)	44,928	3,990,206
Copper (Cu)	17,479	1,552,368
Nickel (Ni)	12	1,066
Selenium (Se)	16,125	1,432,115
Zinc (Zn)	1	89

Source: European Commission (1995).

case study. The major steps followed for the valuation process in value transfer method includes identification of existing studies or values and deciding whether the existing values are transferable and secondly to assess the quality of studies to be transferred and adjusting the existing value to better reflect the value for the case study. According to studies carried out by a foreign researcher, the per kg emission of heavy metals into water can be evaluated and used directly in the value transfer method for calculating the damage cost due to the discharge of heavy metals into water bodies above standard limits. Individual heavy metal costs are shown in Table 16.2.

16.4.8 DISCHARGE BASED METHOD

In this method, economic loss is calculated using the value transfer method and the actual cost method. The economic loss taken into account is the damage caused in water bodies due to the discharge of pollutants from a point source. In this method, much like the value transfer method, a reference study is considered as the beginning study; the potential damage of the concerned pollutant is considered for estimating the cost, and accordingly assessment can be done. The damages are estimated for particular parameter values that violate standards prescribed by authorities of pollution control. In these calculations, greater consideration is given to the additional discharge than to the prescribed standard. The concentration that occurs in the downstream flow when the discharge of a

TABLE 16.3
Damage Cost Assessment Calculation Based on Dilution Model

Year	Parameter (e.g., COD) Exceeded After Dilution in River in mg/l	Unit Damage Cost	Damage Cost in Rupees	Damage Cost Considering Inflation
n^{th}	X	Y	x*y= z	z*sum of inflation rate up to n^{th} year

Source: Bherwani et al. (2019).

pollutant takes place in a water body can be estimated using the Zeroth dimensional model (Bherwani et al., 2019):

$$C_0 = \left(C_s q_s + C_b Q_b\right)/\left(q_s + Q_b\right) \qquad (16.13)$$

where, C_b is Initial concentration in upper stream (mg/l), C_s is Concentration of the pollutant in the waste water (mg/l), Q_b is Initial flow of upstream (m³/s), q_s is the effluent discharge of outfall (m³/s). Damage cost can be calculated as shown in Table 16.3.

The different methods explained above give an idea of the valuation of the cost of water pollution damages for which an on-the-ground study must be done to get original datasets which provide better numbers compared to the numbers determined by shadow pricing or the value transfer method.

16.5 CONCLUSION

The increase in GHGs, air and water pollution, and the extensive deterioration of natural resources directly affect existing ecosystems and human welfare. Raw materials required for different industries consume the available natural resources in the ecosystem. The ultimate aim of public and private players is often solely focused on financial performance, thereby depleting the abundant natural resources to such an extent that disastrous impacts are becoming visible. Measuring the loss of natural capital and analyzing the ecological performance are much needed practices for decision-making and investment planning. Furthermore, the valuation of ecosystem services also nurtures social well-being since vulnerable communities are the ones that are worst affected by environmental catastrophes. For achieving sustainable development economic valuation of environmental impacts needs to exhibit the poignant status of the anthropogenic influence on the ecosystem. Valuation of climate change impacts may become difficult given the number of impacts generated by GHGs, so the use of social costs of carbon is suggested as this encompasses the majority of the impacts generated. The valuation of air pollution and water pollution impacts is based on the disturbance caused in the baseline level of concentration of these pollutants. The additional concentration added by industries should be quantified and a suitable valuation method, such as the ones explained in this chapter, should be utilized to quantify the impacts in monetary terms. The progressive shift toward a sustainable future is mandatory for the next generation, and the financial assessment of the environmental impacts will nurture the conservation of the natural ecosystem. Tools of valuation are necessary with regard to both positive and negative impacts created by industry in order to communicate with various stakeholders and shareholders in a language which is easily understood by all, i.e., in financial terms. Furthermore, valuation helps in understanding the priority areas of action by way of understanding not just the magnitude of emissions but also the monetary value of the impacts generated by emissions. Hence, valuation, once employed, can give insights into the type and magnitude of impacts and help in prioritizing the action steps for sustainability businesses in the interest of shareholders, stakeholders, and the environment at large.

REFERENCES

Bherwani, H., Gupta, A., Nair, M. & Sonwane, H., 2019. *Framework for Environmental Damages Cost Assessment with examples. Special Report on Monetising Damages*, Nagpur: CSIR: National Environmental Engineering Research Institute [NEERI].

Bherwani, H., Nair, M., Kapley, A. & Kumar, R., 2020b. Valuation of ecosystem services and environmental damages: An imperative tool for decision making and sustainability. *European Journal of Sustainable Development Research*, 4(4), em0133. https://doi.org/10.29333/ejosdr/8321.

Bherwani, H., Nair, M., Musugu, K., Gautam, S., Gupta, A., Kapley, A. & Kumar, R., 2020a. Valuation of air pollution externalities: Comparative assessment of economic damage and emission reduction under COVID-19 lockdown. *Air Quality, Atmosphere & Health*, 13, 683–694.

Center for Climate and Energy Solutions, n.d. *Internal Carbon Pricing*. [Online] Available at: https://www.c2es.org/content/internal-carbon-pricing/ [Accessed on 26.09.2021]

Central Pollution Control Board (CPCB), 2013. *Guidelines for the Measurements of Ambient Air Pollutants*, s.l.: s.n.

Dholakia, H., Bhadra, D. & Garg, A., 2014. Short term association between ambient air pollution and mortality and modification by temperature in five Indian cities. *Atmospheric Environment*, 99, 168–174. https://doi.org/10.1016/j.atmosenv.2014.09.071.

European Commission, 1995. *ExternE: Externalities of Energy*. s.l.:s.n.

European Commission, 2020. *Environmental Tax Statistics*. [Online] Available at: https://ec.europa.eu/eurostat/statisticsexplained/index.php/Environmental_tax_statistics#Environmental_taxes_in_the_EU

Fitzgerald, A.J. & Spencer, D., 2020. Governmentality and environmental rights: Regulatory failure and the Volkswagen emissions fraud case. *Critical Criminology*, 28, 43–63.

Friedrich, J., Ge, M. & Pickens, A., 2013. *CAIT Climate Data Explorer*. [Online] Available at: http://cait.wri.org/; https://www.wri.org/blog/2017/04/interactive-chart-explains-worlds-top-10-emitters-and-how-theyve-changed

Gajjar, C. & Adhia, V., 2018. Reducing risk, addressing climate change through internal carbon pricing: A primer for Indian business, *WRI India*, Working Paper, 1–35.

Higgins, P., 2012. Seeding Intrinsic Values: how a law of ecocide will shift our consciousness. *Cadmus, Promoting Leadership in Thought That Leads to Action*, 1(5). Available at: http://cadmusjounal.org/

Hyslop, N.P., 2009. Impaired visibility: the air pollution people see. *Atmospheric Environment*, 43(1), pp. 182–195.

Intergovernmental Panel on Climate Change (IPCC), 2014. *Climate Change 2014: Mitigation of Climate Change,* s.l.: Intergovernmental Panel on Climate Change (IPCC).

Ko, F.W. & Hui, D.S., 2012. Air pollution and chronic obstructive pulmonary disease. *Respirology*, 17(3), pp. 395–401.

Maji, K.J., Dikshit, A.K. & Deshpande, A., 2017. Assessment of city level human health impact and corresponding monitoring cost burden due to air pollution in India taking Agra as a model city. *Aerosol Air Quality Research*, 17(3), pp. 831–842.

Millennium Ecosystem Assessment, 2005. *Ecosystems and Human Well-being*, Washington, DC: Synthesis.

Muller, N.Z. & Mendelsohn, R., 2007. Measuring the damages of air pollution in the United States. *Journal of Environmental Economics and Management*, 54(1), pp. 1–14.

Murty, M. & Kumar, S., 2004. *Environmental and Economic Accounting for Industry.* New Delhi: Oxford University Press.

Nair, M., Bherwani, H., Kumar, S., Gulia, S., Goyal, S., Kumar, R., 2020. Assessment of contribution of agricultural residue burning on air quality of Delhi using remote sensing and modelling tools. *Atmospheric Environment*, 230. https://doi.org/10.1016/j.atmosenv.2020.117504.

Oh, C. & Sverndsen, G.T., 2015. Command-and-control or taxation? The cases of water regulation in California and Denmark. *Environmental Management and Sustainable Development*, 4(2), pp. 141–151. doi:10.5296/emsd.v4i2.8358.

Organization for Economic Co-operation and Development [OECD], 2011. *Organization for Economic Co-Operation and Development [OECD].* [Online] Available at: http://www.oecd.org/env/outreach/2011_AB_Economic%20significance%20of%20NR%20in%20EECCA_ENG.pdf

Pascual, U. et al., 2010. The Economics of Valuing Ecosystem Services and Biodiversity. *The Ecological and Economic Foundations*, Chapter 5, pp. 185–369.

The United Nations, 2015. *World Water Assessment Programme (WWAP)*, Paris: s.n.

United Nations [UN], 2015. *Paris Agreement.* [Online] Available at: https://unfccc.int/files/essential_background/convention/application/pdf/english_paris_agreement.pdf

United Nations Development Programme, 2011. *Managing Natural Resources for Human Development in Low-Income Countries.* s.l.:s.n.

United States Environmental Pollution Agency, 2016. *The Social Cost of Carbon.* [Online] Available at: https://19january2017snapshot.epa.gov/climatechange/social-cost-carbon_.html

United States Environmental Pollution Agency, *n.d. Global Greenhouse Emissions Data.* [Online] Available at: https://www.epa.gov/ghgemissions/global-greenhouse-gas-emissions-data [Accessed on 26.09.2021]

United States Environmental Pollution Agency, n.d. *Greenhouse gas emissions.* [Online] Available at: https://www.epa.gov/ghgemissions/sources-greenhouse-gas-emissions#:~:text=Carbon%20dioxide%20(CO2)%20makes,natural%20gas%2C%20to%20produce%20electricity. [Accessed on 26.09.2021]

Wang, H. & Mullahy, J., 2006. Willingness to pay for reducing fatal risk by improving air quality: A contingent valuation study in Chongqing, China. *Science of the Total Environment*, 367(1), pp. 50–57.

World Economic Forum, 2020. *Environmental Performance Index.* [Online] Available at: https://epi.yale.edu/epi-results/2020/component/epi

Yao, H. et al., 2016. Analysis of surface water pollution accidents in China: Characteristics and lessons for risk management. *Environmental Management*, Volume 57, pp. 868–878.

Yongguan, C., Seip, H.M. & Vennemo, H., 2001. The environmental cost of water pollution in Chongqing, China. *Environment and Development Economics*, 6(3), pp. 313–333.

Zhang, C. et al., 2017. Association between air pollution and cardiovascular mortality in Hefei, China: A Time-Series Analysis. Environmental Pollution, 229, pp. 790–797. doi:10.1016/j.envpol.2017.06.022.

17 Exploring Human-Elephant Conflicts and Mitigation Measures in and Around Rohingya Refugee Camps in Cox's Bazar, Bangladesh

Nafisa Islam
Environmental Sciences Program, Asian University
for Women, Chattogram, Bangladesh

Sayed Mohammad Nazim Uddin
Environmental Sciences Program, Asian
University for Women, and

Center for Climate Change and Environmental Health,
Asian University for Women, Chattogram, Bangladesh

Nazifa Rafa
Environmental Sciences Program, Asian University
for Women, Chattogram, Bangladesh

Mukesh Gupta
Environmental Sciences Program, Asian
University for Women, and

Center for Climate Change and Environmental Health,
Asian University for Women, Chattogram, Bangladesh

CONTENTS

DOI: 10.1201/9781003140382-17

17.1 INTRODUCTION

17.1.1 HUMAN-WILDLIFE CONFLICT

Competition for survival has caused human-wildlife conflicts (HWCs) to persist for millennia. HWC is defined as "any interaction between humans and wildlife that results in negative impacts on human social, economic or cultural life, on the conservation of wildlife populations, or on the environment" (WWF, 2005 as cited in Anthony & Wasambo, 2009, p.5). The impacts are usually massive, where people lose their crops, livestock, property, and sometimes their lives (WWF, 2008). The animals, many of which are already threatened or endangered, are often killed in retaliation, or to prevent future conflicts. Recently, however, the frequency of HWCs has increased due to the expanding human population, harmful anthropogenic activities such as deforestation, and inadequate mediating infrastructures (Anthony & Wasambo, 2009).

Mass deforestation to meet the growing needs of mankind increases the incidence of hostile encounters between the wild animals living there and the people who live in and around those areas. Humans continuously intervene to meet their needs for development by building and constructing on the natural landscape, which fragments and isolates wildlife habitats and decreases the sizes of suitable habitats, fatally threatening many animal species (Li et al., 2018). Crop damage, livestock predation, and human killings by wildlife have negative effects on local attitudes towards the animals (Mir et al., 2015). This has given rise to retaliatory killings and as a result, it has become difficult for humans and wildlife to live in a manner such that they can coexist.

HWC has a prolonged effect on the conservation of threatened and endangered wildlife species and humans. In 1909, the US Forest Service initiated a program of predator elimination in Kaibab Plateau, Arizona, because state and federal agencies deduced that game species were declining due to competition with predator species (Sparling, 2014). They found predator animals to be a form of competition for humans in obtaining natural resources and thus supported human-induced extirpation. Consequently, from 1924, there was massive starvation in which at least 60% of the deer population died in the plateau because the deer population exceeded its ecological carrying capacity and starved in the absence of a predator (Sparling, 2014). Predator and prey species provide balance within the system and therefore the presence of both is necessary. Goswami et al. (2013) explain that the decline in the population of large carnivores in recent times is due to human-animal conflict and the absence of coexistence between humans and animals has been a recurrent theme in conservation biology.

Animal attacks or conflicts can be categorized as direct and indirect (Torres et al., 2018). Direct conflicts occur when animals (for example, tigers, lions, crocodiles, sharks, snakes, etc.) directly attack people. Indirect conflict occurs when animals attack domestic livestock, resulting in damage to crops, houses, fishing gear, and vessels, cause aerial and road accidents, and negatively impact humans in many other ways (Torres et al., 2018). Changes caused by humans in the natural environment can lead to the persecution and exclusion of wild animals from areas where people live, or have settled, and are growing plants or raising animals. As a result, these wild animals are restricted to areas that are inappropriate for agricultural production and unsuitable for the maintenance of wildlife, which might lead to the imminent extinction of wild animals (Gordon, 2009 as cited in Torres et al., 2018).

17.1.2 HWC SPECIES: ASIAN ELEPHANT

Human-elephant conflict (HEC) is one of the most predominant forms of HWCs in areas that host large numbers of elephant populations. Therefore, HEC is particularly pronounced in the Asian and the African continents, which shelter their own unique species of elephants. HEC has remained a consistent problem throughout the years and is gradually increasing due to deforestation. In regions where human populations and economic development are rapidly increasing, humans have been imposing increasing demands for and pressures on wildlife resources and their environments (Li et al., 2018). HEC has often been attributed to several human impacts, such as habitat fragmentation, activities that deteriorate the quality of the living environment of elephants, reduction in forest area, as well as improper management and monitoring of physical barriers (Gubbi, 2012). Elephants tend to tread along the traditional routes and corridors for regular movements and thus try to rid of any obstacles that are present in their paths (IUCN, 2018). Globally, the number of attacks per year by elephants is 250 (Torres et al., 2018).

The situation becomes more precarious when the target animal is endangered, such as Asian Elephants, as classified by IUCN, with less than 50,000 elephants left living in the wild and a third of them in captivity (WWF, 2015). Asian Elephants (*Elephas maximus indicus*), also called Asiatic elephants, are the largest land mammals on the Asian continent. They inhabit dry to wet forest and grassland habitats in 13 countries of South and Southeast Asia (WWF, 2015). Asian elephants eat large amounts of tree roots, barks, stems, and cultivated crops like bananas, rice, and sugarcane, making them serious crop pests. Their actions cause a great economic loss for the owners of nearby croplands and lead the local communities to start viewing these animals in a negative light.

17.1.3 Asian Elephants in Bangladesh: Exploring Vulnerabilities

Bangladesh is a country that is highly dependent on natural resources and its economy is driven by agriculture. The promotion of management of land resources of Bangladesh is done through a number of policies related to land use, agriculture, forestry, water, coastal zone, environment, and fisheries, among others (IUCN & BFD, 2016a). Other policies that are directly and indirectly related to land management are the National Action Programme (NAP) for Combating Deforestation and the Bio-diversity Act. A large area of land in Bangladesh, measuring about 206 million hectares, is forested (IUCN & BFD, 2016a). However, forest resources in Bangladesh continue to be depleted as a result of overexploitation, conversion of forest land for agriculture, fire, and grazing, where Bangladesh lost 2600 hectares of primary forest land between 1990 and 2015 annually (IUCN & BFD, 2016a).

Bangladesh is also a biodiversity hotspot as a consequence of its location at the junction of the Indian and Malayan sub-regions of the Indo-Malayan Realm (Dey & Rabbi, 2015). A large number of native flora species, including 3,000–4,000 species of woody flora, have been recorded in Bangladesh (Khan, 2013 as cited in Dey & Rabbi, 2015). The country also supports 130 species of mammals, 710 species of birds, 164 species of reptiles, and 56 species of amphibians (Khan, 2013 as cited in Dey & Rabbi, 2015). The biodiversity of Bangladesh is threatened by environmental factors such as habitat degradation and fragmentation, change in land-use pattern, change in hydrological regime, pollution, overexploitation of resources, unplanned and uncontrolled tourism, expansion of invasive alien species, and climate change, as well as political factors such as legal and institutional systems that do not curtail unsustainable exploitation, the unsupportive nature of economic systems and policies towards biodiversity conservation, unequal distribution in the sharing of the benefits of biodiversity, lack of knowledge and awareness, and erosion in genetic diversity (IUCN & BFD, 2016a). At the moment, Bangladesh has 38 National Parks and Wildlife Sanctuaries, and the country is committed to conserving biodiversity as agreed to in the Wildlife Conservation and Security Act, 2012 which forbids the cultivation of land, the establishment of industrial operations, disturb or threaten any wildlife, among other initiatives in the sanctuaries. However, the emergence of the Rohingya crisis is making it difficult to follow the Wildlife Conservation and Security Act, 2012, as massive forest coverage within wildlife sanctuaries in Cox's Bazar was cleared by the government for construction of the Rohingya refugee camps. The biodiversity of that area has been highly affected causing habitat loss for many wildlife species and adding to the problem of the existing HECs that are prevalent in that area.

17.1.4   The Rohingya Refugee Crisis: A Brief Overview

The Rohingya are a Muslim minority community in Myanmar, who are mainly settled in the Rakhine State. The Rohingyas have faced decades of discrimination and suppression under successive Burmese governments (HRW, 2017). According to the 1982 Citizenship Law in Myanmar, the Rohingya have been denied citizenship in the country, and as a result represent one of the largest stateless populations in the world (HRW, 2017). Due to the constant severe discrimination they were facing, the Rohingya began a mass exodus to Bangladesh. Subsequently, the government forced the Rohingya to move to the internationally displaced (IDP) concentration camps. These camps were in the border areas of Myanmar and Bangladesh, shown in Figure 17.1.

By late 2017, more than 671,000 Rohingya Muslims fled the Rakhine state and came to Bangladesh to escape the military's extensive campaign of ethnic cleansing (HRW, 2017). From 2015 to 2017, there was a rapid rise in the Rohingya refugee population in Bangladesh, an increase from approximately 225,000 to a staggering 850,000 (Hassan et al., 2018).

17.1.5   Rohingya Refugees, Deforestation, and Wildlife

The mass deforestation that occurred to accommodate the Rohingya refugees resulted in severe environmental damage in Bangladesh. The financial cost of the destruction of forests in Ukhia and Teknaf Upazila is over BDT 2,420 crore as

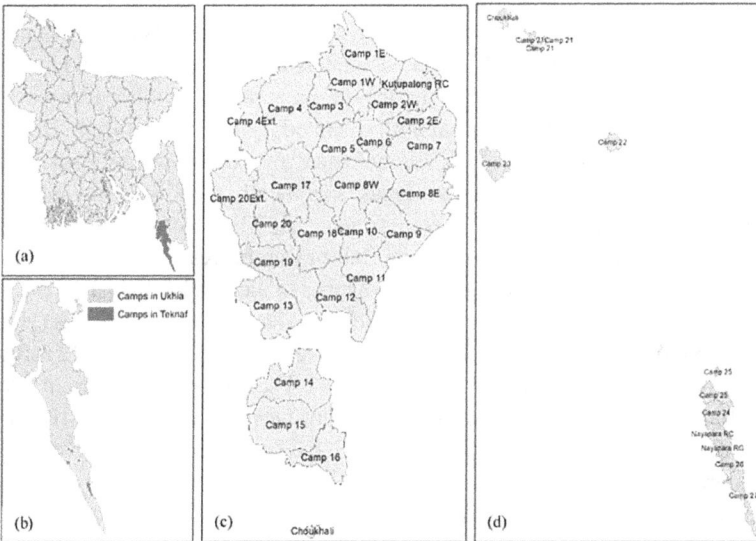

FIGURE 17.1   (a) Map of Bangladesh. The marked red zone is Cox's Bazar, the district where Rohingya refugee camps are present. (b) Map of Cox's Bazar District (marked Rohingya Camps of Ukhia and Teknaf). (c) Rohingya Camps of Ukhia. (d) Rohingya Camps of Teknaf (Map credit: Samiha Nuzhat; samiha.nuzhat@post.auw.edu.bd).

over 8,000 acres of land has been depleted (Hasan, 2019). A study by Hassan et al. (2018) showed that there had been an increase of 175 to 1530 hectares, a net growth of 774%, in three of the Rohingya camps studied between 2016 and 2017 (Hassan et al., 2018). A study by Rahman et al. (2019) focused on changes of vegetation cover in four refugee-occupied Unions of Teknaf and Ukhia Upazila between the pre-and post-2017 influx. Their findings showed that nearly 21,000 acres of dense vegetation had been reduced within the one-year period between 2017 and 2018 in Palong Khali, Whykong, and Nhilla Union. In addition, a reduction in moderate vegetation of more than 1,700 acres occurred in Whykong, Nhilla, and Baharchhara Union within the same period (Rahman et al., 2019). Similarly, Ahmed et al. (2018) found that the dense forest cover had been halved in the span of two years, between 2016 and 2018, while the refugee settlement had increased nine-fold, and concluded that the forced Rohingya migration and deforestation were positively correlated. Khan et al. (2015) explored the main causes of forest destruction in the Teknaf Wildlife Sanctuary and found that the following were responsible: medicinal plant and vegetable collection, hunting, collection of green and dry leaves, grazing, betel leaf cultivation within the forests, sand and stone collection, bamboo and cane extraction, brickfield within the forest, the encroachment of forest land, illicit felling, making fire for better sungrass regeneration, and overexploitation of fuelwood. The Rohingya were involved in eight out of the twelve aforementioned activities and were also shown to be involved in illegal felling, hunting, and fuelwood collection (Mollah 2004, as cited in Khan et al., 2015).

The accelerated deforestation rate in Cox's Bazar, Bangladesh, due to the influx of Rohingya refugees, caused many animals, such as elephants, to lose their natural habitats in the area and expose them to human presence. This exposure to humans has resulted in a higher incidence of hostile encounters between elephants and humans.

All wild elephants are protected under the Bangladesh Wildlife Conservation (Amendment) Act of 1974. They cannot be hunted, killed, or captured. There are 268 wild elephants in Bangladesh as confirmed by IUCN, where resident wild elephants reside only in the southeast areas, specifically the forests in Chittagong, Chittagong Hill Tracts, and Cox's Bazar (Islam et al., 2011). However, the rapid degradation of forested land to accommodate Rohingya refugees has triggered ecological problems in Cox's Bazar, causing disturbances in the wildlife habitat of that area as many of these camps were set up in or near corridors of wild elephants (Hassan et al., 2018). As of 22 February 2018, HECs have caused the death of 12 refugees and one host community member (IUCN, 2019). The Kutupalong camp in Ukhia, Cox's Bazar has completely blocked an elephant movement corridor (Ministry of Environment and Forests, Bangladesh, 2018). It has created a barrier, preventing the free migration of wild elephants between Cox's Bazar and Myanmar.

Such impacts are indications that the relevant stakeholders – governments, local communities, wildlife managers, and scientists – need to recognize this problem and find solutions that will be beneficial for humans, wildlife, and the overall environment. There is a need to adopt measures to form a peaceful coexistence between humans and animals, especially in areas where they share the same landscape and ecosystem services.

17.1.6   HEC MANAGEMENT IN ROHINGYA CAMPS

Currently, IUCN and United Nations High Commissioner for Refugees (UNHCR) are working on a joint project to address HEC in the Rohingya Camps. IUCN has been establishing baselines, conducting pilot interventions, mapping elephant distribution, HEC areas, elephant corridors and paths, and enhancing understanding of HEC issues at the community and decision-maker levels (Islam et al., 2011). In 2013, the Bangladesh Forest Department took three sub-projects under the World Bank-supported "Strengthening Regional Cooperation for Wildlife Protection (SRCWP) Project" to count elephants, identify elephant routes and corridors, and minimize HEC in Sherpur, Chittagong, and Bandarban areas (Wahed et al., 2016). The Bangladesh Forest Department, in collaboration with IUCN, formed ERTs to combat HEC in the central-north and south-eastern areas of Bangladesh. In addition to ERTs, IUCN formed a range of conflict-management techniques. Currently, there are two mitigation methods in place in the camps: Watchtower and Elephant Response Teams (ERTs). Watchtowers are constructed in particular areas of the camp so that the security can see and alert people when the elephants are coming. ERTs are trained to divert the attention of elephants by methods like shouting and using flashlights. IUCN plans on building the capacities of the ERTs and maintain the quality of the constructed watchtowers (IUCN, 2019).

17.1.7   OBJECTIVES

The HEC in and around Rohingya refugee camps serves as, perhaps, the only example of HWC in the refugee context due to the unique locational features of the areas where the refugees have been displaced to. This paper aims to explore the HEC scenario in the Rohingya refugee camps of Cox's Bazar, Bangladesh. It aims to determine how the situation of HEC has changed in the target areas since 2015, the year when the large Rohingya influx began, by investigating the different kinds of impacts elephants have had in those areas. This study also inquires into the perspectives of host community members and Rohingya refugees regarding the protection of elephants. Finally, it attempts to evaluate the effectiveness of the current mitigation methods that are in place in those areas.

17.2   METHODOLOGY

17.2.1   STUDY AREA: KUTUPALONG

The area that was considered for this study is the Kutupalong Camp in Ukhia, Cox's Bazar. The Kutupalong camp is in the vicinity of Cox's Bazar-Teknaf Highway. Currently, the camp has a population of 859,808 with 187,844 households (UNHCR, 2020). The three camps that were chosen for this study were: Camp Four Extension (4E), Camp 13, and Camp 14. The host community areas that were chosen are Thaing Khali and Mosarkola, Palonkhali, where camp 13 is located in the area of Thaing Khali and Camp 14 is located in the area of Hakimpara, as shown in Figure 17.2.

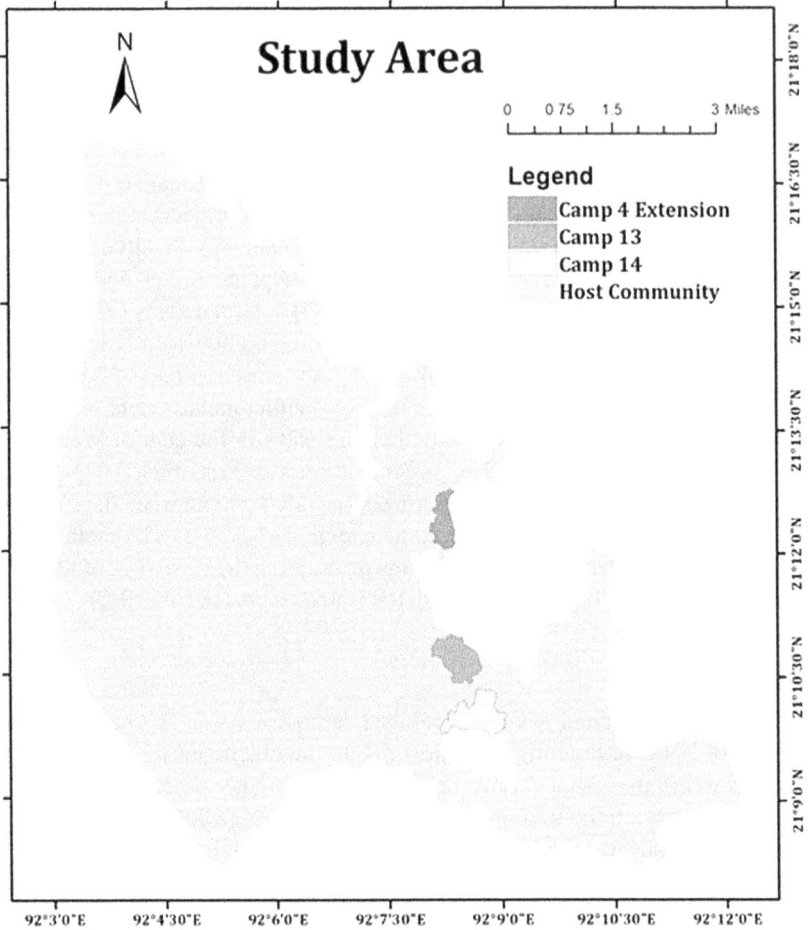

FIGURE 17.2   The study area (Map credit: Samiha Nuzhat; samiha.nuzhat@post.auw.edu.bd).

The campsites were chosen based on findings of the report released by UNHCR and IUCN in 2018, which revealed the vulnerabilities of the study areas to HECs. The chosen camps lie right at the boundaries where the Teknaf Wildlife Sanctuary and the other camps meet, so elephant sightings are higher in these areas, as evidenced by elephant footprints and dung piles around those areas (UNHCR and IUCN, 2018).

17.2.2 Study Design

A cross-sectional study design was used with purposive sampling. The study population was composed of Rohingya refugees in Cox's Bazar living in selected camps close to the elephant corridors and the host community living inside the camps and also in the surrounding areas of the camps. The sample size for this study was 100, which comprised 49 members of the host community and 51 Rohingya refugees. Among the refugees, 20 were part of the ERT; 25% of the respondents were residents of Camp 13 in Thaing Khali, where 15% were host community members, and 10% were refugees; 34% of the participants, all of whom were host community members, lived in Mosarkola, Palonkhali. Of the remaining participants, who were refugees, 21% were situated in Camp 4E and 20% in Camp 14. The majority (79%) of the respondents were male and the remaining were female. From the participants, most of them (61%) belonged to the age group 20–40, followed by 40–60 (17%); 14% were under 20 and the remaining were above 60. Half of the respondents did not have any education, 1% received elementary level education, 26% had primary education, 10% and 8% received junior secondary and secondary level education, and 4% had higher secondary level education. The remaining had undergraduate degrees. Due to time and resource constraints, this study faced limitations in terms of the number and types of participants and camps that could be covered. Moreover, qualitative research tools, such as focus group discussions, and key informant interviews could not be employed even though such methods would have revealed more dimensions to the HEC scenario in the camps.

17.2.3   DATA COLLECTION AND ANALYSIS

The primary data collection was done using semi-structured questionnaire surveys, which included both qualitative and quantitative questions. The first few questions focused on the time (season) and frequency of elephant appearances around the study sites, reports of casualties upon encounter, the economic damage caused by the elephants, and how the frequencies of encounters have changed over the years. The questionnaire also inquired into the perspectives of participants regarding their stance on elephant conservation, and their opinions on the effectiveness of the HEC mitigation strategies. MS Excel 2013 was used for data analysis.

17.2.4   ETHICAL APPROVAL AND CAMP PERMISSION

This research received approval from the Ethical Research Committee (ERC) of the Asian University for Women (AUW), located in Chattogram, Bangladesh, on December 12th, 2019. Permission to conduct research work on the Kutupalong camp and the nearby area was also granted by the Office of the Refugee Relief and Repatriation Commissioner (RRRC), Cox's Bazar, on December 9th, 2019 (No. 2291) for a period of 15 days under the conditions that the concerned Camp-in-Charge was kept informed during research activities and that the research findings are shared with the office for further enhancement of the interventions being applied in the camps.

During data collection, the objectives, procedures, risks, and benefits of participating in the study were explained to the participants prior to the interview.

They were assured that participation was voluntary and that they were free to withdraw at any time without any negative consequences. Before data collection, consent forms were passed out to the participants who either provided a signature or gave their thumbprint to confirm their consent to participating in this research. The confidentiality of all the collected data and the privacy of the participants were ensured and maintained with the highest priority.

17.3 RESULTS AND DISCUSSION

17.3.1 Frequency and Trends of Elephant Encounters

According to the results, the majority (91%) of the participants have seen elephants and they have mostly seen them during the fall and winter (77%) seasons with a few sightings during the monsoon season (22%). Most of the participants (86%) saw elephants fewer than 100 times a year. During the early wet season, elephants are scattered throughout forested areas (IUCN & BFD, 2016b). At this time food and water are more available. One respondent from Camp 4E, who is also an IUCN volunteer, stated, "More elephants come out when it is very cold. They come to eat crops." Another respondent, who lives in Mosarkola, Palonkhali explained, "Elephants come consistently for a time when the crop ripens."

17.3.2 Injuries, Deaths, and Damage From HECs

Overall, 94% and 96% of the participants said there have been no injuries and deaths, respectively, due to elephants. Among the minority of the participants who reported incidents, of those who had been injured, 86% were locals and 14% were refugees, and among those who died, 63% were locals and 38% were refugees. There were no reports of humanitarian workers or government officials being harmed during any of the HECs.

When asked whether the participants suffered any economic harm, the majority (58%) answered "No," while the remaining respondents (42%) answered "Yes." Among the 42% who had sustained economic harm, all were members of the host community. The type of economic harm they mostly suffered from was crop destruction (69%), followed by house destruction (29%), and other types such as medical expenses (2%). The highest amount that a participant had lost due to elephant encounters was BDT 4 lakhs, but most of the participants suffered a loss of BDT 10,000–50,000 (52%); 2%, 21%, and 14% lost BDT 5,000–10,000, BDT 50,000–100,000, and BDT 100,000–200,000 respectively as a result of the conflicts.

17.3.3 Development in the HEC Scenario Since 2015

Sixty-eight percent and 65% of the participants confirmed that HEC has decreased and that the situation has improved, respectively, compared with 2015 when the major Rohingya influx started. Among the participants who held the opposite opinion (21% and 24% respectively), 95% and 88% were members of the host community of Mosarkola, Palonkhali.

None of the participants of the Thaing Khali host community and IUCN volunteers stated that HEC decreased or that the situation had worsened. The IUCN volunteers work directly with HEC so they understand how the situation has changed over the years and are finding improvements. The members of the Thaing Khali host community receive the same protection as the refugees since they live within Thaing Khali (Camp 13) so they are not facing too many HECs. However, the host community members from Mosarkola, Palonkhali are not receiving any protection from elephants, unlike residents in the camps. A host community member from Mosarkola, Palonkhali explained, "The situation of HEC has improved in the camps as they are getting the necessary protection. The mitigation methods are implemented there but not here. As a result, the elephants are getting diverted from the camp area and are coming to our area more often than before."

17.3.4 PERSPECTIVES ON ELEPHANT PROTECTION AND CONSERVATION

More than half (63% and 68%) of the participants said that elephants are not important for the environment and should not be conserved, respectively. Among the participants who agreed that elephants are important and should be conserved (36% and 32%), 50% and 56% were refugees who are also IUCN volunteers and are part of the ERTs. Moreover, 39% and 34% of them were host community members.

Approximately 550 refugees have voluntarily offered to be of service and were trained by IUCN to be part of the ERTs (Irfanullah, 2018). They have been educated about elephants and their importance in our environment and that is why these refugees, who are also part of ERTs, have a positive view of elephants, unlike the other refugees. When asked why they think elephants should be conserved, the majority of the IUCN volunteers said that elephant dung is needed to grow trees, while others said that "more trees means more oxygen" or that they liked seeing elephants. Most of the host community members responded by saying that "elephants are natural assets of our country" and some recognized the elephants' status of being endangered, and therefore highlighted this reason. In response to the question, "Why should elephants not be conserved?" the majority (37%) answered by stating that elephants harm humans and kill people and children, while others highlighted the issue of crop and house destruction.

Participants were also asked to rate the importance of elephants on a scale of 1 to 5 (1 being least important and 5 being most important). The majority of the participants gave a rating of the importance of elephants of 5 (39%), whereas 19% gave a rating of 4, 17% a rating of 3, 14% a rating of 2, and 11% a rating of 1.

17.3.5 EFFECTIVENESS OF MITIGATION EFFORTS

Sixty percent of the participants thought that the current mitigation methods were working and the remaining 40% disagreed. Among the 40% who thought that the mitigation methods were not working, 80% were from Mosarkola, Palonkhali, 13% were refugees, 7% were host community members of Thaing Khali, and none of them were IUCN volunteers. As mentioned above, the host community members of Mosarkola, Palonkhali were not receiving protection, unlike the refugees, so they replied "No."

When asked why the mitigation methods were not working, most of the refugees replied that security personnel do not perform their roles properly because they sleep while on duty. A few stated that the ERT members are not placed in appropriate locations. The host community members responded that they are not getting the same protection as the refugees, so they could not experience the benefits of mitigation methods.

The current mitigation methods are watchtowers and ERTs. There are security members on the watchtowers, and they are tasked with looking out for the elephants. The ERTs are responsible for distracting and scaring the elephants away by shouting, creating noises, and shining lights. When asked to rate watchtowers and ERTs from 1 to 5, 1 being the least effective and 5 being most effective in mitigating HECs, 77% of the participants rated both of the interventions as 5 (most effective).

The participants provided recommendations to enhance the effectiveness of mitigation efforts to improve the current situation of HEC in the Rohingya camps and the host community. A few refugee participants suggested that ERTs be positioned on the paths where elephants emerge from and that barricades be constructed to hinder entrance into the settlements. Some ERT members felt that the watchtowers need to be stabilized and built in a way that is easier to use as currently, the watchtowers are unstable due to the way they are built (often using simple wooden/bamboo poles tied together with ropes) and because of the sloped ground they are constructed on. It is important to remember that the hilly areas were cleared of trees in order to establish the camps.

In addition, the host community members asked to be provided with more lights and recommended that watchtowers be built in host community areas as well. IUCN already has projects being implemented in the Rohingya camps to improve the situation, but attention needs to be provided to the host community too because the elephants are being diverted from the Rohingya camps and towards the host community, as mentioned by the host community respondents. There should be more campaigns to raise awareness about elephants and to build social capital among the host communities and also the refugees. An "us versus them" kind of mentality is often perceived within both the host communities and the refugees, which could hinder not only a harmonious co-existence between the two communities but also between people and the elephants since animosity and conflicts may hinder any fruitful discussion regarding effective HEC interventions. Since this is a joint issue, the two communities should form a truce among themselves and also assist in avoiding HECs and also protecting elephants.

17.4   CONCLUSION AND RECOMMENDATIONS

HEC will remain an ongoing issue in areas that are the habitat of elephants, such as Cox's Bazar. This study showed that the majority of the respondents had a negative view of elephants, but those who had positive views were either part of the ERT and were educated about elephants, or were part of the host community and considered elephants to be natural assets of Bangladesh. The situation of HEC has decreased throughout the years in the Rohingya camps, but many host community members complained that it has increased as a result of the Rohingya crisis. Lastly, the mitigation methods are working quite well in the camps, as confirmed by the refugees.

Considering the current situation of wildlife conservation, and the problems that emerge as a result of it, the government should take more action to mitigate the issue of human-wildlife conflict in general. Greater efforts should be directed at conflicts that are caused by endangered species like Asian elephants. Bangladesh faces imminent long-term, irreparable damage as a result of the loss of forest coverage, in addition to the existing issue of climate change. Therefore, more measures should be taken to focus on wildlife conservation to avoid further impending dangers. The current interventions that are in place were effective in reducing HECs in Rohingya camps, but the initiatives ended up diverting the elephants towards host communities. IUCN-Bangladesh and UNHCR should provide HEC mitigation interventions in host communities as well by recruiting ERTs from the host communities and construct watchtowers in host communities.

Moreover, despite their efficacy, the existing methods have limitations, which can be addressed in the following ways: (i) the recruitment process for ERT workers needs to be improved, as evidenced by the participants' complaints regarding the dedication of these workers towards their work. (ii) Working on shifts and hiring a larger number of volunteers could overcome the issue of being inactive while on duty. (iii) Watchtowers also need to be robust and built using stable materials to make them safer and user-friendly. (iv) Elephants traversing routes must be carefully studied so that ERTs can be strategically placed at elephant entry points. (v) The shortage of flashlights should be addressed. (vi) The relevant actors working in the field can further improve their elephant deterrence strategies by employing acoustic techniques, such as the use of metal banging, tripwire alarms, firecrackers, or recordings of elephant distress calls (Shaffer et al., 2019).

Other diverse forms of HEC prevention techniques that can potentially be operationalized within the study areas should be implemented, which could include, but are not limited to: (i) planting non-palatable, profitable crops for elephants (Wahed et al., 2016), like chamomile, coriander, mint, ginger, onion, garlic, lemongrass, and citrus trees, involving the spatial strategy of interspersing such crops among commonly raided crops (Shaffer et al., 2019). (ii) Chili-based substances have also proven to be good at discouraging elephants from invading croplands and can be applied in various forms, such as chili-grease covering fences, chili dung, or chili ropes (Wahed et al., 2016; Shaffer et al., 2019). Pepper can also be sprayed on crops (Torres et al., 2018). (iii) Because Asian elephants are an endangered species, and Bangladesh is home to fewer than 300, it is important to avoid harmful mitigation methods such as electric fences and barbed wire. HEC mitigation strategies should not only prevent human casualties but elephant casualties as well. Some interventions can include eco-friendly bio-fencing (Wahed et al., 2016) by growing cane/rattan with sharp, spiny stems, improving the elephants' habitat by constructing salt licks and plantations for the elephants (Wahed et al., 2016), and disseminated education related to the importance of elephants beyond ERTs and others working in the field. Initiatives must also be undertaken to improve the social capital between refugees and the host community to make any intervention successful. Both communities must work collectively to tackle HECs. However, this also means that both communities should be equally involved in intervention methods, from the decision-making stage to the implementation stage. That is, participation at all levels is required.

ACKNOWLEDGMENTS

Special thanks to Mr. Mohammed Mizanur Rahman, Additional RRRC, Joint Secretary, Ministry of Public Administration, Government of Bangladesh for his administrative support, and for assisting with the permission process in Kutupalong Camp. We also thank Mr. Shahriar Caesar Rahman, conservation biologist, co-founder, and CEO of Creative Conservation Alliance (CCA) for research development support. The field and translation support provided by Kazi Mohammad Shahadat, Md Tanvir Ahmed, and Mofiz Uddin from the CARE Foundation are highly appreciated. Gratitude is also extended to Samiha Nuzhat for her cartographic support.

BIBLIOGRAPHY

Ahmed, N., Islam, M.N., Hasan, M.F., Motahar, T., & Sujauddin, M. (2018). Understanding the political ecology of forced migration and deforestation through a multi-algorithm classification approach: The case of Rohingya displacement in the southeastern border region of Bangladesh. *Geology, Ecology, and Landscapes*, *3*(4), 282–294.

Anthony, B.P., & Wasambo, J. (2009). Human-wildlife conflict study report: Vwaza Marsh Wildlife Reserve, Malawi. Central European University: Prepared for Malawi Department of National Parks and Wildlife, Budapest. p. 10.

Dey, T.K., & Rabbi, G. (2015). Strengthening regional co-operation for wildlife protection project Bangladesh forest department. In *Guide Book on Wildlife Law Enforcement in Bangladesh*. Dhaka: Bangladesh Forest Department Ministry of Environment and Forests.

Goswami, V.R., Vasudev, D., Karnad, D., Krishna, Y.C., Krishnadas, M., Pariwakam, M., Nair, T., Andheria, A., Sridhara, S., & Siddiqui, I. (2013). Conflict of human–wildlife coexistence. *Proceedings of the National Academy of Sciences*, *110*(2), E108–E108.

Gubbi, S. (2012). Patterns and correlates of human–elephant conflict around a South Indian Reserve. *Biological Conservation*, *148*(1), 88–95.

Hasan, R. (2019, October 18). Rohingya Settlements: 8,000 acres of forests razed. *The Daily Star*. Retrieved from: https://www.thedailystar.net/frontpage/news/rohingya-settlements-8000-acres-forests-razed-1815400.

Hassan, M.M., Smith, A.C., Walker, K., Rahman, M.K., & Southworth, J. (2018). Rohingya refugee crisis and forest cover change in Teknaf, Bangladesh. *Remote Sensing*, *10*(5), 689.

HRW. (2017, October 21). Rohingya Crisis. Retrieved from: https://www.hrw.org/blog-feed/rohingya-crisis.

Irfanullah, H.M. (2018). Elephant Conservation in Bangladesh – Bringing Conservation Effort and Humanitarian Response Together. *Gajah, 49*, 33–35

Islam, M.A., Mohsanin, S., Chowdhury, G.W., Chowdhury, S.U., Aziz, M.A., Uddin, M., Chakma, S., Akter, R., Jahan, I., & Azam, I. (2011). Current status of Asian elephants in Bangladesh. *Gajah, 35*, 21–24.

IUCN & BFD. (2016a). Bangladesh National Conservation Strategy, Part II: Sectoral Profile. Bangladesh Forest Department (BFD), Ministry of Environment and Forests, Government of the People's Republic of Bangladesh, & IUCN, Bangladesh Country Office, Dhaka, Bangladesh. Retrieved from: http://bforest.portal.gov.bd/sites/default/files/files/bforest.portal.gov.bd/notices/c3379d22_ee62_4dec_9e29_75171074d885/Executive%20Summary(NCS).pdf.

IUCN & BFD. (2016b). Status of Asian Elephants in Bangladesh. Bangladesh Forest Department (BFD), Ministry of Environment and Forests, Government of the People's Republic of Bangladesh, & IUCN, Bangladesh Country Office, Dhaka, Bangladesh. Retrieved from: https://portals.iucn.org/library/sites/library/files/documents/2016-085.pdf.

IUCN. (2018). Human-Elephant Conflict Mitigation around the Refugee Camp of Cox's Bazar (Phase 1: 2018). Retrieved from: https://www.iucn.org/asia/countries/bangladesh/human-elephant-conflict-mitigation-around-refugee-camp-coxs-bazar-phase-1-2018.

IUCN. (2019). Human-Elephant Conflict Mitigation around the Refugee Camp of Cox's Bazar (Phase 2: 2019). Retrieved from: https://www.iucn.org/asia/countries/bangladesh/human-elephant-conflict-mitigation-around-refugee-camp-coxs-bazar-phase-2-2019.

Khan, M.A.S.A., Uddin, M.A., & Haque, C. E. (2015). Rural livelihoods of Rohingya refugees in Bangladesh and their impacts on forests: The case of Teknaf Wildlife Sanctuary. In *Counter-Narratives on Rohingya Refugees Issue: Re-look at Migration, Security and Integration.*

Li, W., Liu, P., Guo, X., Wang, L., Wang, Q., Yu, Y, Dai, Y., Li, L., & Zhang, L. (2018). Human-elephant conflict in Xishuangbanna Prefecture, China: Distribution, diffusion, and mitigation. *Global Ecology and Conservation, 16*, e00462.

Ministry of Environment and Forests, Bangladesh. (2018). Bangladesh Elephant Conservation Action Plan. Retrieved from: https://www.asesg.org/PDFfiles/2017/Bangladesh%20Elephant%20Conservation%20Action%20Plan.pdf.

Mir, Z.R., Noor, A., Habib, B., & Veeraswami, G.G. (2015). Attitudes of local people toward wildlife conservation: A case study from the Kashmir Valley. *Mountain Research and Development, 35*(4), 392–400.

Rahman, M., Islam, M.S., & Chowdhury, T.A. (2019). Change of vegetation cover at Rohingya Refugee occupied areas in Cox's Bazar district of Bangladesh: Evidence from remotely sensed data. *Journal of Environmental Science and Natural Resources, 11*(1–2), 9–16.

Shaffer, L.J., Khadka, K.K., Van Den Hoek, J., & Naithani, K.J. (2019). Human-elephant conflict: A review of current management strategies and future directions. *Frontiers in Ecology and Evolution, 6*, 235.

Sparling, D.W. (2014). History of wildlife and natural resource conservation. In *Natural Resource Administration* (1st Edition) (pp. 27–51). Cambridge, MA: Academic Press.

Torres, D.F., Oliveira, E.S., & Alves, R.R.N. (2018). Understanding human–wildlife conflicts and their implications. In *Ethnozoology* (pp. 421–445). Cambridge, MA: Academic Press.

UNHCR & IUCN. (2018). Survey Report on Elephant Movement, Human-Elephant Conflict Situation, and Possible Intervention Sites in and around Kutupalong Camp, Cox's Bazar. Retrieved from: https://www.unhcr.org/5a9946a34.pdf.

UNHCR. (2020). Situation Refugee Response in Bangladesh. Retrieved from: https://data2.unhcr.org/en/situations/myanmar_refugees.

Wahed, M.A., Ullah, M.R., & Irfanullah, H.M. (2016). Human-elephant conflict mitigation measures: Lessons from Bangladesh. IUCN, Bangladesh Country Office, Dhaka. Retrieved from: https://www.researchgate.net/publication/310033702_Human-Elephant_Conflict_Mitigation_Measures_Lessons_from_Bangladesh?fbclid=IwAR3Soy4-aeM1XIsZPCgBzu0sn-L_mwDo6PsykUPPQ4k-STrzH-xnbeV6kOKM.

WWF. (2015). Asian Elephant: WWF Wildlife and Climate Change Series. Retrieved from: https://c402277.ssl.cf1.rackcdn.com/publications/779/files/original/Asian_elephant_-_WWF_wildlife_and_climate_change_series.pdf?1435159896.

WWF. (2008). Common Ground Solutions for reducing the human, economic and conservation costs of human wildlife conflict. Retrieved from: https://wwfeu.awsassets.panda.org/downloads/hwc_final_web.pdf.

18 Transmission Mechanisms of Bioaerosols

An Unseen Threat to Human Health

Sneha Gautam
Department of Civil Engineering, Karunya Institute
of Technology and Sciences, Coimbatore, India

Mohammed Abdus Salam
Department of Environmental Science and
Disaster Management, Noakhali Science and
Technology University, Noakhali, Bangladesh

Mahmud Hossain Sumon
Department of Soil Science, Bangladesh Agricultural
University, Mymensingh, Bangladesh

Muhammad Anwar Iqbal
Department of Environmental Science and
Disaster Management, Noakhali Science and
Technology University, Noakhali, Bangladesh

Bruno Pavoni
Dipartimento di Scienze Ambientali, Informatica e
Statistica, Università Ca' Foscari Venezia,
Mestre, Italy

Md. Badiuzzaman Khan
Department of Environmental Science,
Bangladesh Agricultural University, Mymensingh,
Bangladesh

DOI: 10.1201/9781003140382-18

CONTENTS

18.1 INTRODUCTION

Bioaerosols are airborne biological particulate matter that range in size from nanometers up to about a tenth of a millimeter (Gollakota et al., 2021; Humbal et al., 2019; Humbal et al., 2018). They are comprised of living and dead organisms (algae, archaea, bacteria, and viruses), dispersal units (fungal spores and plant pollen) and various fragments or excretions (plant debris and brochosomes). A considerable variation in size distributions of bioaerosols has been reported: plant pollen grains are 5–100 μm, fungal spores are 1–30 μm, bacteria are 0.1–10 μm, and viruses are generally smaller than 0.3 μm. Bioaerosols are ubiquitous in nature and may contribute up to 25% of atmospheric aerosols. A recent study reported that bioaerosols may contribute up to 34% of indoor air pollution with long-term consequences for human health. The number and mass concentration of bioaerosols throughout vegetative regions are generally ~104 m^{-3} and 1 μg m^{-3} (Després et al., 2012). Table 18.1 shows the emission rate, number, and mass concentration of bioaerosols in near-surface air.

Bioaerosols may have a significant effect on global climate systems (scattering and absorbing radiation), cloud microphysical process (acting as cloud condensation nuclei and ice nuclei), and human health (Huffman et al., 2013). Recently, scientists have expressed concerns about bioaerosols as the prevalence of bioaerosols may be correlated with several human diseases, especially pneumonia, influenza, measles, asthma, allergies, and gastrointestinal illness. However, respiratory diseases caused by bioaerosols are regulated by their particle size distribution and bioaerosols with smaller aerodynamic size generally affect the deeper portion of the respiratory system. There is a link between bioaerosols and human diseases. Various anthropogenic activities are responsible for the generation of bioaerosols. For instance, sneezing/coughing, washing floors/toilet cleaning, and walking/talking (Chen and Hildemann, 2009).

TABLE 18.1

Estimation of Global Emissions, Number and Mass Concentrations in Near-Surface Air for Bioaerosols

	Global Emissions (Tg a⁻¹)	Number Concentration (µg m⁻³)	Mass Concentration
Bacteria	0.4–28	-10^4	-0.1
Fungal spores	8–190	-10^3 to -10^4	$-0.1 - 1$
Fungal hyphal fragments		-10^3	
Pollen	47–84	-10 (up to $- 10^3$)	-1
Plant debris			-0.1 to -1
Algae		-100 (up to -10^3)	-10^{-3}
Fern spores		-10 (up to $- 10^3$)	-1
Viral particles		-10^4	-10^{-3}
Total Primary biological aerosols (PBA)	<10 (dominated by plant debris and fungal spores) to 1000 (includes cellular fragments)		

Source: Després et al. (2012).

The nature, composition, spatio-temporal variation, and distribution of bio-aerosols are affected by various physical and environmental factors, especially air currents, relative humidity, and temperature. Airborne microbial concentrations are influenced by several factors, such as atmospheric temperature, relative humidity, and particulate matter concentrations (Xie et al., 2018).

Following the release of bioaerosols from land and ocean surfaces, they can be transported over long distances and even from one continent to another (Smith et al., 2013) owing to their very small size and light weight. Bioaerosols are the most important medium for the dispersal of reproductive units over long distances such as spores and pollen.

However, there is a lack of information on the transmission of pathogenic microbes via bioaerosols. This chapter explores the components and health impacts of bioaerosols. The results of this review may help to improve our understanding of the transmission mechanisms of pathogenic microbes via bioaerosols, as well as related physical and environmental factors.

18.2 COMPONENTS OF BIOAEROSOLS AND METEOROLOGICAL FACTORS

Bioaerosols consist of different microorganisms and their constituents, such as fungi, bacteria, endotoxin, mycotoxins, and allergens. A description of the main bioaerosol components and related influencing factors are given below:

18.2.1 Fungi

Fungi generally grow on different foods and other organic materials, such as paper, textiles, wood, and damp. Approximately 2.2 to 3.8 million fungal species exist worldwide (Hawksworth and Lücking, 2017). Most of the airborne fungi belong to the division of Ascomycota and Basidiomycota, whereas dominant airborne fungal spores are from the species of *Cladosporium, Alternaria, Penicillium, Aspergillus, Epicoccum* and a variety of yeasts, smuts and rusts, and other basidiomycetes (Després et al., 2012). Fungi are responsible for the production of allergens, enzymatic protein, toxins, and volatile organic compounds that cause various human health problems, including toxic effects, irritation, infections, and allergies. Fungal growth and dispersal showed significant relationships with meteorological parameters. The concentration of fungal bioaerosols ranged from 24 to 654 (mean 177) CFU/m^3 during winter, whereas the value ranged from 60 to 930 (mean 357) CFU/m^3 in summer (Lee et al., 2016) in Seoul Metropolitan area, South Korea.

18.2.2 Bacteria

Globally, 107 to 109 species of bacteria exist, ranging in size from 0.3 μm (mycoplasma) to 0.5 mm (Williams, 2011). However, the size of a single bacterium can be ~μm or even less. Bacteria generally have a long residence time and can be transported over long distances due to their small size. A number of pathogenic bacteria can be disseminated as a bioaerosol, such as *Bordetella pertussis, Bacillus anthracis, Corynebacterium diphtheriae,* and *Neisseria meningitidis,* which are responsible for Pertusis, Anthrax, Diptheriae, and Meningis, respectively (Hendricks et al., 2014). The concentration and composition of bacteria are influenced by anthropogenic activities, as well as altitude, climate, and atmospheric changes, and show diurnal and seasonal variation with higher concentrations in summer and autumn (Kaarkainen et al., 2008).

18.2.3 Endotoxins

Endotoxins are lipopolysaccharides (LPS) in Gram-negative bacteria with very high pro-inflammatory characteristics. Endotoxins comprise a core polysaccharide chain, O-specific polysaccharide side chains (O-angigen), and a lipid component (Lipid A). In addition to their toxic effects, endotoxins are responsible for the reduction of lung diffusion capability, as well as fever, shivering, blood leukocytosis, neutrophilic airway inflammation, arthralgia, dyspnea and chest tightness, and bronchial obstruction. Statistically significant correlations of endotoxin concentrations with relative humidity and temperature have been reported (Traversi et al., 2010).

18.2.4 β Glucans

β glucans are polymers of glucose that are commonly found in the cell walls of different microorganisms, such as bacteria, algae, lichens, fungi, yeasts, and plants (Kurek et al., 2016). β glucans are comprised of β (1→3), β (1→4), or β (1→6) glucose units or mixed linkages. The mean concentrations of β glucans indoors and outdoors in Ohio, USA were 1.0 (0.81–1.2) and 7.34 ng/m³, respectively (Crawford et al., 2009). β-glucans are used to enhance the immune system and are very useful against high cholesterol, diabetes, and cancer.

A study conducted by Hwang et al. (2014) revealed that temperature was significantly correlated with (1→3) β-D-glucan, while a higher concentration was observed during spring (April and May) and a lower level was found in the fall (November–December).

18.2.5 Mycotoxins

Mycotoxins are secondary nonvolatile, low-molecular-weight metabolites produced by fungi and molds such as steroids, carotenoids, alkaloids, cyclopeptides, and coumarins. Approximately 500 different mycotoxins have been identified. Their production is regulated by various environmental drivers, such as temperature, availability of water, nutrients, and presence of microorganisms. Hot and humid climates and poor storage conditions are the most favorable drivers that trigger mould growth and mycotoxin production (Ashiq et al., 2014). Mycotoxins may weaken the immune system, induce allergies or irritation, or cause various identifiable diseases, and can even lead to death (Kim et al., 2017). A list of the commonly observed mycotoxins in food commodities is provided in Table 18.2.

TABLE 18.2
Occurrence of Mycotoxins in Food Commodities

Mycotoxin	Origin of Fungi	Commodities Affected
Aflatoxin	*Aspergillus flavus; A. parasiticus*	Corn, peanuts, tree nuts, cottonseed, cereals, some spices
Citrinin (citreoviridin)	*Penicillium citrinin and other spp. from the Aspergillus and Monascus genera*	Cereals
Ergot alkaloids (ergotamine)	*Claviceps spp.*	Cereals and grasses
Fumonisins	*Fusarium verticillioides; F. proliferatum*	Corn, other cereals
Ochratoxin A	*Aspergillus ochraceus; A. niger; Penicillium verrucosum*	Legumes, grapes, cereals, coffee beans
Patulin	*Penicillium expansum and other spp. from Aspergillus and Byssochlamys genera*	Apples, grapes, pears, other fruits
Penicillic acid	*Penicillium martensii; P. puberulum; P. palitans*	Corn, legumes
Psoralens	*Sclerotiniasclerotiorum, S. rolfsii, Rhizoctiniasolani,*	Celery, figs, parsley, parsnip, lime, cloves
Trichothecenes	*Fusarium graminearum; F. sporotrichioides;*	Wheat, corn, barley, oats
Zearalenone	*Fusarium graminearum, F. culmorum, F. cerealis,*	Corn, sorghum, wheat

Source: Richard and Hayes (1997).

18.2.6 Allergens

The main sources of allergens are fungi (spores and hyphae), arthropods (mites and cockroaches), vascular plants (fern spores, pollen, and soy dust), pet dander, and royal jelly (Jutel et al., 2016). Moreover, humidity and water-damaged items (carpets, ceilings, and walls) are also potential sources of mite and mold allergens. However, exposure of allergens to air is regulated by various factors, such as mechanical disturbance, wind, and rain. Runny and stuffy nose, throat irritation, itchy eyes, and sneezing are typical symptoms associated with allergens.

18.3 THE TRANSMISSION MECHANISM OF PATHOGENIC MICROBES THROUGH BIOAEROSOLS

Bioaerosol transmission is a complex process (Pöschl, 2005). Bioaerosols can be formed by a variety of mechanisms, including bubble bursting, erosion, active dispersal, and mechanical contact between surfaces (Löndahl, 2014). The particles can be dispersed as individual cells, but they can also bind themselves as a biofilm-filled community of cells to other materials, such as pollen, dust, and plant debris. Many microbes have evolved pathways to migrate through the atmosphere, remain viable for a long time in their inhospitable setting, and then be deposited in a new location to begin life. The air provides a good route for rapid transportation of microbes over large areas. Natural surface water is a vital source of environmental bioaerosols as it has an almost inevitable microlayer, which is several hundred μm thick and in contact with the atmosphere (Niazi et al., 2015). It is a unique ecosystem that is enriched with much higher concentrations of microbes, viruses, and other microorganisms, as well as lipids, carbohydrates, polysaccharides, amino acids, and proteins compared with sub-surface waters (Fröhlich et al., 2016). Bioaerosols are formed by precipitation, broken waves, and ship movements in water sources, while in inland areas, these are formed by erosion, wind, or other forms of mechanical activities from the surfaces of plants, grass, soil, and rock. In addition, plants produce bioaerosol particles in the form of pollen and spores for reproduction.

Certain microorganisms can stay biologically active, grow, and sustain reproductive ability in bioaerosols because of ongoing environmental pollution. Numerous pathogens are able to remain alive by developing endospores, while non-sporogenous organisms can also survive through repair and pigmentation mechanisms in bioaerosols. Many microorganisms that do not have a repair function, such as viruses and fungal spores, are sometimes preserved by pollen, dust, biofilm, or other substance attachments in bioaerosols (Löndahl, 2014). Bacteria are also believed to be able to remain viable more than 50 km above the surface of the planet. For certain microorganisms, atmospheric bioaerosols may be habitats of their own as well as a form of transportation.

Bioaerosols typically persist throughout the atmosphere for several weeks before ground deposition, and many of them enter the stratosphere, which increases their atmospheric lifespan. The shape and appearance of bioaerosols are changed in the atmosphere primarily by reformation, coagulation, vaporization, and condensation processes. Bioaerosols combine themselves and shape a larger structure during the coagulation mechanism. Due to increased surface tension or droplet formation, the humidity of a region often affects the transformation of bioaerosols. The rate of coagulation rises significantly with the rise in environmental emissions. Bioaerosols are extracted from the atmosphere by wet or dry ground deposition (Figure 18.1). Wet deposition is the primary sink of bioaerosols as they are washed out from the air by rainfall. During the sedimentation process, bioaerosols interact and adhere to the ground in dry deposition. Bioaerosols with a large droplet diameter are mostly deposited on the ground due to the earth's gravity (Hayleeyesus et al., 2015). Thus, bioaerosols with a large diameter size remain airborne for a short interval, whereas small-sized bioaerosols contaminants stay in the air for a longer duration. Some bioaerosols are found to possess unique aerodynamic structures, such as pollen and fungal spores, which allow them to be transported over long distances. Respiratory diseases, frequently attributed to inhaled microbes and contaminated bioaerosols, are the worst mortality-related

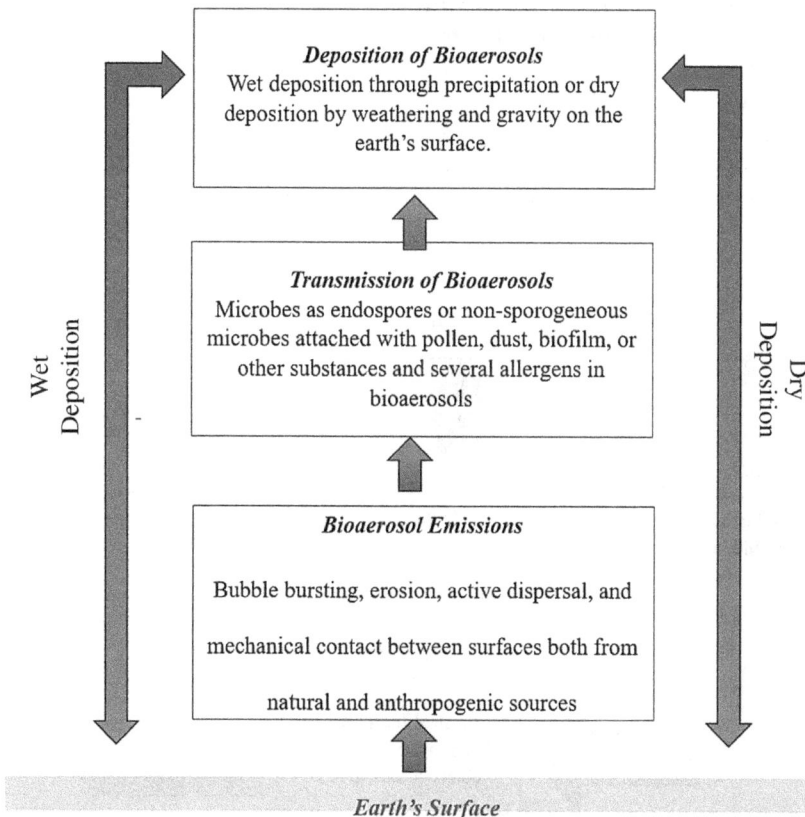

FIGURE 18.1 Transmission mechanism of pathogenic microbes through bioaerosols.

diseases globally as they contain toxic and allergenic materials and can cause infectious diseases from pathogenic biological agents (Leuken et al., 2016). The largest bioaerosols are mainly deposited in the nose and mouth, such as several pollens, fungal spores, and algae, but some bioaerosols containing virus, bacteria, cell fragments, and so on, can go deeper into the respiratory system.

Bioaerosol-related primary infectious diseases are caused by viruses (i.e., COVID-19, pneumonia, influenza, measles, SARS, etc.), bacteria (i.e., tuberculosis, Legionnaire's disease, etc.) or fungal spores (i.e., blastomycosis, aspergillosis) (Heo et al., 2014).

18.4 TRANSMISSION MECHANISM OF COVID-19 VIA BIOAEROSOLS

The COVID-19 pandemic, which is believed to have originated in Wuhan City, China, in late 2019, has become an unprecedented serious concern worldwide due to the rapid rate of human-to-human transmission. Severe acute respiratory syndrome coronavirus 2 (SARS-CoV-2) causes damage to the respiratory system, in much the same manner as MERS-CoV and SARS-CoV (Lai et al., 2020). According to the World Health Organization, this pandemic has caused more than 31,425,029 COVID-19 cases in 235 countries as of September 23, 2020, resulting in more than 967,164 deaths. Although the exact causes of this virus have not yet been established, it is known that this virus causes damage to the respiratory system. It has been postulated that COVID-19 can be transmitted in four transmission pathways (Liu et al., 2020) (Figure 18.2).

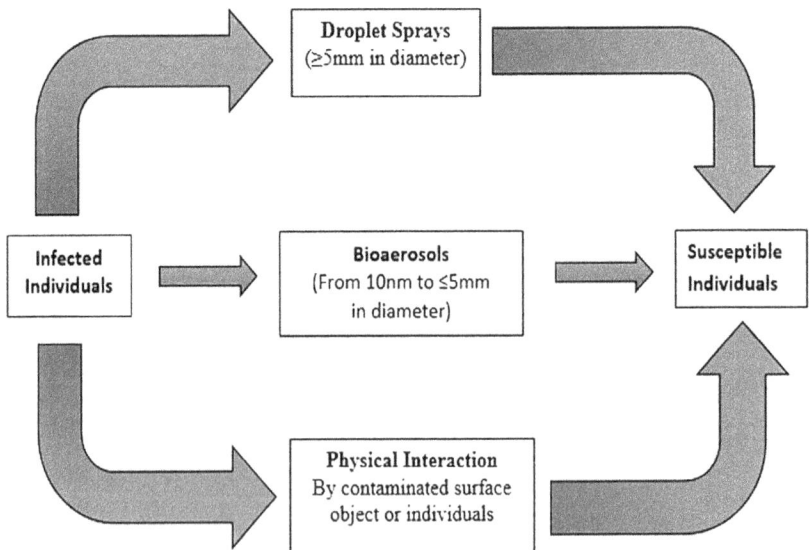

FIGURE 18.2   Transmission mechanism of SARS-CoV-2.

The four transmission pathways are: (1) respiratory droplets emitted by coughing or sneezing into the mucosal surface such as mouth, nose, or eyes, (2) bioaerosols containing viral RNA, (3) close communication as well as touching the hands or other areas of a body affected with contagious respiratory or fecal material, (4) indirect exposure such as physical contact with contaminated objects (fomites) (Liu et al., 2020). Moreover, atmospheric transmission can occur in two different pathways and involves no direct interaction among infected and susceptible individuals. "Droplet sprays" of virus-laden nasal mucosa fluid, which is larger than 5 mm in size, directly influences a susceptible person during a sneeze or a cough. Secondly, bioaerosols, which are composed of the remaining solid components of vaporized coughing and sneezing as well as exhaled respiratory droplets that are sufficiently small (< 5 mm) to stay in the atmosphere for a long time, can be inhaled by a vulnerable person (Guzman, 2020). Infected persons mainly produce infectious bioaerosols when coughing, sneezing, speaking, singing, and exhaling (Ko et al., 2020). Those bioaerosols may further land on the mouth, nose, or eyes of nearby individuals, and they can also potentially be inhaled. However, significantly bigger bioaerosols are deposited on surfaces surrounding the contaminated person, and for several days after deposition, interaction with these bioaerosols could result in infection. Smaller bioaerosols may form nuclei particles in the atmosphere, which can be transmitted far from the source and may readily enter the respiratory system (Sommerstein et al., 2020).

18.5 IMPACT ON HUMAN HEALTH

Over the past few decades, there has been growing interest in bioaerosol exposure to humans and animals. This is largely because of increasing air pollution with bio and inert particulate matter generation and spread in both occupational and residential indoor environments. This form of air pollution is associated with a wide range of adverse health effects that can have a major impact on public health, including respiratory diseases, communicable diseases, and cancer. Table 18.3 shows various bioaerosol-related diseases in humans.

18.6 CONCLUSION

Our review clearly demonstrated that the prevalence of bioaerosols appears to be correlated with several human diseases, especially pneumonia, influenza, measles, asthma, allergies, and gastrointestinal illness. Bioaerosols are extracted from the atmosphere by wet or dry ground deposition, while endospores or non-sporogeneous microbes can be attached to pollen, dust, biofilm, or other substances, as well as several allergens in bioaerosols. It has been well established that COVID-19 can be transmitted through respiratory droplets and physical contact with contaminated objects. Although it is plausible that COVID-19 can be spread through the dispersal of bioaerosols by an infected person, more research is needed in order to establish a definitive conclusion. More work needs to be done with respect to exposure simulation, quantification, and risk assessment of different bioaerosols.

TABLE 18.3
Various Bioaerosol-Related Diseases in Humans

Bioaerosols	Approximate Size	Health Impact	Infection/Transmission
COVID-19	D: 60–140 nm	Severe acute respiratory syndrome	Person-to-person through the air
Mycobacterium tuberculosis	L: 2–4 µm W: 0.2–0.5 um	Tuberculosis	Person-to-person through the air
Bordetella pertussis	L: 40–100 nm D: 2 nm	Whooping cough	Direct contact or inhalation of airborne droplets
Bacillus anthracis	Spore L: 3–5 µm W: 1.0–1.2 µm	Anthrax	Contact with infected animals, flies
Variolavera	L: 220–450nm W: 140–260 nm	Smallpox	Inhalation of airborne variola virus, prolonged face-to-face contact with an infected person
Herpesvirida, HHV-3	D: 150–200 nm	Chickenpox and shingles	Direct contact with fluid from the rash blisters caused by shingles
Morbillivirus measles	L: 125–250 nm D: 21 nm	Measles, mumps, and rubella	Bodily fluids: drops of saliva, mucus from the nose, coughing or sneezing, tears from the eyes, etc.
MicrosporumTrichophyton	L: 5–100 mm W: 3–8 mm	Ringworm	Direct or indirect contact with skin or scalp lesions of infected people and animals
α-amylase	45–49 kDa	Asthma and rhinitis	Direct contact or inhalation of airborne droplets
Mould	D: 10–30 µm	Cough, wheezing, bronchitis, and asthma with dampness	Direct contact or inhalation of airborne mould

REFERENCES

Ashiq, S., Hussain, M., Ahmad, B. 2014. Natural occurrence of mycotoxins in medicinal plants: A review. *Fungal Genetics and Biology.* 66, 1–10.

Chen, Q., Hildemann, L.M. 2009. The effects of human activities on exposure to particulate matter and bioaerosols in residential homes. *Environmental Science and Technology.* 43 (13), 4641–4646.

Crawford, C., Reponen, T., Lee, T., Iossifova, Y., Levin, L., Adhikari, A., et al. 2009. Temporal and spatial variation of indoor and outdoor airborne fungal spores, pollen, and (1→3)-beta-d-glucan. *Aerobiologia*. 25 (3), 147–158.

Després, V.R., Huffman, J.A., Burrows, S.M., Hoose, C., Safatov, A.S., Buryak, G., et al. 2012.. *Tellus Series B-Chemical and Physical Meteorology*. 64 (1), 1–58.

Fröhlich N., Janine K., Christopher W., Bettina H., John P., Christopher A., et al. 20. *Atmospheric Research*. 182 (15), 346–376.

Guzman, M. I. 2020. Bioaerosol Size Effect in COVID-19 Transmission. *Communication*. 10.20944/preprints202004.0093.v1.

Gollakota, A.R.K., Gautam, S., Santosh, M., Sudan, H.A., Gandhi, R., Jebadurai, V.S., Shu, C.M. 2021. Bioaerosols: characterization, pathways, sampling strategies, and challenges to geo-environment and health. *Gondwana Research* 99, 178–203. (10.1016/j.gr.2021.07.003).

Hawksworth, D.L., Lücking, R. 2017. Fungal diversity revisited: 2.2 to 3.8 million species. *Microbiology Spectrum*. 5 (4), 79–95. http://dx.doi.org/10.1128/microbiolspec. FUNK-0052-2016.

Hayleeyesus, S.F., Ejeso, A., Derseh, F.A. 2015. Quantitative assessment of bio-aerosols contamination in indoor air of university dormitory rooms. *International Journal of Health Science*. 9 (3), 249–256.

Hendricks, K.A., Wright, M.E., Shadomy, S.V., Bradley, J.S., Morrow, M.G., Pavia, A.T., et al. 2014. Workgroup on Anthrax Clinical Guidelines, 2014. Centers for disease control and prevention expert panel meetings on prevention and treatment of anthrax in adults. *Emerging Infectious Diseases*. 20 (2), e130687, http://dx.doi.org/10.3201/eid2002.130687.

Heo, K.J., Kim, H.B., Lee, B.U. 2014. Concentration of environmental fungal and bacterial bioaerosols during the monsoon season. *Journal of Aerosol Science*. 77, 31–37.

Huffman, J.A., Prenni, A.J., DeMott, P.J., Pöhlker, C., Mason, R.H., Robinson, N.H., et al. 2013. High concentrations of biological aerosol particles and ice nuclei during and after rain. *Atmospheric Chemistry and Physics*. 13, 6151–6164. http://dx.doi. org/10.5194/acp-13-6151-2013.

Humbal, C., Gautam, S., Joshi, S.K., Trivedi, U.K., 2019. Evaluating the colonization and distribution of fungal and bacterial bioaerosol in Rajkot, western India using multiproxy approach. *Air Quality Atmosphere and Health*. 12 (6), 693–704.

Humbal, C., Gautam, S., Trivedi, U. 2018. A review on recent progress in observations, and health effects of Bioaerosols. *Environmental International*. 118, 189–193.

Hwang, S.H., Yoon, C.S., Park, J.B. 2014. Outdoor (1→3)-β-D-glucan levels and related climatic factors. *Journal of Preventive Medicine & Public Health*. 47 (2), 124–128.

Jutel, M., Agache, I., Bonini, S., Burks, A.W., Calderron, M., Canonica, W., et al. 2016. International consensus on allergen immunotherapy II: Mechanisms, standardization and pharmacoeconomics. *The Journal of Allergy and Clinical Immunology*. 137, 358–368.

Kaarkainen, P., Meklin, T., Rintala, H., Hyvarinen, A., Karkainen, P. et al. 2008. Seasonal variation in the airborne microbial concentration and diversity in landfill, urban and rural sites. *Clean*. 36 (7), 556–563.

Kim, Ki-Hyun. Kabir, E., Jahan, S.A. 2017. Airborne bioaerosols and their impact on human health. *Journal of Environmental Science*. 67, 23–35.

Ko, K., Ellen, S., Manick, M., Kelly, L. 2020. Improving protection from bioaerosol exposure during postoperative patient interaction in the COVID-19 era, a quality improvement study. *American Journal of Otolaryngology*. 41, 102634.

Kurek, M.A., Wyrwisz, J., Wierzbicka, A. 2016. Effect of β-glucan particle size on the properties of fortified wheat rolls. *CyTA-Journal of Food*. 4(1), 124–130.

Lai, C.C., Shih, T.P., Ko, W.C., Tang, H.J., Hsueh, P.R. 2020. Severe acute respiratory syndrome coronavirus 2 (SARS-CoV-2) and coronavirus disease-2019 (COVID-19): The epidemic and the challenges. *The International Journal of Antimicrobial Agents*. 55, 105924.

Lee, B.U., Lee, G., Heo, K.J. 2016. Concentration of culturable 897 bioaerosols during winter. *Journal of Aerosol Science*. 94, 1–8.

Leuken V.J.P.G., Swart A.N., Havelaar A.H., Pul V.A., Hoek V.W., Heederik D. 2016. Atmospheric dispersion modelling of bioaerosols that are pathogenic to humans and livestock- a review to inform risk assessment studies. *Microbial Risk Analysis*. 1, 19–39.

Liu, Y., Ning, Z., Chen Y. 2020. Aerodynamic characteristics and RNA concentration of SARS-CoV-2 aerosol in Wuhan hospitals during COVID-19 outbreak. *Microbiology*. 582, 557–560.

Löndahl, J. 2014. Physical and Biological Properties of Bioaerosols. *Bioaerosol Detection Technologies*. Ch: 3, pp. 33–48. Springer-Verlag, New York. Editors: Per Jonsson, GöranOlofsson, TorbjörnTjärnhage.

Niazi, S., Hassanvand, M.S., Mahvi, A.H., Nabizadeh, R. Alimohammadi, M., Nabavi, S., et al. 2015. Assessment of 967 bioaerosol contamination (bacteria and fungi) in the largest 968 urban wastewater treatment plant in the Middle East. *Environmental Science and Pollution Research*. 22 (20), 16014–16021.

Pöschl U. 2005. Atmospheric aerosols: Composition, transformation, climate, and health effects. *Angewandte Chemie International Edition*. 44 (46), 7520–7540.

Richard, J.L., Hayes, A.W. 1997. Mycotoxins. In: Dulbecco, R. (Ed.), *Encyclopedia of Human Biology*, 2nd ed. Academic Press, San Diego, CA, 6, 5–17.

Smith, D.J., Timonen, H.J., Jaffe, D.A., Griffin, D.W., Birmele, M.N., Perry, K.D., Ward, P.D., Roberts, M.S., 2013. Intercontinental dispersal of bacteria and archaea by transpacific winds. *Applied Environmental Microbiology*. 79(4), 1134–1139.

Sommerstein R.F., Christoph V., Danielle A., Mohamed M., Jonas B., Carlo T., et al. 2020. Risk of SARS-CoV-2 transmission by aerosols, the rational use of masks, and protection of healthcare workers from COVID-19. *Antimicrobial Resistance & Infection Control*. 9 (100), 1–8. 10.1186/s13756-020-00763-0.

Traversi, D., Alessandria, L., Schiliro, T., Piat, S.C., Gilli, G. 2010. Meteo-climate conditions influence the contribution of endotoxins to PM10 in an urban polluted environment. *Journal of Environmental Monitoring*, 12(2), 484–90.

Williams, C., 2011. Who are you calling simple? *The New Scientist*. 211, 38–41.

Xie, Z., Li, Y., Lu, R., Li, W., Fan, C., Liu, P., Wang, W. 2018. Characteristics of total airborne microbes at various air quality levels. *Journal of Aerosol Science*. 116, 57–65.

Index

For Product Safety Concerns and Information please contact our EU
representative GPSR@taylorandfrancis.com
Taylor & Francis Verlag GmbH, Kaufingerstraße 24, 80331 München, Germany

www.ingramcontent.com/pod-product-compliance
Lightning Source LLC
Chambersburg PA
CBHW060809220326
41598CB00022B/2576